MW01007341

Exploring Black Holes

Introduction to General Relativity

Overview

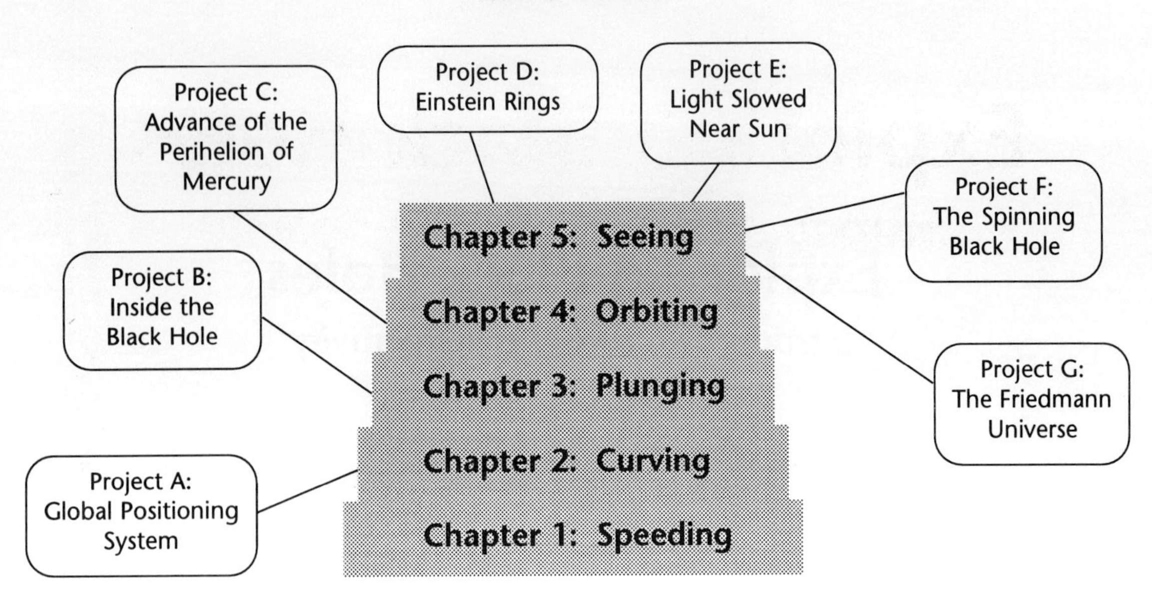

CHAPTERS provide the reader with background needed to carry out exercises and projects.

Chapter 1 Speeding Key ideas from special relativity that are useful in general relativity. We meet the metric for flat spacetime.

Chapter 2 Curving The Schwarzschild metric describes the curvature of spacetime near a non-rotating Earth, Sun, neutron star, or black hole.

Chapter 3 Plunging A stone plunging radially exhibits a constant of the motion: energy.

Chapter 4 Orbiting Orbits of stone and planet derive from two constants of the motion: energy and angular momentum. Predict the shape of an orbit at a glance.

Chapter 5 Seeing What you see when you look at, around, and outward from a black hole. Your view as you plunge through the horizon and approach the crunch point at the center of a black hole.

PROJECTS help the reader explore a topic, fill in the steps, compute physical outcomes, and carry out his or her own investigations.

Project A Global Positioning System General relativity is crucial to its operation.

Project B Inside the Black Hole A one-way trip to the crunch point at the center.

Project C Advance of the Perihelion of Mercury Small changes in the orbit of Mercury showed Einstein that his new general relativity theory was correct.

Project D Einstein Rings Light deflected by dark galactic objects helps us to see them.

Project E Light Slowed Near Sun Light moves slower near Sun. How do we know?

Project F The Spinning Black Hole "Frame dragging" near Earth or black hole.

Project G The Friedmann Universe Simplest model of the evolving Cosmos. Is it correct?

Exploring Black Holes

Introduction to General Relativity

Edwin F. Taylor

Massachusetts Institute of Technology

John Archibald Wheeler

Princeton University

An imprint of Addison Wesley Longman

San Francisco Boston New York
Capetown Hong Kong London Madrid Mexico City
Montreal Munich Paris Singapore Sydney Tokyo Toronto

Acknowledgments

For more than 12 years, students in many classes have read through sequential versions of this text, shared with us their detailed difficulties, and given us advice. (See "Only the Student Knows," *American Journal of Physics*, Volume 60, Number 3, March 1992, pages 201-202.) Many of the "thinker objections" in the text come from these commentators.

Versions of the book were used in classes taught online over the National Teachers Enhancement Network out of Montana State University. Many class members were high school science teachers who brought their professional skill and interest to the educational effectiveness as well as to the physics content of the book.

Nora Thornber has taught several online classes using versions of this book, meticulously read endless drafts, offered a host of suggestions, and improved and simplified a large number of the derivations. She has spent more time and effort on this book than anyone but the authors.

A number of relativity specialists—graduate students, postdoctoral fellows, faculty members, and others—examined parts or all of various versions of this book and made comments that increased clarity, introduced modern results, and corrected many errors. In alphabetical order they are Kashif Alvi, Edmund Bertschinger, Lior M. Burko, Sean M. Carroll, Teviet D. Creighton, Alexei V. Filippenko, Sergio Goncalves, Tom Goodale, William A. Hiscock, Lawrence E. Kidder, Shane L. Larson, Lee Lindblom, Yuk-Tung Liu, Anupam Mazumdar, Jonathan Morris, Mark A. Pelath, Patricia Purdue, Yuri Levin, Kip S. Thorne, Michele Vallisneri, Vrata Venet, Clifford M. Will, and Kazunori Yoshida. The authors regret that they cannot blame remaining errors on this august company.

G. P. Sastry offered much detailed and general advice on the book at several stages and supervised 14 students who read a late version, provided comments, and solved all exercises. He and two students, Manish Niranjan and Saswat Sarangi, turned their solutions into a Solutions Manual for this book.

Philip Morrison made several suggestions and convinced us not to invoke that weird quantum particle, the photon, in a treatment of the classical theory of relativity (except in some exercises).

Eric Sheldon read many drafts of this book; his good judgment and wide acquaintance with the literature have clarified and enriched the presentation. Sanjoy Mahajan significantly improved both the physics and the clarity of expression. Ira Mantz offered a wide spectrum of careful, critical, and useful comments on late versions. Stamatis Vokos and Rachel E. Scherr taught from earlier drafts and offered many helpful suggestions. Rachel Scherr championed inclusion of the questions inside the front cover and revised these questions. Beta Keramati recommended placing questions at the beginnings of chapters and projects. Ewa M. Basinska located many of the references. Peter M. Brown made many suggestions, found quotations, drafted the Glossary of Terms, helped to assemble the reading list, and had the initial idea for the front cover. Michael J. Tsai originated and helped to develop the back cover design. Other reviewers of various drafts: Curt Covey, Sally Ganner, Roy Gould, Vivek Iyer, Roy Lisker, Sergey V. Siparov. Penny Hull's copyediting improved the clarity and simplicity of both the English and the physics.

And then there is Carla . . .

Book design by Karen A. Schriver

Lower front cover Saturn image: Erich Karkoschka, University of Arizona, and NASA.
Upper front cover image: Edmund Bertschinger and Miriam L. Castellano.
(For explanation of these images, see Sections 9 and 10 of Project D, Einstein Rings.)

Published by Addison Wesley Longman, Inc.

Manufactured in the United States of America.

Library of Congress Cataloging-in-Publication Data

Taylor, Edwin F.
Exploring black holes: introduction to general relativity / Edwin F. Taylor,
John Archibald Wheeler.
p.cm.
Includes bibliographical references and index.
ISBN 0-201-38423-X
1. General relativity (Physics) 2. Black holes (Astronomy) I. Wheeler, John Archibald,
II. Title.
QC173.6 .T39 2000
530.11—dc21 00-035551

KEY WORDS: 1. General relativity. 2. Black hole. 3. Relativity. 4. Cosmology. 5. Spinning black hole.

Note: Both females and males make competent observers. Ordinarily we treat as female any observer in free float (whether in orbit or plunging inward) as well as the remote observers called Schwarzschild and Kerr bookkeepers. (See the Glossary of Terms for definitions.) Male observers stand on spherical shells around a nonrotating black hole and ride on circular rings around a spinning black hole. In other settings we try to use gender-neutral language. Where this is not convenient we usually assign gender randomly.

2 3 4 5 6 7 8 9 10—CO—04 03 02 01 00

PREFACE

A Single Goal

This book, Exploring Black Holes, makes a quick, directed thrust through general relativity and black holes. It includes many topics, all with a single goal.

THE GOAL: Power to the Reader! *We provide tools to answer questions and carry out calculations about curved spacetime near Earth and black holes. Topics are limited to those in which you can be an active participant, starting with an elementary knowledge of calculus and special relativity. Tools developed in general relativity then help you to pursue your own investigations.*

—Edwin F. Taylor and John Archibald Wheeler

QUESTIONS *NOT* ANSWERED AND WHAT TO DO ABOUT THEM

Here are some important questions *not* answered in this book.

- What is the full range of phenomena covered by general relativity?
- How did general relativity begin, and who wrestled with the ideas presented in this book?
- What about gravitational waves, pulsars, supernovas, and the formation of stars and black holes?
- Where are the frontiers of the subject, and how far can these frontiers be pushed?
- What are the latest results from observational satellites and the latest theories about the origin of the Universe?

To find the latest observational results, no source is better than the World Wide Web, on which addresses change from month to month, even from hour to hour. Here is a website of Hubble pictures and commentary: http://oposite.stsci.edu/pubinfo/Pictures.html. For more current information, ask your 15-year-old consultant.

To help you engage some of the other questions *not* answered in this book, we know of no better popular source than the following book, which is an almost perfect complement to *Exploring Black Holes:*

Black Holes and Time Warps
Einstein's Outrageous Legacy
Kip S. Thorne
W. W. Norton, New York, 1994
ISBN 0-393-31276-3 paperback
Phone order: 1-800-233-4830

Six hundred pages and 0.9 kilogram of history, people, theory, results, and speculation. It is hard to imagine a more complete, engrossing, or enjoyable survey by a single major participant. The book is thoroughly cross-referenced with notes, list of characters, chronology, glossary, people index, and subject index. We suggest that you start Thorne's book with the Prologue: A Voyage among the Holes, in which the reader, in a science fiction tale, encounters black holes and their strange properties.

One author (EFT) will attempt to provide updates for *Exploring Black Holes* on the web-site http://www.eftaylor.com/

Contents

Chapter 1 Speeding

- *"Spacetime is a unity." Huh?*
- *"Relativity" means everything is relative, right?*
- *What is the farthest galaxy I can possibly visit?*
- *What is the fastest thing ever detected?*

CHAPTER 1

Speeding

The important thing is not to stop questioning. Curiosity has its own reason for existing. One cannot help but be in awe when he [or she] contemplates the mysteries of eternity, of life, of the marvelous structure of reality. It is enough if one tries merely to comprehend a little of this mystery every day. Never lose a holy curiosity.

—Albert Einstein

1 Special Relativity

Key idea: Concepts useful in exploring the very fast help us to examine spacetime near very massive objects.

We use relativity to explore the boundaries of Nature. **Special relativity** describes the very fast. **General relativity**—the **Theory of Gravitation**—describes matter and motion near massive objects: stars, galaxies, black holes. General relativity also describes the Universe as a whole. This chapter discusses a few key concepts of special relativity useful in exploring general relativity. The treatment here is not designed to be an introduction to special relativity; for introductory treatments see Section 11, Readings in Special Relativity, and detailed references to our own introductory treatment at the end of each section.

Special relativity: fast objects
General relativity: spacetime near massive objects

2 Wristwatch Time

Everyone agrees on the wristwatch time between two events.

What is the root of relativity? Is there a single, simple idea that launches us along the road to understanding? Alice's adventures in wonderland begin when a rabbit rushes past her carrying a pocket watch. Our adventure in relativity begins when a small stone flies past us wearing a wristwatch.

Begin relativity with wristwatch time between two ticks.

The wristwatch ticks once at #1 and once at #2 (Figure 1). Wristwatch ticks may be one second apart—or one microsecond. Measure the distance s and time t between these ticks in a particular **free-float** or **inertial** reference frame. (The free-float frame is described in Section 8. Briefly, it is one in which Newton's first law holds: a free particle at rest remains at rest and one in motion continues that motion at constant speed in a straight line.) Special relativity warns us that a different observer passing us in uniform relative motion typically records a different value of spatial separation s and a different value of time lapse t between these two ticks. That is the bad news. The good news is a central finding of special relativity:

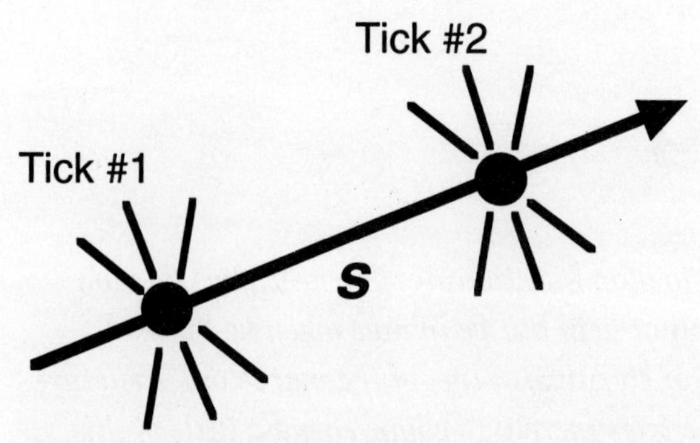

Figure 1 *Straight-line uniform-speed trajectory of a stone through space. The stone wears a wristwatch that ticks and emits a flash at #1 and then ticks again and emits a second flash at #2. These two ticks are a distance s apart and have a time separation t as measured in the frame of reference for which this diagram is drawn.*

All inertial observers, whatever their state of relative motion and whatever values they measure for s and t, agree on the value of the time τ between ticks as recorded on the wristwatch carried by the stone. The formula is simple:

$$\tau^2 = t^2 - s^2 \qquad [1]$$

Define wristwatch time.

We use the Greek letter τ (tau) for the **wristwatch time** between these two watch ticks. The wristwatch time is often called the **proper time** or, more formally, the **timelike spacetime interval** ("timelike" because the time separation t is greater than the space separation s). All observers agree on the value of the wristwatch time between two events. In contrast, the value of t and the value of s between these events will typically differ from frame to frame. Call t the **frame time** and s the **frame distance** between this pair of events. Wristwatch time τ can be used to describe the separation between *any* pair of events for which t is greater than s. It tells the observer in any frame what the time lapse will be on a wristwatch that moves uniformly from one event to the other.

Measure space and time in the same units.

For simplicity, the units of space and time are the same, such as light-years and years, or meters of distance and meters of light-travel time. In both cases the speed of light c is the conversion factor between measures of space and time. For example, the relation between seconds and meters of light-travel time is

$$t(\text{in meters}) = ct_{\text{second}} \qquad [2]$$

The metric: Key to all relativity

Equation [1], which connects the wristwatch time between two adjacent ticks to their space and time separations in a given frame, is called the **metric**. The metric (with a minus sign between squared quantities) tells us the *separation* between events in spacetime, just as the Pythagorean Theorem (with a plus sign between squared quantities) tells us the *distance* between points in a space described by Euclidean geometry. The metric is central in both special and general relativity. In describing physical systems for which it can be derived, the metric provides the answer to every possible question about (nonquantum) features of spacetime. And with a simple extension it also predicts the trajectories of particles and light.

The fact that all free-float observers agree on the wristwatch time τ earns it the label **invariant**. *Invariant* means that all observers calculate the same value, independent of reference frame. In relativity every invariant quantity is a diamond, to be treasured.

Wristwatch time is an INVARIANT.

How fast does the stone travel between ticks? The stone's speed depends on the reference frame. For the frame of Figure 1, the speed (assumed to be constant) is $v = s/t$. Measure distance s and time lapse t in the same unit. For example, a spaceship travels half a light-year of distance during one year of time; its speed is then 0.5 year/year and the units cancel. As another example, if an elementary particle moves 0.7 meter in one meter of light-travel time its speed is 0.7. Hence the speed v has no units. In this book the symbol v represents the speed of an object as a fraction of the speed of light.

Velocity v is a fraction of the speed of light.

Fuller Explanations: *Spacetime Physics*, Chapter 1, Spacetime: Overview; Chapter 3, Same Laws for All; Chapter 6, Regions of Spacetime.

SAMPLE PROBLEM 1 Wristwatch Times

PROBLEM 1A. An unpowered spaceship moving at constant speed travels 3 light-years in 5 years, this time and distance measured in the rest frame of our Sun. What is the time lapse for this trip as recorded on a clock carried with the spaceship?

SOLUTION 1A. The two events that start and end the spaceship's journey are separated in the Sun frame by s = 3 light-years and t = 5 years. Equation [1] gives the resulting wristwatch time:

$$\tau^2 = t^2 - s^2 = 5^2 - 3^2 = 25 - 9 = 16 \text{ years}^2 \qquad [3]$$
$$\tau = 4 \text{ years}$$

which is *less* than the time lapse as measured in the Sun frame.

PROBLEM 1B. An elementary particle is created in the target of a particle accelerator and arrives at a detector 4 meters away and 5 meters of light-travel time later, as measured in the laboratory. The wristwatch of the elementary particle records what time between creation and detection?

SOLUTION 1B. The events of creation and detection are separated in the laboratory frame by s = 4 meters and t = 5 meters of light-travel time. Equation [1] tells us that

$$\tau^2 = t^2 - s^2 = 5^2 - 4^2 = 25 - 16 = 9 \text{ meters}^2 \qquad [4]$$
$$\tau = 3 \text{ meters}$$

Again, the wristwatch time for the particle is less than the time recorded in the laboratory frame.

3 Proper Distance

Everyone agrees on the proper distance between two events.

Two firecrackers explode 1 *meter apart* and *at the same time*, as measured in a particular free-float frame. In this frame these explosions are **simultaneous**. No stone can travel fast enough to be present at both of these explosions without moving at an infinite velocity, which is impossible. Therefore equation [1] is useless to define a wristwatch time τ between these two events.

SAMPLE PROBLEM 2 Speeding to Andromeda

At approximately what constant speed v must a spaceship travel so that the occupants age only 1 year during a trip from Earth to the Andromeda galaxy? Andromeda lies 2 million light-years distant from Earth.

SOLUTION

The word *approximately* in the statement of the problem tells us that we can make some assumptions. We assume that a single free-float frame can stretch all the way from Sun to Andromeda, so special relativity applies. We also predict that the speed v of the spaceship measured in the Sun frame is very close to unity, the speed of light. That allows us to set $(1 + v) \approx 2$ in the last of the following steps:

$$\tau^2 = t^2 - s^2 = t^2\left(1 - \frac{s^2}{t^2}\right) = t^2(1 - v^2) \qquad [5]$$

$$= t^2(1 + v)(1 - v) \approx 2t^2(1 - v)$$

Equate the first and last expressions to obtain

$$1 - v \approx \frac{\tau^2}{2t^2} \qquad [6]$$

Now, we assumed that v is very close to the speed of light. It follows that the time t for the trip in the Sun frame is very close to the time that light takes to make the trip: 2 million years. Substitute this value and also demand that the wristwatch time on the spaceship (the aging of the occupants during their trip) be $\tau = 1$ year. The result is

$$1 - v \approx \frac{\tau^2}{2t^2} = \frac{1\ \text{year}^2}{2 \times 4 \times 10^{12}\ \text{year}^2} \qquad [7]$$

$$= \frac{10^{-12}}{8} = 1.25 \times 10^{-13}$$

Equation [7] expresses the result in sensible scientific notation. However, your friends may be more impressed if you report the speed as a fraction of the speed of light: $v = 0.999999999999875$. This result justifies the assumptions we made about the value of v and the time for the trip as measured in the Sun frame. *Additional question*: What *distance* does the spaceship rider measure between Earth and Andromeda?

Use simultaneous explosions to measure length of a rod.

Simultaneous explosions are thus useless for measuring time. But they are perfect for measuring length. *Question:* How do you measure the length of a rod, whether it is moving or at rest in your frame? *Answer:* Set off two firecrackers at the two ends and *at the same time* ($t = 0$) in your frame. Then *define* the rod's length in your frame as the *distance* s between this pair of explosions.

Special relativity warns us that a different observer passing us in uniform relative motion typically will *not* agree that the two firecrackers exploded at the same time. That is the bad news (and the idea most difficult to understand in all of special relativity). But there is good news: All inertial observers, whatever their state of relative motion, can calculate the distance σ between explosions as recorded in the frame in which they do occur simultaneously. The new metric is a variation of the old metric [1]:

$$\sigma^2 = s^2 - t^2 \qquad [8]$$

Proper distance is an INVARIANT.

The Greek letter σ (sigma) labels what we call the **proper distance** between such events or, more formally, the **spacelike spacetime interval** ("spacelike" because the space separation s is greater than the time separation t). All free-float observers agree on the value of the proper distance—the proper distance is an *invariant*. In contrast, the value of t and the value of s between these events typically differ, respectively, as measured in different frames. Proper distance σ can be used to describe the separation between *any* pair of events for which s is greater than t. It tells the observer in any frame what the distance σ is between the events as measured in a frame in which they occur at the same time.

We attach special significance to the length of a rod measured in the frame in which it is at rest. Let a firecracker explode at each end of a rod at the same time in its rest frame. We call the distance between these explosions the **proper length** of the rod. Any other inertial observer, whatever her state of relative motion, can calculate the proper length of the rod from equation [8] using the time t and distance s that she measures between these particular explosions in her own reference frame.

As in equation [1], the units of space and time in equation [8] are the same, such as light-years and years—or meters of distance and meters of light-travel time.

The name **spacetime interval** is the collective name for the timelike spacetime interval (equation[1]) and the spacelike spacetime interval (equation [8]).

Fuller Explanations: What happens to equations [1] and [8] when s and t have the *same* magnitude? Find the answer in *Spacetime Physics*, Chapter 6, Regions of Spacetime.

4 The Principle of Extremal Aging

The Twin Paradox leads to a definition of natural *motion.*

To get ready for curved spacetime (whatever that may mean), look further at the motion of a free particle in **flat spacetime**, the arena of the free-float frame (Section 8) in which special relativity correctly describes motion.

How does a free particle move in flat spacetime? We say: "What a ridiculous question! Everyone knows that a free particle moves with constant speed in a straight line—at least as observed in a free-float frame." Ah yes, but *why* does a free particle move straight with constant speed? What lies behind this motion? Our answer for flat spacetime will be a trial run for the description of motion in curved spacetime, the arena of general relativity.

A deep description of motion arises from the famous **Twin Paradox**. Recall that one identical twin relaxes on Earth while her twin sister frantically travels to a distant star and returns. When the two meet again, the stay-at-home twin has aged more than her traveling sister. (This outcome can be predicted by extending Sample Problems 1 and 2 to include return of the traveler to the point of origin.) Upon being reunited, the "identical twins" are no longer identical. Very strange! But (almost) no one who has studied relativity doubts the difference in age, and experiments with fast-moving particles verify it.

Twin Paradox predicts the motion of a free particle.

Which twin has the motion we can call *natural*? Isaac Newton has a definition of natural motion. He would say, "A twin at rest tends to remain at

Being at rest is one *natural* motion.

SAMPLE PROBLEM 3 How Slow Is "Speeding"?

A. Answer "yes" or "no" to questions (a) through (e):

Is the stay-at home twin older when they get together again if the traveling twin

(a) streaks to the Andromeda galaxy (2 million light-years distant) and back?

(b) soars to Alpha Centauri (4 light-years distant) and back?

(c) flies to the planet Pluto and back?

(d) hurries to Earth's Moon and back?

(e) strolls next door to the neighbor's house and back?

B. In case (e) of part A, what is the approximate difference in aging between the twins if the traveling twin strolls at 1 meter per second and the next door neighbor's house is 100 meters away?

SOLUTION

A. In principle, one should reply "yes"—the stay-at-home twin will be older—for *all* cases in part A. Part B examines the actual value of the aging difference for small relative velocity.

B. Solve equation [1] for s^2 and apply it to the outward trip from the twins' house to the neighbor's house. The word *approximately* in the statement of the problem gives us permission to make assumptions.

Usually we do not notice results of the Twin Paradox in our everyday lives, so it seems reasonable to assume that the frame time t is very nearly the same as the wristwatch time τ for the stroll next door. This allows us to set $(t + \tau) \approx 2t$ in the following steps. We also set $t = s/v$ in one of the steps.

$$s^2 = t^2 - \tau^2 = (t+\tau)(t-\tau)$$
$$s^2 \approx 2t(t-\tau) = 2\frac{s}{v}(t-\tau) \qquad [9]$$

Equate the first and the last of the expressions in the last line of [9] and multiply through by $v/(2s)$ to obtain

$$t - \tau \approx \frac{sv}{2} \qquad [10]$$

We need to express the velocity v as a fraction of the speed of light. A speed of 1 meter per second is equal to

$$v = \frac{1 \text{ meter/second}}{c} = \frac{1 \text{ meter/second}}{3 \times 10^8 \text{ meter/second}} = 3.3 \times 10^{-9} \qquad [11]$$

Substitute this value of v into equation [10] to yield the time difference for one leg of the round trip:

$$t - \tau \approx \frac{100 \text{ meters} \times 3.3 \times 10^{-9}}{2} = \frac{3.3 \times 10^{-7}}{2} \text{ meters of light-travel time} \qquad [12]$$

The round trip difference will be twice this value, or 3.3×10^{-7} meters of light-travel time. Divide the result by the speed of light to obtain the time difference in seconds:

$$\begin{pmatrix}\text{time difference}\\ \text{for round trip}\end{pmatrix} \approx \frac{3.3 \times 10^{-7} \text{ meter}}{3 \times 10^8 \text{ meter/second}} = 1.1 \times 10^{-15} \text{ second} \qquad [13]$$

(This result justifies our assumption that the two times t and τ are very nearly equal.) So after her stroll next door and back, the traveling twin will be approximately 10^{-15} seconds younger than her stay-at-home sister. To measure this tiny time difference exceeds the sensitivity of even the most accurate atomic clock. That is why we do not notice relativistic effects in our everyday lives! Nevertheless, Nature witnesses the difference by selecting the stay at home twin as the one whose motion (or whose *lack* of motion in this frame) is *natural*.

rest." So it is the stay-at-home twin who moves in the natural way. In contrast, the out-and-back twin suffers the forces required to change her state of motion—from outgoing motion to incoming motion—so that the two sisters can meet again in person. The motion of the traveling twin is forced, *not natural*.

Moving uniformly is another *natural* motion.

Viewed from a second relatively moving free-float frame, the stay-at-home twin moves with constant speed in a straight line. Hers is also *natural* motion. Newton would say, "A twin in motion tends to continue this motion at constant speed in a straight line." So the motion of the stay-on-

Earth twin is also natural from the viewpoint of a second frame in uniform relative motion—or from any frame moving uniformly with respect to the original frame. In *any* such frame, the time lapse on the wristwatch of the stay-at-home twin can be calculated from the metric (equation [1]).

Natural motion in general: Extremal wristwatch time

The lesson of the Twin Paradox is that the natural motion of a free object between two events in flat spacetime is the one for which the wristwatch worn by the object has a maximum time reading between those two events. Purists insist that we say not *maximum* reading but rather *extremal* reading: either maximum or minimum. This book contains only examples of maximum wristwatch time for natural motion. Still, let's try to keep the purists happy! Replace the two words *maximum* and *minimum* with the single word *extremal*. The result is the **Principle of Extremal Aging**.

> ***Principle of Extremal Aging:*** *The path a free object takes between two events in spacetime is the path for which the time lapse between these events, recorded on the object's wristwatch, is an extremum.*

Principle of Extremal Aging: works for general relativity too.

It turns out that the Principle of Extremal Aging describes motion even when spacetime is not flat. The Principle of Extremal Aging accompanies us into curved spacetime, into the realm of general relativity. But for now we stay in flat spacetime and use the Principle of Extremal Aging to derive relativistic expressions for energy and momentum.

5 Energy in Special Relativity

The Principle of Extremal Aging tells us the energy of a free particle.

Derive energy from the metric plus the Principle of Extremal Aging.

Combining the metric (Section 2) with the Principle of Extremal Aging (Section 4) leads to the relativistic expression for energy in flat spacetime—the formula for energy used in special relativity. Here is the plan in outline: A free stone following its natural path carries a wristwatch that emits three flashes. We consider all three flashes to be fixed in space and the emission times for the first and last flashes also to be fixed. We then adjust the time of the middle flash so that the *wristwatch time* from the first flash to the last flash is an extremum. The outcome is the expression for a quantity that is the same along every segment of the path—this quantity is *conserved*. We identify the conserved quantity as the energy. Now fill in some details.

Three flashes: When will the middle flash occur?

Think of a free stone flying along a straight line in space as observed in an inertial frame (Figures 2 and 3). The stone emits three flashes #1, #2, and #3 bracketing two adjacent segments of its trajectory, segments labeled A and B in the figures. These two segments need not be the same length. Fix the *positions* of all three flash emissions in space, fix also the *times* for flash emissions #1 and #3, then ask: At what time t will the free stone pass location #2 and emit the second flash? Find this intermediate time t by demanding that the total wristwatch time from #1 to #3 be an extremum. In other words, use the Principle of Extremal Aging to find the time for the middle flash. The result leads to a conserved quantity, the energy of the stone.

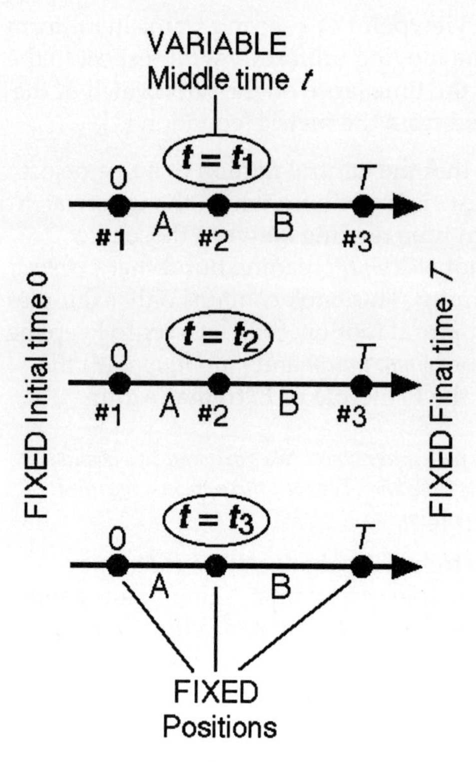

Figure 2 *Three alternative cases of a stone moving along a straight line in space as it emits three flashes, #1, #2, and #3. The space locations of emissions are the same in all three cases, as are the times of first and last emissions #1 and #3. But emission time for the middle flash #2 is different for the three cases. We ask: At what time will a free stone following a natural path pass the intermediate point and emit flash #2? We answer this question by demanding that the total wristwatch time τ from first to last flash emissions be an extremum. From this requirement comes an expression for the energy of the stone as a constant of the motion.*

Now for the full step-by-step derivation.

1. Let t be the frame time between flash #1 and flash #2 and let s be the frame distance between these two flashes. Then the metric [1] tells us that the wristwatch time τ_A along segment A is

$$\tau_A = (t^2 - s^2)^{1/2} \qquad [14]$$

   To prepare for the derivative that leads to extremal aging, differentiate this expression with respect to the intermediate time t:

$$\frac{d\tau_A}{dt} = \frac{t}{(t^2 - s^2)^{1/2}} = \frac{t}{\tau_A} \qquad [15]$$

2. Next, let T be the fixed time between flashes #1 and #3 and S be the fixed distance between them. Then the frame time between flash #2 and flash #3 is $(T - t)$ and the frame distance between them is $(S - s)$. Therefore the wristwatch time τ_B along segment B is

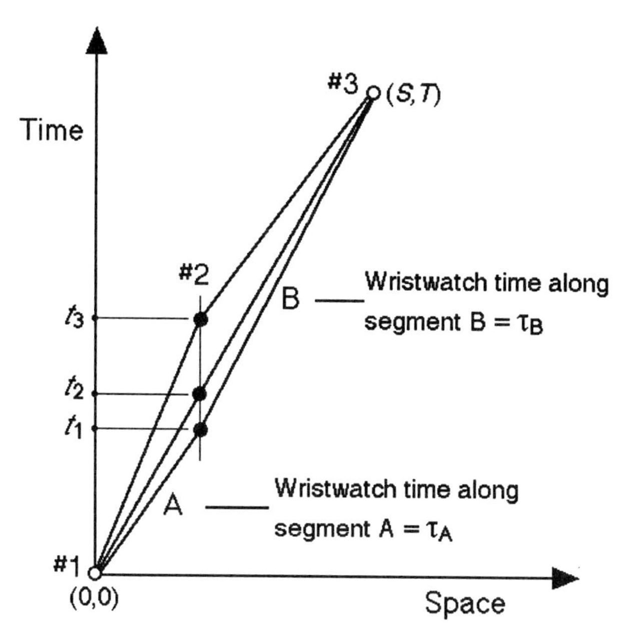

Figure 3 *Three alternative cases of a stone moving along a straight line in space as it emits three flashes, #1, #2, and #3. These are the same three cases shown in Figure 2, but here we plot the stone's path in space* and time. *Such a spacetime plot is called a* **worldline.** *On each of three alternative worldlines, flash emissions #1 and #3 are fixed in space and time. Flash emission #2 is fixed in space (horizontal direction in figure) but its time is varied (up and down in the figure) to find an extremum of the total wristwatch time $\tau = \tau_A + \tau_B$ from #1 to #3. The result is an expression for a quantity that is a constant of the motion: the energy of the stone.*

$$\tau_B = [(T-t)^2 - (S-s)^2]^{1/2} \qquad [16]$$

Again, to prepare for the derivative that leads to extremal aging, differentiate this expression with respect to the intermediate time t:

$$\frac{d\tau_B}{dt} = \frac{-(T-t)}{[(T-t)^2 - (S-s)^2]^{1/2}} = -\frac{T-t}{\tau_B} \qquad [17]$$

3. The total wristwatch time τ from event #1 to event #3 is the sum of the wristwatch time τ_A between events #1 and #2 plus the wristwatch time τ_B between events #2 and #3:

$$\tau = \tau_A + \tau_B \qquad [18]$$

4. Now ask: *When*—at what frame time t—will the a stone, following its natural path, pass the intermediate point and emit the second flash #2? Answer with the Principle of Extremal Aging: Time t will be such that the aging (τ in equation [18]) is an extremum. To find this extremum set the derivative of τ with respect to t equal to zero. Take the derivative of both sides of [18] and substitute from equations [15] and [17]:

Use Principle of Extremal Aging to find the time for the middle flash.

$$\frac{d\tau}{dt} = \frac{d\tau_A}{dt} + \frac{d\tau_B}{dt} = \frac{t}{\tau_A} - \frac{T-t}{\tau_B} = 0 \qquad [19]$$

5. The last equality in equation [19] leads to the equation

$$\frac{t}{\tau_A} = \frac{T - t}{\tau_B} \qquad [20]$$

Quantity whose value is the same for adjacent segments

6. In expression [20] the frame time t is the time for the particle to traverse segment A. Call this time t_A. The time $(T - t)$ is the frame time for the particle to traverse segment B. Call this time t_B. Then equation [20] can be rewritten in the simple form

$$\frac{t_A}{\tau_A} = \frac{t_B}{\tau_B} \qquad [21]$$

7. The locations of segments A and B were chosen arbitrarily along the straight path in space of the particle moving in a region of flat spacetime. Equation [21] holds for *all* pairs of adjacent segments placed *anywhere* along the path. We did not specify where segment A was to begin. Nothing stops us from beginning the analysis with the second segment B and adding to it a third segment C with which to compare it (which may have a different length than either of the first two segments). Then equation [21] applies to the second and third segments. But if the value of the expression is the same for the first and second segments and also the same for the second and third segments, then it must be the same for the first and third segments. Continuing in this way, envision a whole series of adjacent segments, labeled A, B, C, D, . . . , for each of which equation [21] applies, leading to the set of equations

$$\frac{t_A}{\tau_A} = \frac{t_B}{\tau_B} = \frac{t_C}{\tau_C} = \frac{t_D}{\tau_D} = \ldots \qquad [22]$$

In brief, here is a quantity that is a constant of the motion for the free particle—a quantity that has the same value along any segment of the natural path of a free particle moving in flat spacetime. Then equation [22] tells us that

$$\frac{t}{\tau} = \text{a constant of the motion} \qquad [23]$$

$E/m = t/\tau$ is a constant of the motion.

What is this quantity? It is related to the relativistic expression for the total energy of the particle. If we have already studied special relativity, we know that

$$\frac{t}{\tau} = \frac{t}{[t^2 - s^2]^{1/2}} = \frac{t}{t\left[1 - \left(\frac{s}{t}\right)^2\right]^{1/2}} = \frac{1}{(1 - v^2)^{1/2}} = \frac{E}{m} \qquad [24]$$

where m is the mass of the particle. Equation [24] gives the energy per unit mass of a particle that moves with constant speed.

OBJECTION: Baloney! Everyone knows that a free particle moves with constant speed along a straight path in space as observed in a free-float frame. So as this motion proceeds, every possible expression that depends only on $v = s/t$ is also a constant of the

motion, for example the expression v^{12}, which is certainly not the correct expression for energy! Your derivation proves nothing!

RESPONSE: You are *almost* right. Any function of velocity $v = s/t$ is indeed constant for the special case of a free particle in flat spacetime. And if v is constant, so is t/τ, as witnessed by Equation [24]. But notice the priorities used in the derivation: The Principle of Extremal Aging has highest priority; the expression for energy comes out of this principle. Of all the quantities that remain constant because v is constant, the Principle of Extremal Aging picks out $t/\tau = E/m$ as primary. (The following section shows that a similar analysis picks out the relativistic expression for momentum as a constant of the motion.) Chapter 3 contains a new and more general expression for energy in curved spacetime. In that case the velocity is *not* constant—yet that more general expression for energy is correct and a constant of the motion nevertheless. Our derivation of the expression for E/m in flat spacetime is thus a trial run for the derivation of the energy of a particle in the curved spacetime around a center of gravitational attraction.

If the particle changes speed, then it changes energy. In that case it makes sense to talk about *instantaneous speed* and to use calculus notation. Let the pair of flash emissions in Figure 1 be separated by the incremental frame coordinates dt, ds, and incremental wristwatch time $d\tau$. The equation for E/m then becomes

Particle energy in special relativity

$$\frac{E}{m} = \frac{dt}{d\tau} \qquad [25]$$

Ordinarily we use the ratio E/m in equations, instead of E alone. Why? Because it emphasizes two important principles: (1) Only spacetime relations between events appear on one side of equations such as [24] and [25], reminding us that it is *spacetime geometry* that leads to these expressions, not some weird property of matter. (2) The ratio E/m has no units. Therefore, whoever uses these equations has total freedom in choosing the unit of E and m, as long as it is the *same* unit. The same unit in the numerator and denominator of [25] may be kilograms or the mass of the proton or million electron-volts or the mass of Sun. If you insist on using conventional units, such as joules for energy E and kilograms for mass m, then a conversion factor c^2 intrudes into our simple equation:

$$\frac{E_{\text{joules}}}{m_{\text{kg}}c^2} = \frac{dt}{d\tau} \qquad [26]$$

Rest energy: famous formula

Now view the particle from a reference frame in which the particle is at rest. In this rest frame there is zero distance s between sequential flash emissions. Equation [1] says that for $s = 0$ the frame time t and wristwatch time τ have exactly the same value. For a particle at rest, then, equation [26] reduces to the most famous equation in all of physics:

$$E_{\text{joules rest}} = m_{\text{kg}}c^2 \qquad [27]$$

Note that equation [27] describes the *rest energy* of a particle. For a particle in motion, the energy is given by equation [26].

In equation [27], c has the *defined* value 2.99792458×10^8 meters/second. An equation of the same form is correct if E is measured in ergs, m in grams, and c in centimeters/second.

Fuller Explanations: Energy in flat spacetime: *Spacetime Physics*, Chapter 7, Momenergy.

6 Momentum in Special Relativity

The metric plus the Principle of Extremal Aging give us an expression for momentum.

The relativistic expression for momentum is derived by a procedure analogous to the one used to derive the relativistic expression for energy. The figures look similar to Figures 2 and 3, but in this case the *time t* for the intermediate flash emission is *fixed*, while the *position s* for this event is *varied* right and left to yield an extremum for the total wristwatch time from the first flash to the third flash. (You carry out the derivation of momentum in the exercises at the end of this chapter.) The result is a second constant of the motion for a free particle:

$$\frac{s}{\tau} = \frac{s}{[t^2 - s^2]^{1/2}} = \frac{s/t}{[1 - (s/t)^2]^{1/2}} = \frac{v}{(1 - v^2)^{1/2}} = \frac{p}{m} \qquad [28]$$

Equation [28] gives the momentum per unit mass for a particle moving with constant speed. If the particle changes speed, then once again we use calculus notation:

Particle momentum in special relativity

$$\frac{p}{m} = \frac{ds}{d\tau} \qquad [29]$$

Equation [29] has the same form as in Newton's nonrelativistic mechanics, except here the incremental wristwatch time $d\tau$ replaces the Newtonian lapse dt of "universal time."

Fuller Explanations: Momentum in flat spacetime: *Spacetime Physics*, Chapter 7, Momenergy.

7 Mass in Relativity

Everyone agrees on the value of the mass m of the stone.

Find mass from energy and momentum.

An important relation among mass, energy, and momentum follows from the metric and our new expressions for energy and momentum. Suppose a moving stone emits two flashes very close together in space ds and in time dt. Then equation [1] gives the increase of wristwatch time $d\tau$:

$$(d\tau)^2 = (dt)^2 - (ds)^2 \qquad [30]$$

Divide through by $d\tau^2$ and multiply through by m^2 to obtain

$$m^2 = m^2\left(\frac{dt}{d\tau}\right)^2 - m^2\left(\frac{ds}{d\tau}\right)^2 = \left(m\frac{dt}{d\tau}\right)^2 - \left(m\frac{ds}{d\tau}\right)^2 \qquad [31]$$

or, substituting expressions [25] and [29] for energy and momentum,

$$m^2 = E^2 - p^2 \qquad [32]$$

Energy (also momentum) may be different for different observers . . .

In equation [32], mass, energy, and momentum are all expressed in the same units, such as kilograms or electron-volts. In conventional units, the equation has a more complicated form:

$$(mc^2)^2 = E_{\text{conv}}^2 - p_{\text{conv}}^2 c^2 \qquad [33]$$

where the subscript "conv" means "conventional units."

. . . but mass is an invariant, the same for every observer.

Equations [32] and [33] are central expressions in special relativity. The particle energy E will typically have a different value when measured in different frames that are in uniform relative motion. Also the particle momentum p will typically have a different value when measured in different frames that are in uniform relative motion. However, the values of these two quantities in *any* given free-float frame can be used to determine the value of the particle mass m, which is independent of the reference frame. Particle mass m is an *invariant*, independent of reference frame, just as the time $d\tau$ recorded on the wristwatch between ticks in equation [1] is an invariant, independent of the reference frame.

The mass m of key, car, or coffee cup defined in equation [32] is the one we use throughout our study of both special and general relativity. Such a **test particle** responds to the structure of spacetime in its vicinity but has small enough mass not to affect this spacetime structure. (In contrast, the large mass M of a planet, star, or black hole does affect spacetime in its vicinity.) Wherever we are, we can always climb onto a local free-float frame (Section 8) and apply special-relativity expression [32] or some other standard method to measure the mass m of our test particle.

Fuller Explanations: Mass and momentum-energy in flat spacetime: *Spacetime Physics*, Chapter 7, Momenergy.

No Mass Change with Velocity!

The fact that no object moves faster than the speed of light is sometimes "explained" by saying that "the mass of a particle increases with speed." This interpretation can be applied consistently, but what could it mean in practice? Someone riding along with a faster-moving stone detects no change in the number of atoms in the stone, nor any change whatever in the individual atoms, nor in the binding energy between atoms. Our viewpoint in this book is that mass is an *invariant*, the same for all free-float observers when they use equations [32] or [33] to reckon the mass. In relativity, invariants are diamonds. Do not throw away diamonds! For more on this subject, see *Spacetime Physics*, **Dialog: Use and Abuse of the Concept of Mass,** pages 246–251.

8 The Free-Float Frame Is Local

In practice there are limits on the space and time extent of the free-float (inertial) frame.

The free-float (inertial) frame is the arena in which special relativity describes Nature. The power of special relativity applies strictly only in a frame—or in each one of a collection of overlapping frames in uniform relative motion—in which a free particle released from rest stays at rest and a particle launched with a given velocity maintains the magnitude and direction of that velocity.

Limits of local free-float frames imply the need for general relativity.

If it were possible to embrace the Universe with a single free-float (inertial) frame, then special relativity would describe that universe and general relativity would not be needed. But general relativity *is* needed precisely because typically inertial frames are inertial in only a limited region of space and time. Inertial frames are **local**. The free-float frame can be realized, for example, inside various "containers," such as (1) an unpowered spaceship in orbit around Earth or Sun or (2) an elevator whose cables have been cut or (3) an unpowered spaceship in interstellar space. Riding in these free-float frames for a short time, we find no evidence of gravity.

Free-float frame cannot be too large.

Well, *almost* no evidence. The enclosure in which we ride cannot be too large or fall for too long a time without some unavoidable changes in relative motion being detected between particles in the enclosure. Why? Because widely separated test particles within a large enclosed space are differently affected by the nonuniform gravitational field of Earth—to use the Newtonian way of speaking. For example, two particles released side by side are both attracted toward the center of Earth, so they move closer together as measured inside a falling long narrow horizontal railway coach (Figure 4, left). Moving toward one another has nothing to do with gravitational attraction between these test particles, which is entirely negligible.

As another example, think of two test particles released far apart vertically but one directly above the another in a long narrow vertical falling railway coach (Figure 4, right). For vertical separation, their gravitational accelerations toward Earth are in the same direction, according to the Newtonian analysis. However, the particle nearer Earth is more strongly attracted to Earth and gradually leaves the other behind: the two particles move far-

Elevator Safety

Could the cables snap and send an elevator plummeting down the shaft?

This is every rider's worst fear, but experts say there's no need to worry. You're being supported by four to eight cables, each of which could support the weight of the car by itself. In fact, the only time an elevator has been known to go into freefall—with all of its cables cut—was during World War II, when an American bomber accidentally hit the Empire State Building [in New York City]. The plane's crew died, but the lone elevator passenger survived.

—*Good Housekeeping Magazine*, February 1998, page 142.

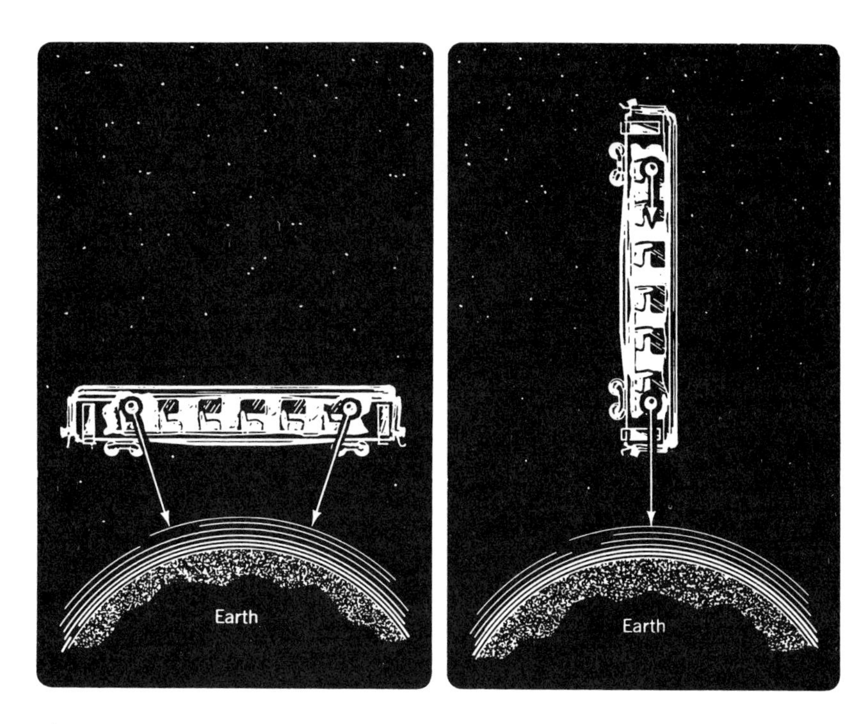

Figure 4 *Einstein's old-fashioned railway coach in free fall. Left: horizontal orientation. Right: vertical orientation.*

ther apart as observed inside the falling coach. Conclusion: The large enclosure is not a free-float frame.

A rider in either railway car shown in Figure 4 sees the pair of test particles *accelerate* toward one another or away from one another. These relative motions earn the name **tidal accelerations,** because they arise from the same kind of nonuniform gravitational field—this time the field of Moon—that account for ocean tides on Earth.

Now, we want the laws of motion to look simple in our free-float frame. Therefore we want to eliminate all relative accelerations produced by external causes. "Eliminate" means to reduce them below the limit of detection so that they do not affect measurements of, say, the velocity of a particle in an experiment. We eliminate the problem by choosing a room that is sufficiently small. Smaller room? Smaller relative motions of objects at different points in the room!

Reduce space or time extension to preserve free-float frame.

Let someone have instruments for detection of relative motion with any given degree of sensitivity. No matter how fine that sensitivity, the room can always be made so small that these perturbing relative motions are too small to be detectable in the time required for the experiment. Or, instead of making the room smaller, shorten the time duration of the experiment to make the perturbing motions undetectable. For example, very fast particles emitted by a high-energy accelerator on Earth traverse the few-meter span of a typical experiment in so short a time that their deflection in

Earth's gravitational field is negligible. The result: The frame of the laboratory at rest on Earth's surface is effectively free-float for purposes of analyzing these experiments.

Both space and time enter into the specification of the limiting dimensions of a free-float frame. Therefore—for a given sensitivity of the measuring devices—a reference frame is free-float only within a limited region of *spacetime.*

Test for free-float property within the frame itself.

An observer tests for a free-float frame by releasing particles from rest throughout the space and noting whether they remain effectively at rest during the time set aside for our particular experiment. Wonder of wonders! Testing for free float can be carried out entirely within the frame itself. The observer need not look out of the room or refer to any measurements made external to the room. A free-float frame is "local" in the sense that it is limited in space and time—and also "local" in the sense that its free-float character can be determined from within, locally.

One way to get rid of "gravitational force" is to jump from a high place toward a trampoline below. That is to say, a locally free-float frame is always available to us. But no contortion or gyration whatsoever will eliminate the *relative* accelerations of test particles that indicate the limits of the free-float frame. These relative accelerations are the central indicators of the *curvature of spacetime*. They stand as warning signs that we are reaching the limits of special relativity.

General relativity requires more than one free-float frame.

How can we analyze a pair of events widely separated near Earth, near Sun, or near a neutron star, events too far apart to be enclosed in a single free-float frame? For example, how do we describe the motion of an asteroid whose orbit completely encircles Sun, with an orbital period of many years? The asteroid passes through many free-float frames but cannot be tracked using a single free-float frame. Special relativity has reached its limit! To describe accurately motion that oversteps a single free-float frame, we must turn to general relativity—the Theory of Gravitation—as we do in Chapter 2.

Fuller explanations: *Spacetime Physics*, Chapter 2, Floating Free, and Chapter 9, Gravity: Curved Spacetime in Action.

9 The Observer

Ten thousand local witnesses

Detect each event locally, using a latticework of clocks.

How, in principle, do we record events in space and time? Nature puts an unbreakable speed limit on signals—the speed of light. This speed limit causes problems with the recording of widely separated events, because we do not *see* a remote event until long after it has occurred. To avoid the light-velocity delay, adopt the strategy of detecting each event using equipment located right next to that event. Spread event-detecting equipment over space as follows. Think of assembling metersticks and clocks into a cubical latticework similar to a playground jungle gym (Figure 5). At every intersection of the latticework fix a clock. These clocks are identical and measure time in meters of light-travel time.

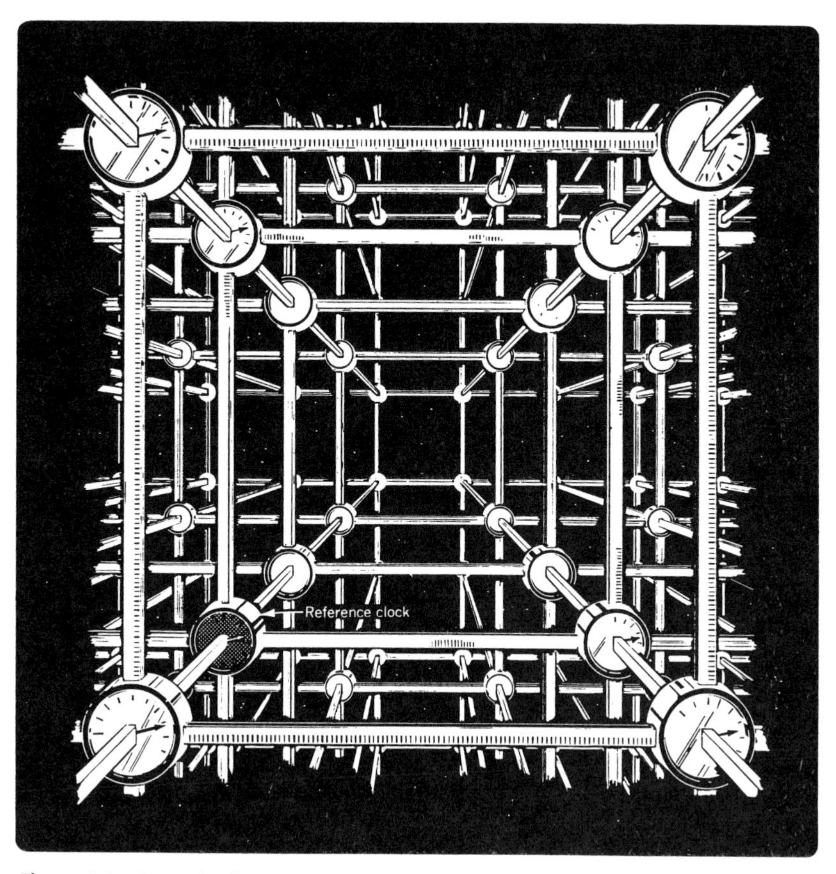

Figure 5 *Latticework of metersticks and clocks*

These clocks should read the *same time*. That is, the clocks need to be **synchronized** in this frame. There are many valid ways to synchronize clocks. Here is one: Pick one clock as the standard, the **reference clock**. At midnight the reference clock sends out a **synchronizing flash** of light in all directions. Prior to emission of the synchronizing flash, every other clock in the lattice has been stopped and set to a time (in meters) later than midnight equal to the straight-line distance (in meters) of that clock from the reference clock. Each clock is then started when it receives the reference flash. The clocks in the latticework are then said to be *synchronized.*

Synchronize clocks in the lattice.

Use the latticework of synchronized clocks to determine the location and time at which any given event occurs. The spatial position of the event is taken to be the location of the clock nearest the event and the time of the event is the time recorded on that clock. The location of this nearest clock is measured along three directions, northward, eastward, and upward from the reference clock. The spacetime location of an event then consists of four numbers, three numbers that specify the space position of the clock nearest the event and one number that specifies the time the event occurs as recorded by that clock.

Measuring the space and time location of an event

Specify the location of an event as the location of the clock nearest to it. With a latticework made of metersticks, the location of the event will be uncertain to some substantial fraction of a meter. For events that must be located with greater accuracy, a lattice spacing of 1 centimeter or 1 millimeter would be more appropriate. To track an Earth satellite, lattice spacing of 100 meters might be adequate.

The lattice clocks, when installed by a foresighted experimenter, will be **recording clocks**. Each clock is able to detect the occurrence of an event (collision, passage of light flash or particle). Each reads into its memory the nature of the event, the time of the event, and the location of the clock. The memory of all clocks can then be read out and analyzed later at some command center.

The "observer" is all the recording clocks in one frame.

In relativity we often speak about the **observer**. Where is this observer? At one place or all over the place? Answer: The word *observer* is a shorthand way of speaking about the whole collection of recording clocks associated with one free-float frame. This is the sophisticated sense in which we hereafter use the phrase "the observer measures such-and such."

What happens to our latticework of clocks in the vicinity of Earth or Sun or neutron star or black hole? Suppose one of these centers of attraction is isolated in space and we stay far away from it. Then there is no problem in setting up an extensive latticework that starts far from the center and stretches even farther away in all directions. Such an extensive *far-away lattice* can represent a single valid free-float frame. And in studying general relativity we often speak of a **far-away observer**.

The far-away lattice is not free float when extended to near Earth or black hole.

But there are problems in extending the far-away latticework of clocks down toward the surface of any of these structures. A free particle released from rest near that center does not remain at rest with respect to the far-away lattice. A single free-float frame no longer provides a simple description of motion.

Many local frames are required near Earth or black hole.

To describe motion near a center of gravitational attraction we must give up the idea of a single global free-float frame, one that covers all space and time around Earth or black hole. Replace it with many local frames, each of which provides only a small part of the global description. A world atlas binds together many overlapping maps of Earth. Individual maps in the atlas can depict portions of Earth's surface small enough to be essentially flat. Taken together, the collection of maps bound together in the world atlas correctly describes the entire spherical surface of Earth, a task impossible using a single large flat map for the entire Earth. For spacetime near nonrotating Earth or black hole, the task of binding together individual localized free-float frames is carried out by the *Schwarzschild metric*, introduced in Chapter 2. The Schwarzschild metric frees us from limitation to a single free-float frame and introduces us to curved spacetime.

Fuller Explanations: *Spacetime Physics*, Chapter 2, Section 2.7, Observer.

10 Summary

The wristwatch time τ between two events, the time recorded on a watch that moves uniformly from one event to the other, is related to the separation s between the events and the time difference t between them as measured in a given frame. For space and time measured in the same units, this relation is given by the equation

$$\tau^2 = t^2 - s^2 \qquad [1]$$

The wristwatch time τ is an *invariant*, the same calculated by all observers, even though t and s may have different values, respectively, as measured in different reference frames. Equation [1] is an example of the *metric.*

Of all possible paths between an initial event and a final event, a free particle takes the path that makes the wristwatch time along the path an extremum. This is called the *Principle of Extremal Aging.*

From the metric and the Principle of Extremal Aging one can derive two quantities that are constants of the motion for a free particle. One constant of the motion is the energy per unit mass E/m:

$$\frac{E}{m} = \frac{dt}{d\tau} \qquad [25]$$

The second constant of the motion is the momentum per unit mass p/m:

$$\frac{p}{m} = \frac{ds}{d\tau} \qquad [29]$$

The spacetime arena for special relativity is the *free-float (inertial) frame,* one in which a free test particle at rest remains at rest and a free test particle in motion continues that motion unchanged. We call a region of spacetime *flat* if a free-float frame can be set up in it.

In principle one can set up a latticework of synchronized clocks in a free-float frame. The position and time of any event is then taken to be the location of the nearest lattice clock and the time of the event recorded on that clock. The *observer* is the collection of all such recording clocks in a given reference frame.

Most regions of spacetime are flat over only a limited range of space and time. Evidence that a frame is not inertial (so that its region of spacetime is not flat) is the relative acceleration ("tidal acceleration") of a pair of free test particles with respect to one another. If tidal accelerations affect an experiment in a region of space and time, then we say that spacetime region is *curved,* and special relativity cannot validly be used to describe this experiment. In that case we must use *general relativity, the theory of gravitation,* which correctly describes the relations among events spread over regions of space and time too large for special relativity.

Note on terminology: In this book we use the convention recommended by the International Astrophysical Union that names for objects in the solar system be capitalized and used without the article. For example, we say "orbits around Sun" or "the mass of Moon." This provides a consistent convention; one would not say "orbits around *the* Mars." We also capitalize the words *Nature* and *Universe* out of respect for our cosmic home.

11 Readings in Special Relativity

Spacetime Physics, Introduction to Special Relativity, Second Edition, Edwin F. Taylor and John Archibald Wheeler, W. H. Freeman and Co., New York, 1992, ISBN 0-7167-2327-1. Our own book, to which reference is made at the end of several sections in Chapter 1 and elsewhere in the present book.

Special Relativity, A. P. French, W. W. Norton & Co., New York, 1968, Library of Congress 68-12180. An introduction carefully based on experiment and observation.

A Traveler's Guide to Spacetime, An Introduction to the Special Theory of Relativity, Thomas A. Moore, McGraw-Hill, Inc., News York, 1995, ISBN 0-07-043027-6. A concise treatment by a master teacher.

Flat and Curved Space-Times by George F. R. Ellis and Ruth M. Williams, Clarendon Press, Oxford, 1988, ISBN 0-19-851169-8. A leisurely, informative, and highly visual trip through special relativity is followed by treatment of curved spacetime. See more on this book in the section Readings in General Relativity at the end of the present book.

Space and Time in Special Relativity, N. David Mermin, Waveland Press, Inc., Prospect Heights, IL, 1989, ISBN 0-8813-420-0. Rigorous and mildly eccentric.

Understanding Relativity: A Simplified Approach to Einstein's Theories, Leo Sartori, University of California Press, Berkeley, 1996, ISBN 0-520-20029-2. Thoughtful and complete.

Relativity, The Special and General Theory, Albert Einstein, Crown Publishers, New York, 1961, ISBN 0-517-025302. A popular treatment by the Old Master himself. Published originally in 1916. Enjoyable for the depth of physics, the humane viewpoint, and the charm of old-fashioned trains racing past embankments.

Relativity Visualized, Lewis Carroll Epstein, Insight Press, San Francisco, 1997, ISBN 0-953218-05-X. An enjoyable and eccentric presentation of special and general relativity, done primarily with figures and graphics. Available in some bookstores, or send $19.95 plus $2 handling to Insight Press, 614 Vermont Street, San Francisco, CA 94107-2636, USA.

Of historical interest

Relativity and Its Roots, Banesh Hoffmann, Scientific American Books, New York, 1983, ISBN 0-7167-1510-4. History of the subject by one of Einstein's collaborators.

The Principle of Relativity, A. Einstein, H. A. Lorentz, H. Weyl, H. Minkowski, Dover Publications, Inc., New York, 1952, Standard Book Number 486-60081-5. Translations of many of the original papers. See the following reference for a more recent translation of Einstein's special relativity paper.

Albert Einstein's Special Theory of Relativity: Emergence (1905) and Early Interpretation (1905–1911), Arthur I. Miller, Addison-Wesley Publishing Co., Inc., 1981, ISBN 0-201-04680-6. Careful historical analysis of Einstein's original special relativity paper "On the Electrodynamics of Moving Bodies," the setting in which it was produced, and early consequences for the scientific community. Includes a modern, corrected translation of the paper itself.

12 Reference

Initial quote: Personal memoir of William Miller, an editor of *Life* magazine, quoted in the issue of May 2, 1955. See *The Quotable Einstein,* edited by Alice Calaprice, Princeton University Press, 1996, page 199.

Chapter 1 Exercises

1. Spatial Separation I

Two firecrackers explode at the same place in the laboratory and are separated by a time of 3 years as measured on a laboratory clock.

A. What is the spatial distance between these two events in a rocket in which the events are separated in time by 5 years as measured on rocket clocks?

B. What is the relative speed of the rocket and laboratory frames?

2. Spatial Separation II

Two firecrackers explode in a laboratory with a time difference of 4 years and a space separation of 5 light-years, both space and time measured with equipment at rest in the laboratory. What is the distance between these two events in a rocket in which they occur at the same time?

3. Super Cosmic Rays

The Akeno Giant Air Shower Array of detectors spread over 100 square kilometers in Japan detects the energy of individual cosmic ray particles indirectly by the resulting shower of particles these cosmic rays create in the atmosphere. This array has detected a few cosmic ray particles with an energy as high as 10^{20} electron volts.

A. A regulation tennis ball has a mass of 57 grams. If this tennis ball is given a kinetic energy of 10^{20} electron volts, how fast will it move, in meters per second? (*Hint:* Try Newtonian mechanics first.)

B. Research workers find no upper limit on cosmic ray energies. The proposed Pierre Auger Cosmic Ray Observatory will consist of detectors spread over 3000 square kilometers at each of two sites: Utah USA in the northern hemisphere and Argentina in the southern hemisphere. Suppose the new arrays detect a cosmic ray proton of energy 10^{21} electron-volts, ten times more energetic than those so far observed. How long would it take this proton to cross our galaxy (take the galaxy diameter to be 10^5 light-years) as measured on the proton's wristwatch? Give your answer in seconds. (The answer is not zero!)

4. Mass-Energy Conversion

A. How much mass does a 100-watt bulb dissipate (in heat and light) in one year?

B. Pedaling a bicycle at full throttle, you generate 1/2 horsepower of *useful* power (1 horsepower = 746 watts). The human body is about 25 percent efficient; that is, 25 percent of the food burned can be converted to useful work. How long a time will you have to ride your bicycle in order to lose 1 kilogram by conversion of mass to energy? Express your answer in years. (Conversion factor inside back cover.) How can reducing gymnasiums stay in business?

C. One kilogram of hydrogen combines chemically with 8 kilograms of oxygen to form water; about 10^8 joules of energy is released. A very good chemical balance is able to detect a fractional change in mass of 1 part in 10^8. By what factor is this sensitivity more than enough—or insufficient—to detect the fractional change of mass in this reaction?

5. Units and Conversions

A. Show that for a particle of nonzero rest mass, the speed (as a fraction of the speed of light) is given by the expression

$$v = \frac{ds}{dt} = \frac{p}{E}$$

B. What value does the speed v derived in part A take when the mass of the particle is zero, as is the case for a flash of light? Is this result the one you expect?

C. The mass and energy of particles in beams from accelerators is often expressed in GeV, that is billions of electron-volts. Journal articles describing these experiments refer to particle momentum in units of GeV/c. Explain.

6. The Pressure of Light

A flash of light has zero mass. Use equation [33], in conventional units, to answer the following questions.

A. You can feel on your hand an object with the weight of 1 gram mass. You detect an equal force on a black piece of wood that you hold in your hand as it absorbs a steady laser beam. What power does the laser beam deliver, in watts?

B. The block of wood described in part A absorbs the energy of the laser beam. Will the block burst into flame?

7. Derivation of the Expression for Momentum

A. Carry out the derivation of the relativistic expression for momentum described in Section 6. Draw figures for this case similar to Figures 2 and 3. *Hint:* Follow steps 1 through 7 on pages 1-7 through 1-11, but take derivatives with respect to s rather than with respect to t.

B. Write an expression for p in conventional units, similar to equation [26] for energy.

Chapter 2 Curving

- *What does "curvature of spacetime"* **mean**?
- *How can I observe this curvature?*
- *Does light "get tired" as it moves away from Earth?*
- *General relativity describes only tiny effects, right?*

CHAPTER 2

Curving

It is not my purpose in this discussion to represent the general theory of relativity as a system that is as simple and as logical as possible, and with the minimum number of axioms; but my main object here is to develop this theory in such a way that the reader will feel that the path we have entered upon is psychologically the natural one, and that the underlying assumptions will seem to have the highest possible degree of security.

—Albert Einstein

1 "Distances" Determine Geometry

Describe an object with a table of distances between points.
Describe spacetime with a table of intervals between events.

Nothing is more distressing on first contact with the idea of curved spacetime than the fear that every simple means of measurement has lost its power in this unfamiliar context. One thinks of oneself as confronted with the task of measuring the shape of a gigantic and fantastically sculptured iceberg as one stands with a meterstick in a tossing rowboat on the surface of a heaving ocean.

Reproduce a shape using nails and string.

Were it the rowboat itself whose shape were to be measured, the procedure would be simple enough (Figure 1). Draw it up on shore, turn it upside down, and lightly drive in nails at strategic points here and there on the surface. The measurement of distances from nail to nail would record and reveal the shape of the surface. Using only the table of these distances between each nail and other nearby nails, someone else can reconstruct the shape of the rowboat. The precision of reproduction can be made arbitrarily great by making the number of nails arbitrarily large.

It takes more daring to think of driving into the towering iceberg a large number of pitons, the spikes used for rope climbing on ice. Yet here too the geometry of the iceberg is described—and its shape made reproducible—by measuring the distance between each piton and its neighbors.

The event is a nail driven into spacetime.

But with all the daring in the world, how is one to drive a nail into spacetime to mark a point? Happily, Nature provides its own way to localize a point in spacetime, as Einstein was the first to emphasize. Characterize the point by what happens there: firecracker, spark, or collision! Give a point in spacetime the name *event*.

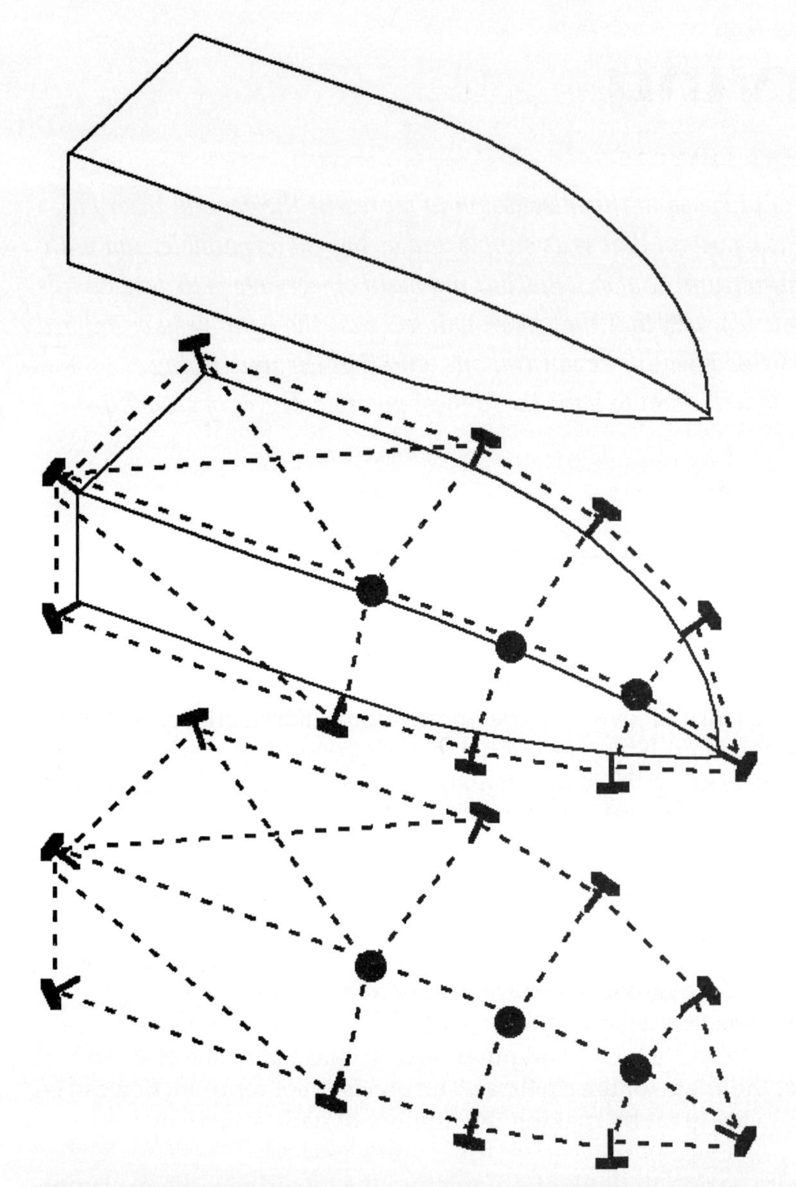

Figure 1 *Reproducing the shape of an overturned rowboat (top) by driving nails around its perimeter, then stretching strings between each nail and every nearby nail (middle). The shape of the rowboat can be reconstructed (bottom) using only the lengths of string segments—the distances between nails. To increase the precision of reproduction, increase the number of nails, the number of string segments, the table of distances.*

Interval: Separation between events in spacetime

Events are the nails, the pitons, the steel surveying stakes of spacetime. How can events describe the geometry of spacetime? Measure the "distance" between each event and every one of its neighboring events. We already know that for spacetime the "distance" between each pair of events means the *spacetime interval* between them (Chapter 1). The table of *distances* between points in space becomes a table of *intervals* between events in spacetime.

The table of distances between points allows us to describe and reproduce the spatial geometry of a surface—whether plane or curved. The table of spacetime intervals between events allows us to describe and reproduce the "shape," the geometry of spacetime—whether the flat spacetime geometry described by special relativity or the curved spacetime geometry described by general relativity.

Events and intervals reproduce "shape" of spacetime.

2 Reference Frames Are Secondary

Lab and rocket frames give different viewpoints on flat spacetime.
Different reference frames give different viewpoints on curved spacetime.

Events themselves are the nails on which science hangs. Spacetime intervals between events evidence the geometry of spacetime, its curvature. This geometry, this curvature from point to point, exists whether one or another competing reference frame is used to describe it. Spacetime geometry exists—and can be described—even when no frame of reference is used at all!

Curvature exists with or without reference frames.

We may—and will—choose to use several different reference frames to describe the same events near a gravitating star or planet. One frame spans the interior of an unpowered spaceship orbiting a spherically symmetric center of attraction. Another occupies the inside of a second unpowered spaceship plunging radially toward that center. A third reference frame consists of the inside of a powered spaceship, rockets blasting, that stands at rest outside the same heavenly body. (Or save rocket power by constructing and standing on a stationary spherical shell concentric to the star or planet.) There are many other possible frames. A central idea of general relativity is that reference frames are not fundamental—all are equally valid. People who use general relativity as a tool change reference frames more often than they change clothes. Each different frame illuminates some features of curved spacetime geometry, but rarely does any single reference frame reveal every important feature of that geometry.

Special relativity uses laboratory and rocket frames as different vantage points to get an insight into flat spacetime that exists independent of any reference frame. In the same way we use alternative reference systems around a star to get insight into curved spacetime—a curved geometry that exists independent of any frame of reference. By using different frames for different purposes, we glimpse the spacetime geometry that lies behind all frames of reference.

Different frames offer different "vantage points" to study spacetime.

You keep talking about "curvature" of spacetime. What is curvature?

The word *curvature* is an analogy, a visual way of extending ideas about three-dimensional space to the four dimensions of spacetime. Travelers detect curvature—in both three and four dimensions—by the gradual increase or decrease of the "distance" between "straight lines" that are initially parallel. In three space dimensions, the actual paths in space converge or diverge. Think of two travelers who start near one another at the equator of Earth and march "straight north." Neither traveler deviates to the right or to the left, yet as they continue northward they discover that the distance between them decreases, finally reaching zero as they arrive at the north pole. They can use this deviation to describe the curved spherical surface on which they travel. Similarly, in four-dimensional spacetime, travelers detect the deviation from parallelism of nearby worldlines of free particles, each of which follows an ideally straight spacetime path, often called a ***geodesic***. This curvature can be measured by the travelers and varies from place to place in spacetime.

Einstein: Coordinate systems are not fundamental.

We use frames of reference for our own convenience, for concreteness and economy of thought. But reference frames and their coordinates are not fundamental to Nature. Geometry is fundamental. It took Einstein seven years to achieve this basic insight. In a few sentences he summarizes the transition from special relativity to general relativity:

> *Now it came to me: . . . the independence of the gravitational acceleration from the nature of the falling substance, may be expressed as follows: In a gravitational field (of small spatial extension) things behave as they do in a space free of gravitation. . . . This happened in 1908. Why were another seven years required for the construction of the general theory of relativity? The main reason lies in the fact that it is not so easy to free oneself from the idea that coordinates must have an immediate metrical meaning.*

3 Free-Float Frame

Our old, comfy, free-float (inertial) frame carries us unharmed to the center of a black hole. Well, unharmed almost *to the center!*

No escape from inside the horizon of a black hole

We want to experience the spacetime geometry around a black hole, a star that has collapsed "all the way," without limit. General relativity predicts this fate for any too-massive collection of matter. General relativity predicts further that nothing, not even light, can escape from a black hole if the emitting satellite gets closer to the black hole than what is called the *horizon* (the radius of no return, defined more carefully in Section 9). If light cannot escape from an object, this object appears black from the outside. Hence the name "black hole."

"Capsule of flat spacetime"

No one can stop us from observing a black hole from an unpowered spaceship that drifts freely toward the black hole from a great distance, then plunges more and more rapidly toward the center. Over a short time the spaceship constitutes a "capsule of flat spacetime" hurtling through

Escape from a Black Hole? Hawking Radiation

Einstein's equations predict that nothing escapes from the so-called "horizon" of a black hole. In 1973, Stephen Hawking demonstrated a contrary conclusion using quantum mechanics. For years quantum mechanics had been known to predict that particle-antiparticle pairs, such as electrons and positrons, are continually being created and recombined in undisturbed space, despite the frigidity of the vacuum. These processes have, indirectly, important and well-tested observational consequences. Never in cold flat spacetime, however, do such events ever present themselves to direct observation. For this reason the pairs receive the name "virtual particles." When such a particle-antiparticle pair is produced near the horizon of a black hole, Hawking showed, one member of the pair will occasionally be swallowed by the black hole, leaving the other one to escape. Escaped particles form what is called **Hawking radiation**. The energy of the escaping particle comes from the black hole. Over time this loss of energy causes the black hole to "evaporate." The final stage may be a super-H-bomb explosion. For a black hole of several solar masses, however, the time required to achieve this explosive state exceeds the age of the Universe by a fantastic number of powers of ten. For this reason we ignore such emissions here. (See also the box on page 5-27.)

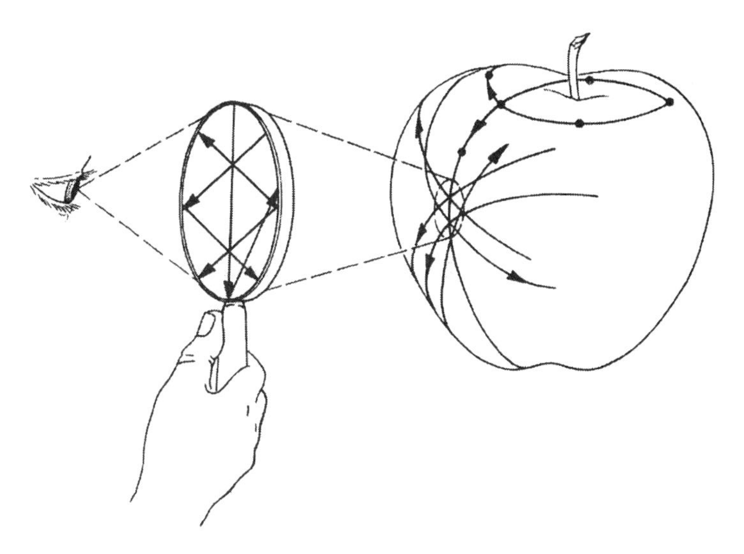

Figure 2 *The curved spacetime geometry of general relativity symbolized by the two-dimensional geometry of the surface of Newton's apple. The locally straight* **(geodesic)** *tracks followed by ants crawling on the apple's surface symbolize the tracks followed through spacetime by free particles. In any sufficiently localized region of spacetime, the geometry can be idealized as flat, as symbolized on the apple's two-dimensional surface by the straight-line course of the tracks viewed in the magnifying glass. In a region of greater extension, the curvature (curved two-dimensional space in the case of the apple; four-dimensional spacetime in the case of the real physical world) makes itself felt. On a larger scale, two tracks originally diverging from a common point later approach, cross, and go off in very different directions. In Newtonian theory this effect is ascribed to gravitational force acting at a distance from a massive body, symbolized here by the stem of the apple. According to Einstein, a particle gets its moving orders locally, from the geometry of spacetime right where it is. Its instructions are simple: "Go straight! Follow the straightest possible worldline (geodesic)." Physics is as simple as it could be locally. Only because spacetime is curved in the large do the tracks diverge, converge, and cross ("tidal accelerations").*

curved spacetime. It is a free-float frame like any other. Special relativity makes extensive use of such frames, and special relativity continues to describe Nature correctly for an astronaut in a local free-float frame, even as she falls through curved spacetime, through the horizon, and into a black hole. Keys, coins, and coffee cups continue to move in straight lines with constant speed in such a local free-float frame. (Figure 2 illustrates, by analogy, that paths curved in three space dimensions appear straight when we view small enough portions of these paths.) Collisions, creations, and annihilations of particles continue to follow the special relativity law of conservation of momentum-energy. What could be simpler?

Unavoidable relative accelerations near a star

However, as we approach the black hole the dimensions of our frame must be progressively constricted if we are to verify that it is free-float. In free fall near Earth, relative accelerations change the separation between two test particles, thus restricting the size of the spacetime region in which both are observed to be in free float (Chapter 1, Section 8). In imagination we can extend our near-Earth experience to regions exterior to more massive spherically symmetric objects: our Sun, a similar star, a white dwarf, a neutron star, a black hole. As we get closer and closer to each of these more and more compact spherically symmetric bodies, greater and greater become the relative accelerations between test particles. Near the center of a black hole these relative accelerations become lethal.

Relative accelerations are called **tidal accelerations**, because they are similar to the difference of our Moon's gravitational attraction on opposite sides of Earth that lead to tides. (See Section 2.3 of *Spacetime Physics*.)

Lethal effects of relative accelerations near black hole

Consider, for example, the plight of an experimental astrophysicist freely falling feet first toward a black hole. As the trip proceeds, various parts of the astrophysicist's body experience different gravitational accelerations. His feet are accelerated toward the center more than his head, which is farther away from the center. The difference between the two accelerations (the tidal acceleration) pulls his head and feet apart, growing ever more intense as he approaches the center of the black hole. The astrophysicist's body, which cannot withstand such extreme tidal accelerations, suffers drastic stretching between head and foot as the radial distance drops to zero.

But that is not all. Simultaneous with this head-to-foot stretching, the radial attraction toward the center funnels the astrophysicist's body into regions of space with ever-decreasing circumferential dimension. Tidal gravitational accelerations *compress* the astrophysicist on all sides as they *stretch* him from head to foot. The astrophysicist, as the distance from the center approaches zero, is crushed in width and radically extended in length. Both lethal effects are natural magnifications of the relative motions of test particles released from rest at opposite ends of free-float frames near Earth (Chapter 1, Section 8).

How small can a free-float frame be?

Confronted by tidal accelerations, how can we define a free-float frame falling into a black hole? At the center of the black hole we cannot; general relativity predicts infinite tidal accelerations there. However, short of the center, we employ the strategy used in free-float near Earth (Section 8 of Chapter 1): Limit the space and the time—the region of *spacetime*!—in which experiments are conducted. Very near the center we restrict ourselves to an ever smaller and more pinched local region of spacetime in which to define a free-float frame and in which to employ special relativity. How small can a free-float frame be? A single radioactive atomic nucleus can emit a detectable signal, for example a high-energy flash ("gamma ray"). In principle a reference frame can have space dimensions as tiny as that of the nucleus and time dimension equal to the emission time of the gamma ray. If gravitational tidal accelerations do not distort the nucleus "too much" within this spacetime region, the laws of special relativity accurately describe the nucleus in such a frame—the frame is effectively free-float for purposes of this experiment.

The constant, ever-present "force of gravity" that we experience on Earth is gone, eliminated as we step into a free-float frame. What remains of "gravity"? Only curvature of spacetime remains. What is this curvature?

Nothing but tidal acceleration. Curvature is tidal acceleration and tidal acceleration is curvature. Kip Thorne says it clearly: "Einstein and Newton, with their very different viewpoints on the nature of space and time, give very different names to the agent that causes test particles to accelerate toward or away from one another in a frame that is not quite free-float. Einstein calls it spacetime curvature; Newton calls it tidal acceleration. But there is just one agent acting. Therefore, *spacetime curvature and tidal accelerations must be precisely the same thing, expressed in different languages."* (Quotation slightly edited; original in the references.)

Need global coordinates

One limitation of a free-float frame near a black hole is the tidal accelerations experienced by test particles as the frame falls toward zero radius. Another limitation is the large-scale consequence of tidal acceleration: No single free-float frame is large enough to describe relations between two events that occur on opposite sides of the central mass. Two such events might be the emission of two flashes at different times by an object whose orbit girdles a black hole. To relate such widely separated events, we need a global rather than a local coordinate system. Karl Schwarzschild provided the basis for such a global coordinate system around a spherical, nonspinning center of attraction. Schwarzschild coordinates apply approximately to slowly spinning bodies such as Earth and Sun and to nonspinning or slowly spinning neutron stars and black holes. But these coordinates also have limitations. Points of view provided by free-float and Schwarzschild-related coordinate systems—and by still other coordinate systems—probe deeply the geometry of empty spacetime around a star. We now begin the study of the coordinates used by Schwarzschild.

4 The *r*-coordinate: Reduced Circumference

How to measure the radius while avoiding the trap in the center

Spherically symmetric centers of attraction

Matter has, by virtue of gravitation, a marvelous ability to agglomerate into spherical centers of attraction. Nicolaus Copernicus is credited with the insight that replaced Earth as the only assumed center of attraction with multiple centers of gravity. Standing as witness to the simplicity of the spherical shape are Earth, Moon, planets, Sun, and stars. Each of these structures is compressed—more dense—near its center and less dense near its surface. But this density changes with radius only, not with angle around the center. Such structures earn the label **spherically symmetric**. (Strictly speaking, an astronomical object can be spherically symmetric only if it does not rotate on its axis. For our Sun and planets this rotation rate is small enough so that departures from spherical symmetry can be neglected in the interpretation of many observations.)

The closer the distribution of mass to exact spherical symmetry, the better the spacetime geometry around such a structure conforms to the wonderfully simple solution to the equations of general relativity discovered by Karl Schwarzschild in 1915. Schwarzschild's solution describes spacetime external to *any* isolated spherically symmetric body in the Universe.

Nicolaus Copernicus *Born Torun, Poland, February 19, 1473; died Frombork, Poland, May 24, 1543. The flower is an old symbol for medicine, which Copernicus learned at Padua. His medical skill was always at the service of the poor.*

The Dictionary of Scientific Biography says*: Whereas the pre-Copernican cosmos had known only a single center of gravity or heaviness, the physical universe acquired multiple centers of gravity from Copernicus, who thus opened the road that led to universal gravitation . . . [H]e put forward a revised conception of gravity, according to which heavy objects everywhere tended toward their own center—heavy terrestrial objects toward the center of the earth, heavy lunar objects toward the center of the Moon, and so on. [Copernicus wrote:]*

> *"For my part, I think that gravity is nothing but a certain natural striving with which parts have been endowed . . . so that by assembling in the form of a sphere they may join together in their unity and wholeness. This tendency may be believed to be present also in the sun, the moon, and the other bright planets, so that it makes them keep that roundness which they display.'"*

Schwarzschild's simple solution

What does this "Schwarzschild geometry" around Earth, star, or black hole look like? "What a nonsensical question!" we say at first. Whoever looks *at* space? We look *through* space. Or we thrust skeleton skyscrapers out into space, we push out into space the Buckminster Fuller framework of a great spherical building (Figure 3). Ha! Just such a Buckminster Fuller construction gives us at last a way to "see" what space looks like, as described in what follows.

To be specific, take the center of attraction to be a black hole. Let it have the same mass as Sun. Build around it, in imagination, an open spherical shell of rods fitted together in a mesh of triangles (Figure 3) similar to hemispherical jungle gyms found on playgrounds. This spherical shell, this scaffolding, is an alternative to our latticework of rods and clocks in local free-float frames. Mount clocks on this shell. The rods and clocks of this shell provide one system of spacetime coordinates to locate events.

Spherical shell of rods and clocks

We say to build this shell "in imagination," because neither steel nor tantalum nor any modern wonder material has a ratio of strength to weight adequate to support such a structure against the inward pull of gravity. However, the surface of a planet, moon, or star has itself the character of a shell. We walk around on such a shell every day: Earth's surface! In the absence of an actual spherical shell, we can use a spaceship that stands still above the surface by blasting its rockets inward.

We cannot measure radius directly.

How shall we define the size of the sphere formed by this latticework shell? Shall we measure directly its distance from its center? That won't do. Yes, in imagination we can stand on the shell. Yes, we can lower a plumb bob on a "string." But for a black hole, any string, any tape measure, any steel wire—whatever its strength—is relentlessly torn apart by the unlimited pull the black hole exerts on any object that dips close enough to its center. Even for Earth or Sun, the surface keeps us from lowering our plumb bob directly to the center.

Derive radius from measurement of circumference.

Then try another way to define the size of the spherical shell. Instead of lowering a tape measure from the shell, run a tape measure around it. Call the distance so obtained the *circumference* of the sphere. Divide this circumference by $2\pi = 6.283185\ldots$ to obtain a distance that would be the directly measured radius of the sphere if the space inside it were flat. But it isn't flat. Yet this procedure yields the most useful known measure of the size of the spherical shell.

The "radius" of a spherical object produced by this method of measuring has acquired a name, the **coordinate radius**, despite its being no true radius. We call it also the **reduced circumference**, to remind us that it is derived ("reduced") from the circumference:

$$\begin{aligned}\text{coordinate radius} &= r = \text{reduced circumference}\\ &= (\text{circumference})/2\pi \\ &= r\text{-coordinate}\end{aligned} \qquad [1]$$

The phrases *coordinate radius* and *reduced circumference* are such mouthfuls that we usually call it simply the ***r*-coordinate** and represent it by the symbol r. The r-coordinate is the radius computed from the sphere's circumference. This value of r is stamped on every shell for all to see.

Having constructed—in imagination—one spherical shell around our black hole and found its coordinate radius, its reduced circumference r, we construct inside it a second such framework of rods and likewise determine its radius. We find the reduced circumference r of the inner sphere to

Figure 3 Geodesic globe named Spaceship Earth, the symbol of Disney Epcot Center in Orlando, Florida. Fifty meters in diameter, it contains a ride highlighting the history of communication from cave dwellers to the present. The spherical shells surrounding our black hole are openwork lattices, not a closed surface as shown here.

be 1 kilometer less than that of the first one—based on tape-measure determinations of distance around the two spheres.

Directly measured separation between nested shells is **greater** than the difference in r-value.

Now, finally, we lower a plumb bob from the outer sphere and for the first time measure directly the true radial distance perpendicularly from the outer sphere to the inner one. Will we find a 1-kilometer radial distance between the two spheres? We would if space were flat. But it is *not* flat. Schwarzschild geometry tells us that the directly measured radial distance between the two nested spheres is *more* than 1 kilometer. That increase over the expectations of Euclidean geometry provides the most striking evidence in principle one can easily cite for the curvature of space we call gravitation. To examine such discrepancies is to see what space looks like around a black hole.

Small effect near Sun

Built around our Sun, the inner sphere cannot lie inside Sun's surface. Its *r*-coordinate can be no less than that of Sun's surface, which is approximately 695 980 kilometers. Around this inner shell we erect a second one—again in imagination—of *r*-coordinate 1 kilometer greater: 695 981 kilometers. The directly measured distance between the two would be not 1 kilometer, but 2 millimeters more than 1 kilometer.

How can we get closer to the center of a stellar object with mass equal to that of our Sun—but still be external to that object? A white dwarf and a neutron star each has roughly the same mass as our Sun, but each is much smaller. Therefore we can—in principle—conduct a more sensitive test of the nonflatness of space much closer to the centers of these objects while staying external to them. The effects of the curvature of space are much greater near the surface of a white dwarf or neutron star than near the surface of our Sun.

Huge effect near a black hole

Turn attention now to a black hole of one solar mass. Close to it the departure from flatness is much larger than it is anywhere in or around a white dwarf or a neutron star. Construct an inner sphere having an *r*-value, an *r*-coordinate, a reduced circumference of 4 kilometers. Let an outer sphere have an *r*-coordinate of 5 kilometers. In contrast to these two distances, defined by measurements around the two spheres, the directly measured radial distance between the two spheres is 1.723 kilometers, compared to the Euclidean-geometry figure of 1 kilometer (Sample Problem 2, page 2-28). At this location the curvature of space results in measurements quite different from anything that textbook Euclidean geometry would lead us to expect!

WHY is the directly measured distance between spherical shells greater than the difference in r-coordinates between these shells? Is this discrepancy caused by gravitational stretching of the measuring rods?

No, the quoted result assumes infinitely rigid measuring equipment. In practice, of course, a measuring rod held by the upper end will be subject to gravitational stretching. So think of flinging the rod up from below so that it comes to rest temporarily with its two ends next to the two shells and thereby measures the separation directly while in free float. Even in this case there will still be tidal forces on the rod. Strain gauges affixed along the rod can permit us to "calculate away" this stretching. For smaller and smaller separation between the shells the stretching can be reduced below any specified limit.

Don't avoid the issue! You have not answered the question: What CAUSES the discrepancy, the fact that the directly measured distance between spherical shells is greater than the difference in r-coordinates between these shells? WHY this discrepancy?

A deep question! Fundamentally, this discrepancy is evidence of space curvature resulting from the mass contained in the center of attraction. External to this center, the fabric of spacetime does not tear but transmits the ever-diluted curvature outward to influence locally every spherical shell, every test particle, every satellite in the surroundings.

5 Gravitational Red Shift

Rising light shows fatigue by increasing its period of oscillation.
Light rising from the horizon has infinite period—so it does not exist!

Time also enters into the (space*time*) curvature around a black hole. In no way is "time curvature" more apparent than in the behavior of a signal emitted from a clock bolted to a spherical shell near a gravitating body. Let this clock tick by emitting light in a radially outward direction. The emitted light increases its period of vibration as it climbs up out of the gravitational field.

The period of light increases as it climbs.

How does light increase its period of vibration? Every period (every back-and-forth undulation in the wave) of the light can be considered a measure of time, a "tick of the clock." Suppose that the light has a short period when emitted by the clock on the shell. The shell observer records that the emitting clock ticks rapidly; for him time is short from one tick to the next tick. When the light finally arrives at a remote observer, its period is longer. The received clock-tick signals are observed to be farther apart in time than the sent clock-tick signals. Light emitted from a shell clock still closer to the black hole suffers an even greater increase in its period—a greater "time between ticks"—when this light has climbed to infinity.

The period of the received light increases more and more as the emitter stands closer and closer to the black hole. Details of this increased period imply curvature not only of space but of time—curvature of *spacetime*! The increased period means also that the time dt_{shell} between two events—such as clock ticks—measured by an observer standing on a shell (or occupying a spaceship at rest, rockets blasting inward) will be different from the "far-away time dt" between these events as transmitted to and recorded by a clock remote from the gravitating body.

Gravitational red shift

Visible light with the longest period is red. The remote observer sees light emitted by the close-in clock to be "redder"—that is, of longer period—than it was at the point of emission. This effect thus earns the name **gravitational red shift.**

"Blackness" of a black hole

Why is a black hole black? Why cannot light escape from a black hole? After all, light cannot stop moving! Every local observer records the speed of light to be unity as it passes on its upward journey. The gravitational red shift result allows us to give a meaning to the phrase "cannot escape." Light of any period emitted from near the horizon (the threshold radius of no return) suffers a gravitational red shift to a very long period. The closer the clock is to the horizon, the farther toward infinity the period grows as the light climbs out of the black hole to a great distance. But a light signal with infinite period is no light signal at all! It cannot be detected. In this case almost no light has escaped from near the horizon of the black hole. For a clock *at* the horizon, as a limiting case, *no* light escapes to even 1 centimeter above the horizon (Chapter 5 exercises). Light is red shifted all the way to infinite period. This crisp result accounts for the blackness of a black hole (which is black except for Hawking radiation, a quantum phenomenon described in the box on page 2-4).

The gravitational red shift occurs between two clocks that are at different radii and both at rest with respect to the black hole or other center of gravitational attraction. Another and different red shift occurs due to the Doppler effect when two clocks move away from one another. An example is the red shift of light that we receive from nearby galaxies outside our own, thought to be due to the recession of these galaxies from us. (A generalization of the Doppler shift to curved expanding spacetime is the reddening of light from distant galaxies as the Universe expands—see Project G, The Friedmann Universe.) For observers on Earth this recessional red shift is in principle partly canceled by the **gravitational blue shift** of the light as it drops into the gravitational well surrounding Earth. However, for many everyday purposes the gravitational blue shift for Earth is negligible. (See exercise at the end of this chapter.)

The gravitational red shift is different from the Doppler shift due to relative motion.

We have described two consequences of spacetime curvature: the augmentation of distance between adjacent spherical shells and the increase in the period of light escaping outward from one of these shells. How these effects come about, and why they become so impressive at the horizon of a black hole, shows on an examination of the expressions describing the Schwarzschild solution to Einstein's great and still standard 1915 equation for the bending of spacetime geometry. Before we can write down these expressions in simple form, we need to describe the mass of the central body, not in the unperceptive conventional units of kilograms, but rather in the same geometric units we use to measure distance: meters or kilometers.

6 Mass in Units of Length

Want to make everything geometry? Then measure mass in meters!

Descriptions of spacetime near any gravitating body are simplest when the mass M of that body is expressed in units of distance—in meters or kilometers. This section is devoted to finding the conversion factor between, say, kilograms and meters.

Measure mass in meters.

Earlier when we wanted to measure space and time in the same units (Chapter 1, Section 2), we used the conversion factor c, the speed of light. The conversion from kilograms to meters is not so simple. Nevertheless, here too Nature provides a conversion factor, a combination of the speed of light and the **universal gravitation constant** G that characterizes the gravitational interaction between bodies.

Newton's theory of gravitation predicts that the gravitational force between two spherically symmetric masses M and m is proportional to the product of these masses and inversely proportional to the square of the distance r between their centers:

$$F = \frac{GM_{\mathrm{kg}}m_{\mathrm{kg}}}{r^2} \qquad \text{[2. Newton]}$$

Subscripts tell us that in this equation the masses M_{kg} and m_{kg} are in units of kilograms. In this equation G is the "constant of proportionality." The numerical value of this constant depends on the units with which mass

Numerical values of G and c: historical accident

and distance are measured. Historically the units of mass and the units of distance were developed independently, without appreciation of their relationship. The numerical value of G was not built into Nature by law but arose by accident of human history, as the numerical value of the speed of light c likewise arose from historical accident alone. When we measure mass in kilograms and distance in meters, then G has the experimentally determined value

$$G = 6.6726 \times 10^{-11} \frac{\text{meter}^3}{\text{kilogram second}^2} \qquad [3]$$

Divide G by the square of the speed of light, c^2, to find the conversion factor that translates the conventional unit of mass, the kilogram, into what we have already found to be the natural geometric unit, the meter:

$$\frac{G}{c^2} = \frac{6.6726 \times 10^{-11} \dfrac{\text{meter}^3}{\text{kilogram second}^2}}{8.9876 \times 10^{16} \dfrac{\text{meter}^2}{\text{second}^2}} \qquad [4]$$

$$= 7.424 \times 10^{-28} \frac{\text{meter}}{\text{kilogram}}$$

Now convert from mass M_{kg} measured in conventional units of kilograms to mass M in units of length—meters—by multiplication with this conversion factor:

$$M = \frac{G}{c^2} M_{kg} = \left(7.424 \times 10^{-28} \frac{\text{meter}}{\text{kilogram}}\right) M_{kg} \qquad [5]$$

Mass in units of meters unclutters equations.

Why make this conversion? First, it is an elegant way to proclaim that mass is fundamentally tied to geometry. Second, it allows us to get rid of the factors G and c^2 that would otherwise clutter up the equations to follow.

Wait a minute! Stars and planets are not the same as space. No twisting or turning on your part can make mass and distance the same. Therefore mass cannot be measured in units of distance. How can you possibly propose to measure mass in units of meters?

True, mass is not the same as distance. Neither is time the same as space: Clock ticks are different from meterstick lengths! Nevertheless, we have learned to measure both time and space in the same units: light-years of distance and years of time, for example, or meters of distance and meters of light-travel time. Using the same units for both space and time helps us to get rid of people-made complications and to recognize the unity we call spacetime. The conversion factor between time in seconds and space in meters is the speed of light c.

The same comments hold in the present case for measuring mass in units of length. Mass is not the same as length; no one claims it is. But we gain insight when we measure both in geometric units. When we express the mass of a star in meters, we can convert this figure to any other measure we want: grams, kilograms, or number of solar masses. For the translation from kilograms to meters, the conversion factor is not a mere power of the speed of light but includes the gravitational constant G. The factor that converts kilograms to meters is G/c^2. And the payoff of this conversion is similar to earlier payoffs: we see more simply how Nature works and we arrive more quickly at correct predictions. Mass, and therefore gravitation, is elevated (not reduced!) to geometry.

All right. Wonderful! Now go one step further and make the definition of the kilogram in terms of the meter an official international standard. Since 1983 the official international standard for the meter is the distance light travels in 1/299,792,458 second, thus tying the meter to a measurement of time. By defining (at some more enlightened future date) the kilogram in terms of the geometric unit meter, we link it also to a measurement of time. All other physical units—energy, momentum, electric charge—have long been defined in terms of time, length, and mass. By officially defining the kilogram in units of length, and therefore ultimately in units of time, we unify the world of measurement to a single quantity.

Your proposed unification is a good idea in principle but not yet satisfactory in practice. Measurement of mass is very precise. So is measurement of length and time. However, the conversion factor between mass and length, G/c^2, is not known with corresponding precision. The fault lies with the gravitational constant G, which is difficult to measure—presently accurate to 5 digits at most. Compare that with the nine-digit accuracy of the speed of light that allowed a redefinition of length in terms of time.

Why wait until G is known more accurately? Why not just define the kilogram in the unit of length using the conversion factor 7.424×10^{-28} meter/kilogram, this figure taken to be exact by definition?

There is no logical reason why G cannot be *defined* to have an exact value right now. However, convenience and accessibility are no less important criteria for standards than logical simplicity. The present standard of mass—a particular chunk of metal—can be accurately duplicated, providing secondary standards for calibration of the scales used in science and commerce. This standard is unlikely to be replaced until a way is discovered to measure the gravitation constant G much more accurately, with apparatus available in any well-equipped laboratory

Table 1 displays in both kilograms and meters the mass of Earth, the mass of Sun, and the mass of the huge spinning black hole believed to explain the activity observed at the center of our galaxy and a similar black hole in one other galaxy. (See the references.) Thus does the geometric language of relativity cut the stars down to size.

7 Satellite Motion in a Plane

Once moving in a plane, always moving in that plane

Orbits stay in a plane.

An isolated satellite zooms around a spherically symmetric massive body. Our very first look shows that this motion lies in a plane determined by the satellite's position, its direction of motion, and the center of the attracting body. We know that forever afterward the motion will remain confined to that same plane. Why? The reason is simple: symmetry! No distinction between "up out of" and "down below" that plane, so the satellite cannot choose either. Such a rise would provide immediate evidence that there is some further force at work beyond any exerted by the spherically symmetric body—evidence, in other words, that the satellite's environment is not spherically symmetric with respect to that center of attraction.

Table 1 Masses of some astronomical objects

Object	Mass in kilograms	Geometric measure of mass	Equatorial radius
Earth	5.9742×10^{24} kilograms	4.44×10^{-3} meters or 0.444 centimeters	6.371×10^{6} meters or 6371 kilometers
Sun	1.989×10^{30} kilograms	1.477×10^{3} meters or 1.477 kilometers	6.960×10^{8} meters or 696 000 kilometers
Black hole at center of our galaxy	5.2×10^{36} kilograms (2.6×10^{6} Sun masses)	3.8×10^{9} meters (see references)	
Black hole in center of Virgo cluster of galaxies	6×10^{39} kilograms (3×10^{9} Sun masses)	4×10^{12} meters	

Locate satellite using r and ϕ.

The satellite moves in a plane, so we need two quantities, and only two, to specify its location at any instant. Adopt for one the *r*-coordinate, the reduced circumference of a circle cutting through the satellite. For the second coordinate take the **azimuthal angle** ϕ of the satellite's progression in the plane around the center of attraction (Figure 4).

Every astronaut, every satellite, every light pulse independently orbiting around a spherically symmetric body will remain in its own plane of motion, each position in the plane described by the reduced circumference *r* and the azimuthal angle ϕ in that particular plane. This limitation to a plane greatly simplifies the analysis of physical events described in the remainder of this book.

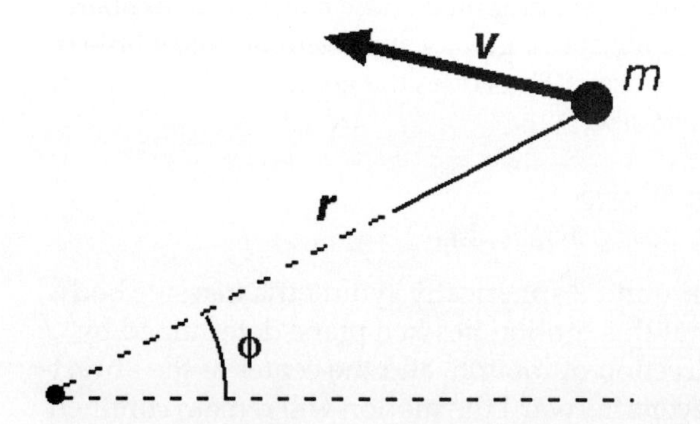

Figure 4 *A satellite moves in an orbit with instantaneous velocity v around a spherically symmetric body. This orbit lies in a plane and remains in that plane for all time. Satellite position on the plane is specified uniquely by two coordinates: we choose the r-coordinate and the azimuthal angle ϕ with respect to some arbitrary initial direction in the plane (horizontal dashed line in the figure). The inner part of the r-line is also dashed, because in the case of a black hole the radius cannot be surveyed directly.*

8 Metrics for Flat Spacetime

Rectangular space coordinates or polar space coordinates: Either can appear in a metric of flat spacetime.

What gives us security as we move from flat spacetime geometry to Schwarzschild geometry? On what can we depend? What can we trust? Answer: *Events!* Events are the nails of reality on which all of science hangs (Section 1). And between event and event we seek the basic relation, the basic separation, the four-dimensional "distance" between firecracker explosion and firecracker explosion. We seek the *spacetime interval* between any pair of events.

Events and intervals form a universal language.

When no large mass is in the vicinity we say that spacetime is *flat*. In flat geometry the expression for the wristwatch time τ between two events can be written in the usual rectangular coordinates described by Descartes ("Cartesian coordinates"). Let t, x, and y mark the separation between two events on a spatial plane when this separation is timelike (time separation greater than space separation). Then τ^2, the square of the wristwatch time between them, is given by the expression

$$\tau^2 = t^2 - x^2 - y^2 \qquad \text{[6. flat spacetime]}$$

Timelike spacetime interval

When, instead, the separation between the two events is spacelike, that is, when the space part of the separation predominates over the time part, we reverse the signs of the terms on the right of [6] to keep the combination positive. Give the resulting squared quantity the Greek letter σ (sigma):

$$\sigma^2 = -t^2 + x^2 + y^2 \qquad \text{[7. flat spacetime]}$$

Spacelike spacetime interval

The corresponding equations [1] and [8] of Chapter 1 for the spacetime interval earned the name *metric*; equations [6] and [7] are metrics too. A *metric* provides the method by which we *meter* or *measure* spacetime.

For describing the linear motion of one rocket with respect to another in flat spacetime, the Cartesian system of coordinates was perfect. Direction of relative motion: x. Direction transverse—perpendicular—to that relative motion: y. The Cartesian rectangular system is not so convenient as we prepare to describe spacetime around a spherically symmetric gravitating mass. Here the preeminent dimension is radial, toward and away from the center of attraction, with angle ϕ describing the location of an event on an imaginary circle of given radius r lying on a plane through that center. Rewrite the expression for the timelike interval (equation [6]) in polar coordinates. The resulting metric is

Polar coordinates are convenient when there is a center of attraction.

$$(d\tau)^2 = (dt)^2 - (dr)^2 - (rd\phi)^2 \qquad \text{[8. flat spacetime]}$$

The box on page 2-18 presents a derivation of the space separation part of this expression, namely $(dr)^2 + (rd\phi)^2$.

Spatial Separation in Flat Spacetime, Expressed in Polar Coordinates

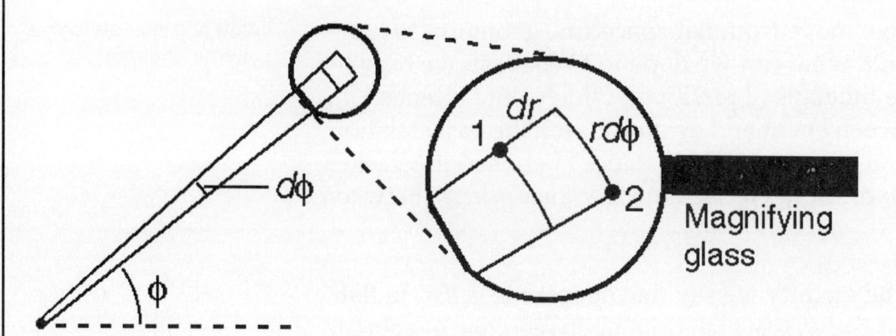

Figure 5 *Spatial separation between two points in polar coordinates.*

Let the coordinate separations between two events near one another be dx and dy in the x and y directions, respectively. Then the square of the spatial separation between two events is written

$$(\text{spatial separation})^2 = (dx)^2 + (dy)^2$$

Look for a similar expression for two events numbered 1 and 2 separated by the spherical polar coordinate increments dr and $d\phi$. See Figure 5.

Draw little arcs through events 1 and 2 to form a tiny rectangle, as shown in the magnified inset. The squared distance between events 1 and 2 is—approximately—the sum of the squares of two adjacent sides of the little rectangle. Each complete circle, if drawn here, would run through a total arc of 2π and possess a circumference of $2\pi r$. It was this circumference that was the starting point for our very definition of the reduced circumference r. The portion of each arc that is depicted in the figure extends only over the angle $d\phi$. It comprises only the fraction $d\phi/2\pi$ of the whole circle. By proportion, its length is $(d\phi/2\pi)$ times $2\pi r$, or $r d\phi$. This arc is so short that its length closely approximates the length of the corresponding straight line. We spell out this part of the reasoning because it goes over unchanged to the curved space geometry around a spherically symmetric body. Not so for the distance dr! Consider two points that lie at the same azimuth but have r-coordinates r and $r + dr$. Only in flat space is the distance between them equal to dr. Therefore only for flat space are we entitled to figure the distance in space between event 1 and event 2 by the formula

$$(\text{spatial separation})^2 = (dr)^2 + (r d\phi)^2$$

This squared spatial separation is the space part of the squared interval for flat spacetime, equation [8]. Notice that this derivation depends on $d\phi$ being small, so the small segment of arc $r d\phi$ is indistinguishable from a straight line.

Equation [8] is still true only for flat spacetime—the domain of special relativity. Why? Because the everyday world is still the everyday world, whether you view it while standing on your feet or standing on your head! Similarly, flat spacetime is flat whether the interval between events is expressed in Cartesian (rectangular) coordinates or in polar (spherical) coordinates. In brief, no massive body is yet positioned at the origin of this coordinate system

Now place the origin of the spherical coordinate system at the center of a nonspinning spherical object, approximated by Earth or Moon, Sun or white dwarf, neutron star or black hole. Examine the new spacetime geometry external to such a body. This new geometry is described by the Schwarzschild metric, introduced in Section 9.

9 The Schwarzschild Metric for Curved Spacetime

Spherically symmetric massive center of attraction?
Then the Schwarzschild metric describes curved spacetime around it.

The metric for the proper time between two timelike events in a plane in *flat* spacetime is given by equation [9]:

$$d\tau^2 = dt^2 - dr^2 - r^2 d\phi^2 \qquad \text{[9. flat spacetime]}$$

$r^2 d\phi^2$ term is still OK to describe spacetime near Earth.

How is this metric altered for two nearby events on a plane that passes through the center of a spherically symmetric massive body? The last term on the right stays the same because of the way we chose the r-coordinate. The r-coordinate is *defined* so that $2\pi r$ is the measured distance around a circle centered on the attracting mass; hence its name, *reduced circumference.* Measurement of the total circumference $2\pi r$ is the sum of measured distances ($r\, d\phi$) along many small segments of the circle. As a result, the last term on the right, $(r\, d\phi)^2$, remains correct for the Schwarzschild metric.

Timelike form of Schwarzschild metric

What about the time term and the radial term? How will they change near a black hole—or near Earth? The answer is embodied in the **Schwarzschild metric**. For two events close to one another the Schwarzschild metric introduces us to *curved empty spacetime* on a spatial plane through the center of a spherically symmetric (nonspinning) center of gravitational attraction:

$$d\tau^2 = \left(1 - \frac{2M}{r}\right)dt^2 - \frac{dr^2}{\left(1 - \frac{2M}{r}\right)} - r^2 d\phi^2 \qquad \text{[10. timelike form]}$$

The coordinates r, ϕ, and t appear in this equation. The angle ϕ has the same meaning in Schwarzschild geometry as it does in Euclidean geometry. We have defined r, the *reduced circumference,* so that $r d\phi$ is the incremental distance measured directly along the tangent to the shell. The time t is called **far-away time** and is measured on clocks far away from the center of attraction, as discussed in detail in Section 11.

The timelike Schwarzschild metric is so important that we write it for reference as equation [A] on the last page of this book.

Equation [10] is the timelike form of the Schwarzschild metric, for events in which the time separation predominates. In contrast, the spacelike form, describing a pair of events in which the space separation predominates, equation [10] is replaced by the equation

Spacelike form of Schwarzschild metric

$$d\sigma^2 = -\left(1 - \frac{2M}{r}\right)dt^2 + \frac{dr^2}{\left(1 - \frac{2M}{r}\right)} + r^2 d\phi^2 \qquad \text{[11. spacelike form]}$$

Equation [11] is placed for reference on the last page of this book as equation [B].

Sloppy Use of Differentials in Relativity

In going from equation [8] to equation [9], we have begun using differentials dr, $d\phi$ to describe the space separation between events and dt for the separation in time. Where did these differentials come from, and why do we suddenly start to use them? The analysis in the box on page 2-18 makes the approximation that the sides dr and $r d\phi$ of the little rectangle are *straight*. But the inner and outer sides are *not* straight: each is a portion of a circular arc. The approximation is sensible only if the little arc "looks like" a straight line, only if the angular separation $d\phi$ is a very small fraction of 2π, the angle for a complete circle. Our mathematician friends insist that the approximation is "correct" only in the limit of zero angle. Physicists tend to be a bit sloppy about applying mathematical differentials to nonzero (but still small) separations in real space and time.

But sloppy use of differentials by physicists goes farther than this. Equation [8] is usually written in the even more irresponsible form of equation [9]:

$$d\tau^2 = dt^2 - dr^2 - r^2 d\phi^2 \qquad \text{[9. flat spacetime]}$$

Compare equation [9] with [8]. Equation [9] is squeezed into the compact algebraic notation that by now has become standard. Legalistically it is wrong. On the left should appear $(d\tau)^2$, as it does in equation [8]. If we were credulous enough to take it seriously, $d\tau^2$ would give us not the square of the change in proper time, but rather the crazy idea of the small change in the *square* of proper time.

How did this sloppiness come about? Pure laziness. People got tired of writing down those extra parentheses, left them out, whispered a warning to their friends to write them back in—mentally at least—when putting the metric formula to use, and by now we're all in on the little secret. The same with the terms on the right-hand side, which should read $(dt)^2$, $(dr)^2$, and $(r\,d\phi)^2$, respectively, as they do in equation [8].

In late 1915, within a month of the publication of Einstein's general theory of relativity and just before his own death from battle-induced illness, Karl Schwarzschild (1873 –1916) derived this metric from Einstein's field equations. Einstein wrote to him, "I had not expected that the exact solution to the problem could be formulated. Your analytic treatment of the problem appears to me splendid."

The Schwarzschild metric appears here out of thin air. Where does it come from?

The Schwarzschild metric derives from Einstein's field equations for general relativity, equations that relate the "warping" of spacetime across a spacetime region to the mass and pressure in that region. Different distributions of mass lead to different metrics in the vicinity of the mass. Deriving a metric from the field equations is a major professional accomplishment. Einstein himself did not think it possible that anyone could carry out the task, even for a nonrotating, uncharged, spherically symmetric structure. The metric for a spinning black hole was not published until 1963, almost 50 years later. (See Project F, The Spinning Black Hole.)

Einstein's field equations themselves are not "derived," any more than Newton's laws of motion are derived. Indeed, a Newtonian prediction of the existence of a "horizon," the radius from which only light can escape, is given in the box on page 2-22. The field equations are, as Einstein was fond of saying, "the free invention of the human mind." This invention rests on Einstein's deep intuition for physical reality and symmetry—how Nature *must* behave. Of course the results must lead to correct predictions of experimental results, as they have repeatedly. In this book we start with the metric around each center of attraction. Each of these metrics is one step removed from the (underived) field equations. For a brief account of Einstein's development of the field equations and a description of their content, see Kip Thorne, *Black Holes and Time Warps*, pages 113–120.

The Schwarzschild description is **complete**.

Further investigation has shown that the Schwarzschild metric gives a *complete* description of spacetime external to a spherically symmetric, nonspinning, uncharged massive body (and everywhere around a black hole but at its central crunch point). *Every* (nonquantum) feature of spacetime around this kind of black hole is described or implied by the Schwarzschild metric. This one expression tells it all! Moreover, the vast majority of experimental tests of general relativity have been tests of the Schwarzschild metric. All test results have been consistent with Einstein's theory.

Different terms for "horizon"

At the radius $r = 2M$ something strange happens to the Schwarzschild metric. The time term goes to zero and the radial term increases without limit as r approaches the value $2M$ in both the timelike and spacelike versions, equations [10] and [11]. This value of r marks the location of the one-way surface through which anything may pass inward but nothing passes outward. This special value of the radial coordinate is given various names: the **Schwarzschild radius** or the **event-horizon radius**. The "membrane" at $r = 2M$ is called the **Schwarzschild surface** or the **Schwarzschild sphere**, the **Schwarzschild horizon**, the **event horizon**, or simply the **horizon**. (*Caution:* Some workers in the field refer to the geometric measure of mass M as the *gravitational radius*. Others reserve this name for $2M$. That is why we avoid the term in this book.)

Ways the Schwarzschild metric makes sense:

For us the Schwarzschild metric—one step from the field equations—is not derived but given. However, we need not accept it uncritically. Here we check off the ways in which it makes sense.

1. Depends only on r-coordinate.

First, the **curvature factor** $(1 - 2M/r)$ that appears in both the dt term and the dr term depends only on the r-coordinate, not on the angle ϕ. How come? Because we are dealing with a spherically symmetric body, an object for which there is no way to tell one side from the other side or the top from the bottom. This impossibility is reflected in the absence of any direction-dependent curvature factor multiplying dt^2 or dr^2.

2. Goes to flat spacetime metric for large r.

Second, as the r-coordinate increases without limit, the curvature factor $(1 - 2M/r)$ approaches the value unity, as it must. Why must it? Because an observer far from the center of attraction can carry out experiments in her vicinity without noticing the presence of the distant object at all. For her spacetime is locally flat. In other words, for large r the Schwarzschild metric [10] must go smoothly into the metric for flat spacetime [9].

3. Goes to flat spacetime metric for zero M.

Third, as the mass M goes to zero, the curvature factor $(1 - 2M/r)$ approaches the value unity, as it must. Why must it? Because a center of attraction with zero mass is the same as the absence of a massive body at that center, in which case equation [10] becomes equation [9], the expression for the interval in flat spacetime.

4. Confirms dr is less than the directly measured distance between shells.

Fourth, consider the factor for dr^2, namely $1/(1 - 2M/r)$. For $r > 2M$ this factor has a value greater than one, which is consistent with our first "experiment in principle" around a black hole (Section 4). The directly measured separation $d\sigma$ is larger than that calculated from the difference dr in r-values between two adjacent Buckminster Fuller shells. Think of a rod held vertical to the shell, spanning the radial separation between two nested spherical shells. Set off two firecrackers, one at each end of this rod, at the same time, $dt = 0$. Take these explosions to be the two events whose separation is described by the metric [11]. The two explosion events have zero separation in azimuth, so $d\phi = 0$. Then the *proper distance* between the explosions is the distance that the shell observer measures directly; call it $dr_{shell} = d\sigma$. The spacelike equation [11] leads to

$$d\sigma = dr_{shell} = \frac{dr}{\left(1 - \frac{2M}{r}\right)^{1/2}} \qquad \text{[12. radial rod at rest on shell]}$$

Here dr is the difference in reduced circumference between two shells. Prior knowledge of the factor $(1 - 2M/r)^{1/2}$ in the denominator was used in describing our first "experiment in principle" that dr_{shell} is greater than

Newton Predicts the Horizon of a Black Hole?

A stone far from a black hole and initially at rest with respect to it begins to move toward the black hole. Gradually the stone picks up speed, finally plunging to the center. With what speed v does this stone pass a spherical shell at radius r? For low velocities and weak gravitational fields the speed is easily derived from Newtonian conservation of energy. In conventional units, the potential energy $V(r)$ of a particle of mass m_{kg} (measured in kilograms) in the gravitational field of a spherical body of mass M_{kg} is given by the expression

$$V(r) = -\frac{GM_{kg}m_{kg}}{r} \qquad \text{[13. Newton]}$$

Here G is the gravitational constant, and the zero of potential energy is taken to be at infinite radial distance r. A particle that starts at that great distance with zero velocity and therefore zero kinetic energy has a total energy zero for all later times and positions r given by the expression

$$E = 0 = \frac{1}{2}m_{kg}v_{conv}^2 - \frac{GM_{kg}m_{kg}}{r} \qquad \text{[14. Newton]}$$

where v_{conv} is velocity measured in the conventional units meters/second. From this equation,

$$v_{conv} = \left(\frac{2GM_{kg}}{r}\right)^{1/2} \qquad \text{[15. Newton]}$$

The particle moves radially inward at this speed. Divide through by c to give speed dr/dt with distance and time in the same units:

$$\frac{v_{conv}}{c} = v = \left(\frac{2GM_{kg}}{c^2 r}\right)^{1/2} \qquad \text{[16. Newton]}$$

But $(G/c^2)\,M_{kg} = M$, the central attracting mass expressed in units of length (equation [5]). The resulting speed is

$$v = \left(\frac{2M}{r}\right)^{1/2} \qquad \text{[17. Newton]}$$

Surprisingly, equation [17] is correct in general relativity too, but only when the speed is interpreted as the speed of the in-falling object *as measured by the shell observer* (Chapter 3, Section 5).

Escape Velocity

Equation [17] provides a prediction for the "radius of a black hole." Think of hurling a stone radially outward from radius r with the speed given by equation [17]. Then Newtonian mechanics, which runs equally well both forward and backward in time, predicts that this stone will coast to rest at a great distance from the center of attraction. Thus equation [17] tells us the **escape velocity**—the minimum velocity needed to escape from the gravitational attraction—for a stone launched outward from radius r. What is the maximum possible escape velocity? Here we elbow Newton aside and give the relativistic answer: The maximum escape velocity is the speed of light, $v = v_{conv}/c = 1$. Place this value in equation [17] to find the minimum radius from which an object can escape—the Newtonian prediction for the radius of the Schwarzschild horizon:

$$r_{horizon} = 2M \qquad \text{[18. Newton]}$$

According to general relativity this is the correct value—provided r is the reduced circumference! The physical interpretation, however, is quite different in the two theories. Newton predicts that a stone launched from the horizon with a speed less than that of light will rise some radial distance, slow, stop before escaping, and fall back. In contrast, Einstein predicts that nothing, not even light, can be successfully launched outward from the horizon (exercise in Chapter 5), and that light launched outward EXACTLY at the horizon will never increase its radial position by so much as a millimeter. (For historical details, see the box on page 3-3.)

dr, as this equation affirms. The change of this factor from place to place implies space curvature.

Equation [12] is so useful that we place it for reference as equation [D] in Selected Formulas at the end of this book.

Fifth, the curvature factor $(1 - 2M/r)$ in the numerator of the dt^2 term also has a value less than one, which is consistent with the gravitational red shift (Section 5). Think of a clock bolted to the shell at radius r. Choose two events to be two sequential ticks of this shell clock. Call dt_{shell} this wristwatch time $d\tau$ between ticks of the shell clock. Between these two ticks the coordinate separations dr and $d\phi$ are both zero. The timelike equation [10] leads to

5. Confirms gravitational red shift.

$$d\tau = dt_{shell} = \left(1 - \frac{2M}{r}\right)^{1/2} dt \qquad \text{[19. clock at rest on shell]}$$

Here dt is the corresponding lapse of far-away time. From our second "experiment in principle" we know that the time dt_{shell} between pulses emitted by the clock is smaller at emission than their red-shifted value dt when received at a great distance. In brief, dt_{shell} is less than dt. This result is consistent with the less-than-one value of the curvature factor $(1 - 2M/r)$ in the time term of the Schwarzschild metric (equation [10]).

Equation [19] is placed for reference as equation [C] in Selected Formulas at the end of this book.

The Schwarzschild metric, equation [10], governs the motion of a free test particle external to any spherically symmetric, nonspinning, uncharged massive body. It applies with high precision to slowly spinning objects such as Earth or an ordinary star like our Sun. For the motion of a particle outside such an object, it makes no difference what the coordinates are inside the attracting sphere because the particle never gets there; before it can it collides with the surface of the star—collides with the fluid mass in hydrostatic equilibrium. The more compact the configuration, however, the greater the region of spacetime the test particle can explore. Our Sun's surface is 695 980 kilometers from its center. A white dwarf with the mass of our Sun has a radius of about 5000 kilometers, approximately that of Earth. The Schwarzschild metric describes spacetime geometry in the region external to that radius. A neutron star with the mass of our Sun has a radius of about 10 kilometers, so the test particle can come even closer and still be "outside," that is in a region described correctly by the Schwarzschild metric if the neutron star is not spinning.

The Schwarzschild metric applies only outside the surface of an object.

The ideal limit is not a star in hydrostatic equilibrium. It is a star that has undergone complete gravitational collapse to a black hole. Then the Schwarzschild metric, equation [10], can be applied almost all the way down to zero radius, $r = 0$. The wonderful thing about a black hole is that it has no surface, no structure, and no matter with which one will collide. A test particle can explore *all* of spacetime around a black hole without bumping into the surface—since there is no surface at all.

Black hole has no "surface."

How can a black hole have "no matter with which one will collide"? If it isn't made of matter, what is it made of? What happened to the star or group of stars that collapsed to form the black hole? Basically, how can something have mass without being made of matter?

The mass is all still there, inducing the curvature of adjacent spacetime. It is just crushed into a singularity at the center. How do we know? We don't! It is a prediction from the Schwarzschild metric. Can you verify this prediction? Only if you drop inside the horizon, perform experiments, and make measurements as you approach the singularity—and then neither you nor your reporting signal can make it back out through the horizon. Startling? Crazy? Absurd? Welcome to general relativity!

What is this "singularity" business, anyway? I've heard the term before, but I don't know what it means.

A singularity is a "nobody knows" phenomenon. Coulomb's law of electrical force between point charges has a $1/r^2$ factor in it, which goes to infinity at $r = 0$. But it doesn't really, because there is no such thing as a point charge in structures described by classical (non-quantum) physics. In the Schwarzschild metric the curvature factor $(1 - 2M/r)$ goes to zero at $r = 2M$, leading one term in the metric to blow up. However, it was discovered after long study that this singularity in the metric is due to the choice of coordinates and is not "real." Someone falling inward in free float feels no jolt as she passes $r = 2M$. (More on this smoothness at $r = 2M$ in Chapter 3 and Project B, Inside the Black Hole.) On the other hand, the singularity at $r = 0$ appears to be "real." That is, anything falling to the center of a black hole is crushed to zero volume—to a single point. That is the prediction of general relativity, which is a classical (non-quantum) theory. In contrast, quantum theory predicts that nothing—not even a single electron—can be confined to a point. So what's the truth? The truth is, nobody has figured it out yet! No one has developed a **theory of quantum gravity** that combines quantum mechanics and general relativity. Anyway, Nature has hidden away the singularity inside a one-way surface at $r = 2M$, so we cannot find out while remaining outside. This situation is often described by saying that all real singularities are "clothed," as if there is cosmic censorship. Are there any "naked" (uncensored) singularities not hidden by a one-way surface? None that we yet know about.

One cannot predict the future

If there are non-trivial singularities which are naked, i.e., which can be seen from infinity, we may as well all give up. One cannot predict the future in the presence of a spacetime singularity since the Einstein equations and all the known laws of physics break down there. This does not matter so much if the singularities are all safely hidden inside black holes but if they are not we could be in for a shock every time a star in the galaxy collapsed.

— Stephen Hawking

10 Picturing the Space Part of Schwarzschild Geometry

Freeze time; examine curved space.

How can one visualize the geometry around a black hole? In general relativity, every coordinate system is partial and limited, correctly representing one or another feature of curved spacetime and misrepresenting other features. Figures and diagrams that display these coordinate systems embody the same combination of clarity and distortion.

One partial visualization displays the *spatial* part of the Schwarzschild metric. Freeze time (set $dt = 0$) and limit ourselves to a spatial plane passing through the center of the black hole. Then the spacelike form of the Schwarzschild metric [11] becomes

Visualize the spatial part of the metric.

$$d\sigma^2 = \frac{dr^2}{1 - \frac{2M}{r}} + r^2 d\phi^2 \qquad [20.\ dt = 0]$$

Figures 6 and 7 represent this special case. The radius r of each circle is the r-coordinate, the reduced circumference, locating the intersection of a spherical shell with a spatial plane through the center of the black hole. The differential dr is the difference in reduced circumference between adjacent circles. We have *added* the vertical dimension in the diagram and scaled it so that the slanting distance upward and outward along the surface represents $d\sigma$, the proper distance between adjacent circles measured directly with a plumb bob and tape measure. The "funnel" surface resulting from this scaling condition is called a *paraboloid of revolution*, and the heavy curved line in Figure 7 is a *parabola*—actually half a parabola.

Figure 7 embodies the fact that $d\sigma$ is greater than dr, the demonstration in principle that evidenced the curvature of space around a black hole in the first place (Section 4). The figure further shows that the ratio $d\sigma/dr$ increases without limit as the radial coordinate decreases toward the critical value $r = 2M$ (vertical slope of the paraboloid at the throat of the funnel).

These figures embed curved-space geometry in the flat Euclidean three-space geometry perspective shown on the printed page. Therefore these figures are called **embedding diagrams**. But flat Euclidean geometry is *not* curved space geometry. Therefore we expect embedding diagrams to misrepresent curved space in some ways. They lie! For example, the vertical dimension in Figures 6 and 7 is an artificial construct. It is *not* an extra dimension of spacetime. *We* have added this Euclidean three-space dimension to help us visualize Schwarzschild geometry. In the diagram, only the paraboloidal surface represents curved-space geometry. Observers posted on this paraboloidal surface must stay on the surface, not because they are physically limited in any way, but because locations off the surface simply do not exist in spacetime.

"Embedding diagrams" help visualize space curvature.

Observers constrained to the paraboloidal surface cannot measure directly the radius of any circle shown in Figures 6 and 7. They must derive this radius—the reduced circumference—indirectly by measuring the distance around the circle and dividing this circumference by the quantity 2π. From the circumference of an adjacent circle they derive its different radius and calculate the difference dr in the reduced circumferences of the two circles. In contrast, they can directly measure $d\sigma$, the proper distance between these adjacent circles, and compare their result with the computed difference in reduced circumference dr. Result: $d\sigma$ is greater than dr. The ratio $d\sigma/dr$ becomes infinite at $r = 2M$ (where parabola is vertical in diagram).

Observers cannot measure dr directly.

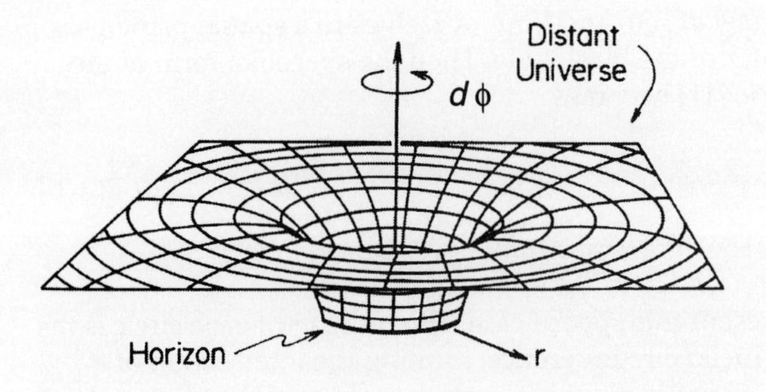

Figure 6 *Space geometry for a plane sliced through the center of a black hole, the result "embedded" in a three-dimensional Euclidean perspective. All of the curvature of empty space (space free of any mass–energy whatsoever) derives from the mass of the black hole. Circles are the intersections of the spherical shells with the slicing plane. WE add the vertical dimension to show that dσ is greater than dr in the spatial part of the Schwarzschild metric, as shown more clearly in Figure 7.*

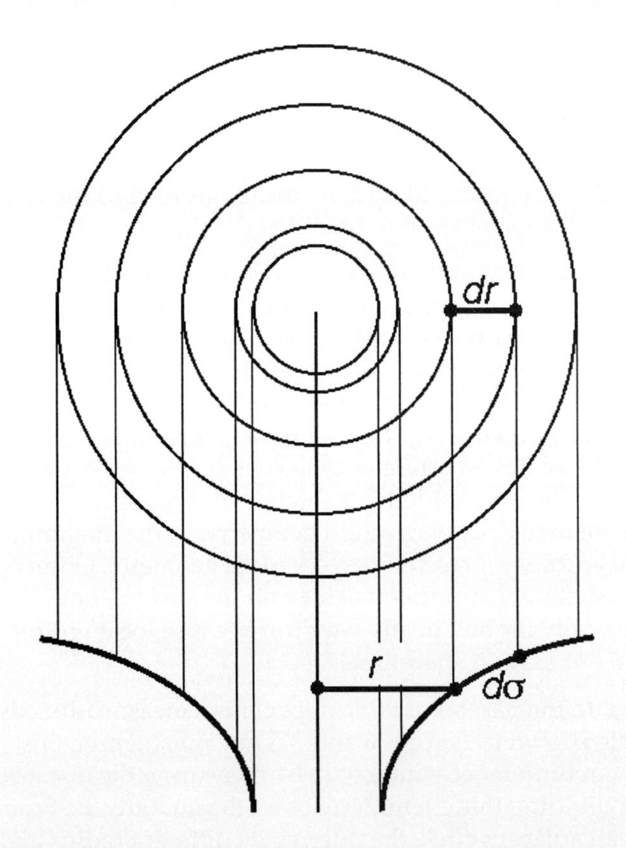

Figure 7 *Projections of the embedding diagram of Figure 6, showing how the directly measured radial distance dσ between two adjacent spherical shells is greater than the difference dr in r-coordinates. Real observers exist only on the paraboloidal surface (shown edge-on as the heavy curved line). They can measure dσ directly but not r or dr. They derive the r-coordinate (the* **reduced circumference)** *of a given circle by measuring its circumference and dividing by 2π. Then dr is the computed difference between the reduced circumferences of adjacent circles.*

SAMPLE PROBLEM 1 Limits of Small Curvature

The curvature factor $(1 - 2M/r)$ in the Schwarzschild metric marks the difference between flat and curved spacetime. How far from a center of attraction must we be before this curvature becomes extremely small?

A. As a first example, find the value of the radius r from the center of our Sun ($M \approx 1.5 \times 10^3$ meters) such that the curvature factor differs from the value unity by one part in a million. Compare the value of this radius with the radius of Sun ($r_S \approx 7 \times 10^8$ meters).

B. As a second example, find the radial distance from Sun such that the curvature factor differs from the value unity by one part in 100 million. Compare the value of this radius with the average radius of the orbit of Earth ($r \approx 1.5 \times 10^{11}$ meters).

SOLUTION

A. We want $1 - \frac{2M}{r} \approx 1 - 10^{-6}$, which yields

$$r \approx \frac{2M}{10^{-6}} = 2 \times 1.5 \times 10^3 \times 10^6 \text{ meters} = 3 \times 10^9 \text{ meters} \qquad [21]$$

This radius is approximately four times the radius of Sun.

B. This time we want $1 - \frac{2M}{r} \approx 1 - 10^{-8}$, so

$$r \approx \frac{2M}{10^{-8}} = 2 \times 1.5 \times 10^3 \times 10^8 \text{ meters} = 3 \times 10^{11} \text{ meters} \qquad [22]$$

which is approximately twice the radius of Earth's orbit.

Curvature of spaceTIME is needed to describe orbits.

The embedding diagrams, Figures 6 and 7, represent one cut through the spatial part of the Schwarzschild geometry. Time does not enter, since $dt = 0$. There being no place on this surface for changing time, it depicts nothing moving. Therefore this representation has nothing to tell us directly about the motion of particles and light flashes through the spacetime of Schwarzschild geometry (in spite of all the steel balls you have seen rolling on such surfaces in science museums!). In Chapters 3 and 4 we describe trajectories near a black hole, including trajectories that plunge through the Schwarzschild surface at $r = 2M$ "into" the black hole. But first, Section 11 describes the meaning of "far-away time" t in the Schwarzschild metric.

11 Far-Away Time

Freeze space; examine curved spacetime.

It is not enough to know the geometry of space alone. To know the grip of spacetime that tells planets how to move requires knowing the geometry of spacetime. We have to know not merely the distance between two nearby points, P, Q, in space but the interval between two nearby *events*, A, B, in spacetime.

Far-away time t measured at large r.

The Schwarzschild metric uses what we call **far-away time *t***. There can be many remote clocks recording far-away time t. These remote clocks form a latticework that extends in all directions from the isolated black hole. Far from the influence of the black hole, these clocks are in a region of flat spacetime, so they can be synchronized with one another using light flashes similar to the synchronization pulse for free-float frames described in Chapter 1 (Section 9). However, in the present case the synchronizing

SAMPLE PROBLEM 2 Sample of "Radial Stretching"

Verify the statement at the end of Section 4 that for a black hole of one solar mass, the directly measured radial distance calculates as 1.723 kilometers between a shell at $r = 4$ kilometers and a shell at $r = 5$ kilometers. In Euclidean geometry, this measured distance would be 1 kilometer.

SOLUTION

The mass of Sun to four significant figures is $M = 1.477$ kilometers. Express all masses and distances in kilometers. Use the increments of the Schwarzschild metric to obtain

$$dr_{\text{shell}} = \frac{dr}{\left(1 - \frac{2M}{r}\right)^{1/2}} = \frac{1 \text{ kilometer}}{\left(1 - \frac{2.954 \text{ kilometer}}{r}\right)^{1/2}} \qquad [23]$$

Which radius r do we use in the denominator of the right-hand expression? If we use $r = 4$ kilometers, the result is

$$dr_{\text{shell}} = 1.956 \text{ kilometer} \qquad [24.\ r = 4 \text{ km}]$$

On the other hand, if we use $r = 5$ kilometers, the result is

$$dr_{\text{shell}} = 1.563 \text{ kilometer} \qquad [25.\ r = 5 \text{ km}]$$

The trouble here is that the term $2M/r$ changes significantly over the range $r = 4$ kilometers to $r = 5$ kilometers. The radial stretch factor differs from radius to radius. The results in equations [24] and [25] bracket the answer. An exact calculation requires that we sum all the increments of dr_{shell} from $r_1 = 4$ kilometers to $r_2 = 5$ kilometers. This "summation" is an integration. The result of the integration will be Δr_{shell} between the values 1.563 kilometer and 1.956 kilometer:

$$\Delta r_{\text{shell}} = \int_{r_1}^{r_2} \frac{dr}{\left(1 - \frac{2M}{r}\right)^{1/2}} = \int_{r_1}^{r_2} \frac{r^{1/2}\,dr}{(r - 2M)^{1/2}} \qquad [26]$$

This integral is not in a common table of integrals. So make the substitution $r = z^2$, from which $dr = 2z\,dz$. Then the integral and its solution become

$$\int_{z_1}^{z_2} \frac{2z^2\,dz}{(z^2 - 2M)^{1/2}} = z(z^2 - 2M)^{1/2} + 2M \ln\left|z + (z^2 - 2M)^{1/2}\right| \Big|_{z_1}^{z_2} \qquad [27]$$

Here ln is the natural logarithm (to the base e) and | | stands for absolute value. Substitute the values (units omitted)

$$2M = 2.954$$
$$z_1 = \sqrt{4} = 2 \qquad [28]$$
$$z_2 = \sqrt{5} = 2.236$$

and recall that in general ln(B) – ln(A) = ln(B/A). The result is

$$\Delta r_{\text{shell}} = 1.723 \text{ kilometer} \qquad [29]$$

This value, given at the end of Section 4, lies between the bracketing values in equations [24] and [25] for the fixed choices r_1 and r_2.

pulse or pulses must stay in the remote region, not travel through regions where the value of the curvature factor $(1 - 2M/r)$ differs significantly from the value unity. We call the time *far-away time* as read on these clocks at rest with respect to the attracting body. The technical term is **ephemeris time**. Often we say ***t*-coordinate** and give it the symbol t. The increment dt of far-away time appears on the right side of the Schwarzschild metric.

By definition, the time lapse dt between two events is that recorded on a remote clock by an observer far from the attracting mass.

The relation between dt_{shell} and dt can be read directly from the Schwarzschild metric. Think of an Earth clock mounted in a fixed position on the surface of Earth, which we consider to be nonrotating for purposes of this example. The spatial position of the Earth clock does not change between

Your Own Personal Far-Away Clock

If he wishes, an observer at rest on a spherical shell deep in the gravitational pit of a black hole (but outside the horizon!) can have, in addition to his regular shell clock, a second clock that reads far-away time t directly. To this end he needs to carry out two tasks: (1) adjust the rate at which his personal far-away clock runs and (2) synchronize his personal far-away clock with a remote clock that really is far away.

1. **Rate adjustment.** By turning the fast-slow screw on his personal far-away-time clock, he adjusts it to run fast by the factor $1/(1 - 2M/r)^{1/2}$ compared with his regular proper clock, this factor reckoned using the known mass M of the black hole and his measured reduced circumference r. No such rate adjustment is required by the British resident of New York City who always carries a second wristwatch (far-away-time clock) set to Britain's Greenwich time.

2. **Synchronization.** He synchronizes his personal far-away-time clock by some such procedure as the following: (a) Send radially outward to a remote clock an "inquiring" light pulse requesting the time. (b) Upon receiving the inquiry, the remote clock immediately sends a reply flash that encodes its time. (c) When he receives the reply flash, the inner observer assumes that the encoded time is halfway in time between the events of emission of the inquiry flash and reception of the reply flash—and sets his personal far-away-time clock accordingly.

By placing personal far-away clocks on all shells, one can in effect extend the far-away latticework of rods and clocks down to the horizon of a black hole (or down to the surface of a nonrotating star, planet, white dwarf, or neutron star). The far-away time of any event is recorded by these clocks, and the value of the r-coordinate is stamped on every shell.

Relation between dt_{shell} and dt

ticks. Hence dr and $d\phi$ are both equal to zero. Both ticks occur at the clock. Therefore the interval between the ticks on the same clock is the proper time $d\tau$ read on the clock: $dt_{shell} = d\tau$. Two events that occur at the same place evidently have a timelike relation, so choose the timelike version of the Schwarzschild metric. The result was displayed in equation [19]:

$$dt_{shell} = \left(1 - \frac{2M}{r}\right)^{1/2} dt \qquad [19]$$

Equation [19] tells us that an observer remote from Earth records a time separation dt between the arrival of the two pulses that is different from the time recorded on the Earth surface clock that emits the two pulses. In the Schwarzschild metric the curvature factor for time is identical with the curvature factor for space except for one circumstance. In the case of space, *divide* by the curvature factor less than one and get outward distances greater than expected from radial coordinates. In the case of time, *multiply* by the same less-than-one curvature factor and get time lapses near a black hole *less* than expected from the readings on far-away clocks.

Hold it! How can the time dt between two events always be the time lapse as recorded "on a remote clock by an observer far from the center of attraction"? What about two events that occur close to the center of attraction? For example, suppose a clock at rest on Earth's surface ticks twice and we on Earth read off the change in clock time. How is the time lapse between these ticks to be recorded by your remote observer?

There are two equivalent ways to determine far-away time lapse between two events occurring on Earth's surface: (1) Compare the reading of a clock on Earth's surface to the reading of a far-away clock by sending a light signal between them. The Earth-surface clock sends a light signal outward with each tick. The two signals are separated by time dt_{shell} as recorded on the shell clock. An observer remote from Earth receives the two signals and times their separation dt using her clock. (Since time *sep-*

aration is being measured, the flight time of the signals cancels out.) (2) Alternatively, have a far-away clock on the shell, as described in the box on page 2-29.

Gravitational red shift of "climbing light"

Instead of using two separate pulses to make the comparison of shell time with far-away time (equation [19]), use a light wave. Every period (every back-and-forth undulation of the wave) of the light emitted by the close-in clock can be considered as a measure of the time dt_{shell} between its ticks. When this signal is received by a remote observer, the period dt is longer, as given in equation [19]. Visible light with a longer period is more red. As described in Section 5, the *gravitational red shift* is named after this 1915 prediction—and 1960 finding —that the remote observer sees light emitted by the close-in clock to be redder than it was at the point of emission.

Gravitational blue shift of "falling light"

If the signal originates farther from the center of gravity and is sent inward toward the center, the received period decreases. The receiver detects a shorter period than the "proper period" of the sender. The light is shifted toward the blue. We call this the **gravitational blue shift**.

Near a black hole such effects are very much greater than they are near Earth. When the light originates at the black hole horizon ($r = 2M$) and is sent outward radially, the square root of the curvature factor, $(1 - 2M/r)^{1/2}$, becomes zero. Far from the black hole the period dt of the received light measured by the far-away observer becomes infinite, no matter how short is the period dt_{shell} of the emitted light measured by the emitting shell observer. But a light signal with an infinite period is no light signal at all! As described earlier, this is the sense in which no light can escape from the horizon of a black hole—and makes the name "black hole" so descriptive.

Come on! A clock is a clock. You say a lot about exchanging signals between clocks, but nothing about the real time recorded on a real clock. Which observer's clock records the REAL time between a pair of events?

We learned in special relativity that there are measured and verified differences in the time between two events as recorded in different frames in uniform relative motion. Similarly, in general relativity there are measured and verified differences in the time between two events as recorded by different observers near a black hole, even when these observers are relatively at rest. In both special and general relativity *you cannot tell by observation* whether these differences are due to the method of exchanging signals or due to clock rates themselves. The phrase "real time" does not have a unique meaning independent of the means by which that time is measured.

Every schoolboy in the streets of Göttingen

Many not close to his work think of Einstein as a man who could only make headway by dint of pages of complicated mathematics. The truth is the direct opposite. As the great mathematician of the time, David Hilbert, put it, "Every schoolboy in the streets of Göttingen understands more about four-dimensional geometry than Einstein. Yet . . . Einstein did the work and not the mathematicians." The amateur grasped the simple central point that had eluded the expert.

—John Archibald Wheeler

SAMPLE PROBLEM 3 Shining Upward

What happens when light emitted from one shell is absorbed at another shell? In particular, let light be emitted from the shell at $r_1 = 4M$ and absorbed at the shell $r_2 = 8M$. By what fraction is the period of this light increased by the gravitational red shift?

SOLUTION

Equation [19] relates the period dt_{shell} of light measured by a shell observer at r to the period dt measured by a remote observer. But we want the period measured by a second shell observer at a different radius. One way to find the period at the second shell is to use equation [19] twice, once for each observer, and make the remote time lapse dt equal in both cases. Ask the remote observer to hold up a mirror that reflects the light from the inner shell back down to the second shell. This procedure must give the same result as direct transmission between the two shells. Use equation [19] twice.

$$\frac{dt_{\text{shell 1}}}{\left(1-\frac{2M}{r_1}\right)^{1/2}} = dt = \frac{dt_{\text{shell 2}}}{\left(1-\frac{2M}{r_2}\right)^{1/2}} \qquad [30]$$

From which

$$\frac{dt_{\text{shell 2}}}{dt_{\text{shell 1}}} = \frac{\left(1-\frac{2M}{r_2}\right)^{1/2}}{\left(1-\frac{2M}{r_1}\right)^{1/2}} \qquad [31]$$

Substitute $r_1 = 4M$ and $r_2 = 8M$ to yield

$$\frac{dt_{\text{shell 2}}}{dt_{\text{shell 1}}} = \frac{\left(1-\frac{1}{4}\right)^{1/2}}{\left(1-\frac{1}{2}\right)^{1/2}} = \frac{0.866}{0.707} = 1.22 \qquad [32]$$

The period of the light is increased (redshifted) by the factor 1.22 as it climbs from $r = 4M$ to $r = 8M$. This factor would shift, for example, spectral yellow light to deep red.

12 Three Coordinate Systems

(1) Free-float, (2) Spherical shell, (3) Schwarzschild bookkeeping.
Live locally in the first two; span spacetime with the third.

Many possible reference frames

Ride in an unpowered satellite as you fall toward a black hole. Or stand on the scaffolding of a spherical shell and observe this satellite up close as it streaks past. Or analyze the satellite motion using the reduced circumference r, angle ϕ, and far-away time t. Each of these observations requires a different set of spacetime coordinates, a different point of view from which to examine and analyze the motion of the satellite and the structure of spacetime around the black hole. Can a person exist in each of these frames, and if so what is this existence like? How do we describe satellite motion in each one of these frames? And how is the description in one frame related to the description in another frame? We conclude this chapter with brief answers to these questions.

Free-float frame

Nowhere could life be simpler or more relaxed than in a free-float frame, such as an unpowered spaceship falling toward a black hole. The speed of this spaceship increases with time as observed by a sequence of shell observers past which it plunges. For those of us who ride inside, however, the spaceship serves as a special-relativity capsule in which we can be oblivious to the presence of the black hole. Up-down, right-left, back-forth: every direction is the same. We observe that keys, coins, and

The Metric as Micrometer

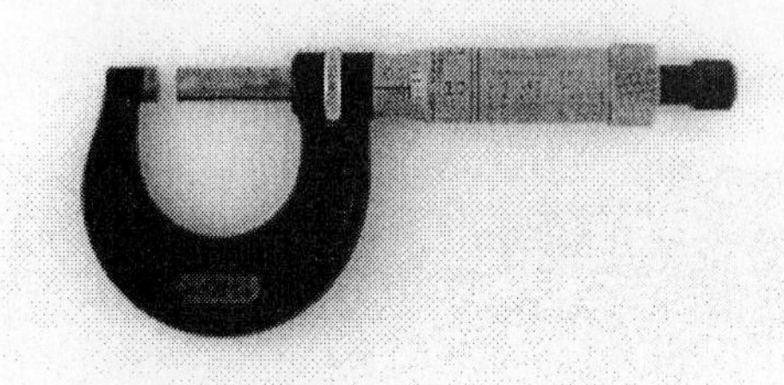

Figure 8 *A micrometer caliper, used to measure small distances, such as the thickness of metal sheet. A calibrated screw on the right meters the gap between cylinders at the left.*

What *is* the metric? What is it *good* for? Think of a **micrometer caliper** (Figure 8), a device used by metalworkers and other practical workers to measure small distances. The worker *owns* the caliper and *chooses* which distance to measure.

The metric is a "four-dimensional micrometer" for measuring the small spacetime separation between a chosen pair of events. You *own* the metric. You *choose* the events whose separation you wish to measure with the metric. The "metric micrometer" translates bookkeeper coordinate increments dr, $d\phi$, and dt into proper time $d\tau$ or proper distance $d\sigma$ between the pair of events *you* choose.

I. *One possible choice for two events*: Two sequential ticks of a clock bolted to a spherical shell. Then $dr = d\phi = 0$ and the proper time lapse $d\tau$ is that read directly on the shell clock, dt_{shell}. The result is equation [19], page 2-23, for the relation between shell time and far-away time.

II. *A second possible choice of two events:* Events at the two ends of a stick held at rest radially between two adjacent shells. Choose $dt = d\phi = 0$, then the proper distance is the directly measured length of the stick dr_{shell}. The result is equation [12], page 2-22, for the relation between directly measured distance between shells and their radial separation dr in the Schwarzschild coordinate.

III. *A third possible choice:* Two ticks on the wristwatch of a particle in free fall inward along a radius. Then $d\phi = 0$ and the proper time is read directly on this wristwatch, which leads to several important results in Chapter 3.

And so on. There is an infinite set of event pairs near one another that you can choose for measurement using your four-dimensional micrometer—the metric.

What advice will the "old spacetime machinist" give to her younger colleague about the practical use of the metric micrometer? She might say the following:

1. Think *events* and separations between pairs of events, not fuzzy concepts like "length" or "location."
2. Do not confuse results from one pair of events with results from another pair of events.
3. Whenever possible, choose the pair of events so that the differential of one or more coordinates is zero.
4. Whenever possible, identify the proper time or proper distance with someone's direct measurement.
5. When a light flash can move directly from one event to another event in the time between them, then the proper time between the events is zero: $d\tau = 0$.

coffee cups remain at rest or if pushed move with constant speed in a straight line. In this free-float frame we use special relativity to compute the spacetime interval between nearby events and analyze collisions as if we were in interstellar space devoid of gravity.

Free-float frame is only local.

However, the simplicity of our free-float frame is only local. We detect curvature of spacetime by the tide-producing relative accelerations among two or more free test particles situated far enough apart or observed for a long enough time to reveal the nonflat nature of nonlocal spacetime. Tidal accelerations drive toward one another objects that lie separated along some directions in the free-float frame; tidal accelerations drive apart particles that lie separated along another direction (Figure 4 of Chapter 1). Detecting the presence of tidal accelerations identifies our reference frame as not free float. To make these tidal accelerations undetectable—by instruments of given sensitivity—we either narrow the spatial extent of our free-float frames or limit the time duration of any particular experiment or both. These constraints are minor in free-float frames situated near Earth or far from any gravitating body. They become progressively and inexorably confining as we approach the center of a black hole. Near this center, general relativity predicts that tidal accelerations tear apart every physical object.

The free-float frame is familiar, simple, and universal. It is the only one of our three frames in which humans can exist near a black hole. Even a body made of steel would be crushed by the "gravitational force" while standing on a spherical shell near the horizon. In contrast, for a large enough black hole the tidal forces can be tolerated by the human body even inside the horizon, at least at a sufficiently great distance from the central singularity. (See Project B, Inside the Black Hole.)

The spherical shell

We live on a (nearly) spherical shell: the surface of Earth. This shell is our home. We habitually construct latticeworks—called buildings—with mutually perpendicular axes on which we mount synchronized clocks. To sit or stand on our spherical shell—Earth's surface—forces us away from the natural motion of a free particle. This departure from natural motion we experience as a "force of gravity" pointing toward the center of Earth. In everyday life we simply include this "force" with other forces in order to get on with the practical analysis of events around us. This approximation works admirably well for the small space and time regions of everyday experience. It works also for high-speed particle interactions in a laboratory, which are over so quickly that the "force of gravity" has little effect. For such experiments the Earth frame is effectively free-float, and analysis using special relativity gives good results even for observations from our shell frame. (See Chapter 1, Section 8.)

Earth's surface is a "spherical shell."

Is special relativity sufficient for the shell observer? Yes, at least locally in space and time. This conclusion is supported by the form of the metric for a shell observer. Substitute into the Schwarzschild metric (equation [10]) the expression for dr_{shell}^2 from equation [12] and the expression for dt_{shell}^2 from equation [19]. The result is

Shell observer uses special relativity.

$$d\tau^2 = dt_{\text{shell}}^2 - dr_{\text{shell}}^2 - r^2 d\phi^2 \qquad [33]$$

The right side of equation [33] contains only coordinate increments measured directly by the shell observer. (Recall that radius r is *defined* so that $rd\phi$ is the directly measured distance along the surface of the shell—Section 4 and the box on page 2-18.) The metric [33] *looks* like that of flat spacetime. But spacetime is *not* flat on a shell near a black hole, and this limits the usefulness of equation [33] to *local* measurements. After all, dt_{shell} and dr_{shell}, along with $rd\phi$, are all functions of the radius r. Still, equation [33] has its uses. For example, it implies that the shell observer measures light ($d\tau = 0$) to have the speed unity—most easily seen by substituting the incremental distance squared ds_{shell}^2 (with a minus sign) for the last two terms in on the right of [33]. Of course every experiment that

Shell frame is also local.

takes place on the shell is influenced by the apparent "gravitational force" due to the fact that the shell is not a free-float frame.

Free-float and shell observers use special relativity to exchange data.

Except for this gravitational force, the **local shell frame** implied by equation [33] is similar to the local free-float frame of a passing plunging observer. In both frames the local speed of light is unity and special relativity correctly describes brief, local experiments. Moreover, the shell observer and a passing free-float observer can use the Lorentz transformation of special relativity to exchange data on local events that are close together in spacetime. We shall use this ability to exchange data many times in the chapters and projects that follow.

In two important ways the local free-float frame is more general than the local shell frame: (1) Special relativity can be made to work well for a longer time in a free-float frame by making the spatial extent smaller, whereas on the shell direct effects of the gravitational force cannot be reduced by any such ruse. (2) The free-float observer can cross the horizon and continue her experiments without interruption, at least until tidal forces overwhelm her. In contrast, inside the horizon neither shells nor shell observers can exist, and equation [33] is useless there.

Different rates for clocks separated vertically

Less familiar than "gravitational" effects to most Earth inhabitants is the difference in rates between clocks separated vertically in the gravitational field, an effect that is a daily experience for anyone designing or predicting the performance of the Global Positioning System, which uses atomic clocks in satellites (see Project A, Global Positioning System).

Vertical separations are affected by curvature.

Least familiar of all effects of general relativity for the shell observer is the difference between the radius of Earth, directly measurable in principle, and the reduced circumference obtained by dividing the circumference by 2π. For two concentric spherical shells near a black hole, the directly measured radial distance between them is greater than the difference in r-values of their reduced circumferences.

These effects in clock rates and vertical separations limit the region of spacetime—space *and* time—in which to analyze experiments using special relativity expressed in shell coordinates. Now we move to a set of coordinates that are global in extent but farther removed from the reality of most experiments carried out near Earth, Sun, or black hole.

Bookkeeper coordinates r, ϕ, and t

Schwarzschild coordinates span full region of spacetime.

A free-float observer makes observations that span only a little patch of spacetime. A local shell observer has similar limitations. In contrast, the coordinates r, ϕ, and t, called **Schwarzschild coordinates**, satisfy the need for a global description of events, a description that encompasses, for example, two events located so far apart in space that they lie on opposite sides of the black hole. The reduced circumference r, the azimuthal angle ϕ, and the far-away time t span all of spacetime surrounding a black hole. They report measurements made in a distant frame at rest with respect to the center of gravitational attraction. Latin was the international language of medieval Europe; the coordinates r, ϕ, and t form the international language for describing events that take place near a black hole.

The **Schwarzschild observer** is a bookkeeper, an archivist, a top-level accountant who rarely measures anything herself. Instead she spends her time examining reports from local shell and free-float observers and combining them to describe events that span spacetime around a black hole. Before accepting a report, this perfectionist demands that coordinate separations between events described in the report be translated into her language: increments in reduced circumference r, azimuthal angle ϕ, and far-away time t. Therefore we call her the **Schwarzschild bookkeeper.**

Schwarzschild observer is a bookkeeper.

An orbiting satellite rapidly emits two sequential flashes as it streaks past two shells concentric to a black hole. The local shell observer measures directly the small separations between the emissions of these flashes: time separation dt_{shell} measured by nearby clocks bolted to the shell and vertical separation dr_{shell} measured with a tape measure. The shell observer also measures the change in azimuthal angle $d\phi$ in the plane of the orbit and verifies by direct meterstick measurement that this increment of angle corresponds to the tangential separation $rd\phi$, where r is the reduced circumference stamped on every shell by the original builders. The shell observer converts dt_{shell} to dt using equation [19] and dr_{shell} to dr using equation [12]—both special cases of the Schwarzschild metric. The shell observer then reports the resulting separations dt, dr, and $d\phi$ to the Schwarzschild bookkeeper.

Now the Schwarzschild bookkeeper swings into action. She knows the space and time coordinates r, ϕ, and t at the beginning of this increment of time. To these coordinates she adds increments dr and $d\phi$ for each lapse of far-away time dt reported by a local shell observer. The result is a table, a diagram, or what we call a **Schwarzschild map** that traces the satellite through spacetime as expressed in her coordinates r, ϕ, and t. Such a map is shown in Figure 9.

Schwarzschild bookkeeper traces out orbit.

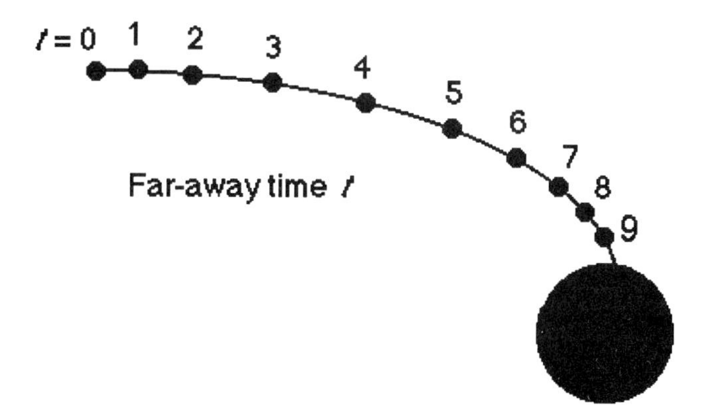

Figure 9 *Schematic Schwarzschild map of the trajectory of an object that plunges into a black hole. Only every hundredth flash is numbered and shown; adjacent flashes are very close together in space and time so they can be observed directly by one or another free-float observer or shell observer. NO ONE observes directly the trajectory shown on this map. Question: Why are numbered event dots closer together near both ends of the trajectory than in the middle of the trajectory? (Answer in Chapter 3.)*

Schwarzschild map is a bookkeeping device.

Notice that the Schwarzschild map of Figure 9 is a summary, an artifact, a bookkeeping device. It indicates events each of which is recorded locally. Nobody observes this entire trajectory directly. The price paid for the universal language of r-coordinate and t-coordinate is the loss of direct experience. No one lives in or on a road map, but we use road maps to describe the territory and plan our trips. Similarly, coordinates r, ϕ, and t are calibrations on a Schwarzschild map of spacetime. These coordinates simply and precisely locate events in the entire spacetime region outside the surface of any spherically symmetric gravitating body. The Schwarzschild map guides our navigation near a black hole.

What if an astronaut riding in the satellite wants to transmit to the Schwarzschild bookkeeper data about separation between events she observes inside her unpowered spaceship? She begins by using special relativity to transform her coordinate separations to values on the passing shell. Then the shell observer can transmit the results to the far-away bookkeeper, as before.

r, ϕ, t are bookkeeping coordinates.

In summary, the full range of coordinates r, ϕ, and t are primarily for bookkeeping. A computer can replace the bookkeeper. Then nobody lives in the coordinates r, ϕ, and t, nobody works there, nobody takes data directly using the wide span of these three coordinates. They form an accounting system, a bookkeeping device, a data bank, a spreadsheet, a tabulating mechanism, an international language, the basis for a spacetime map that describes events and motions in the entire spacetime region surrounding Earth, Moon, Sun, or black hole. For this reason, we often call r, ϕ, and t **bookkeeper coordinates**. The strength of bookkeeper coordinates is universality; their weakness is isolation of most data entries from direct experience.

People can live on a shell and in a free-float frame.

In contrast, people can, in principle, live and work in free-float frames and on spherical shells, taking and analyzing data as if they were in flat spacetime, but unfortunately they can do so only for limited patches of spacetime.

Don't tell me I cannot experience directly the entire trajectory shown in Figure 9! Station me a great distance from a black hole. Then I can view the satellite directly with my eyes as it orbits the black hole or plunges toward it.

True, you see the entire orbit—at least until the satellite reaches the horizon of the black hole. But what you see by eye are not the coordinates r, ϕ, and t of this trajectory. First of all, there is a time delay between emission of a flash by the satellite and the instant at which you see this flash with your eye. The relative delay increases as the satellite moves farther from you or deeper into the gravitational pit.

Second, there is an effect we have not yet mentioned (see Figure 10): Light is deflected in a gravitational field. In 1919 Arthur Eddington verified Einstein's prediction for starlight passing Sun, making Einstein an instant worldwide celebrity. The tiny deflection near Sun becomes dramatic near a black hole (Chapter 5 and Project D, Einstein Rings). As a result, you typically do not see the satellite where it was but in some other direction.

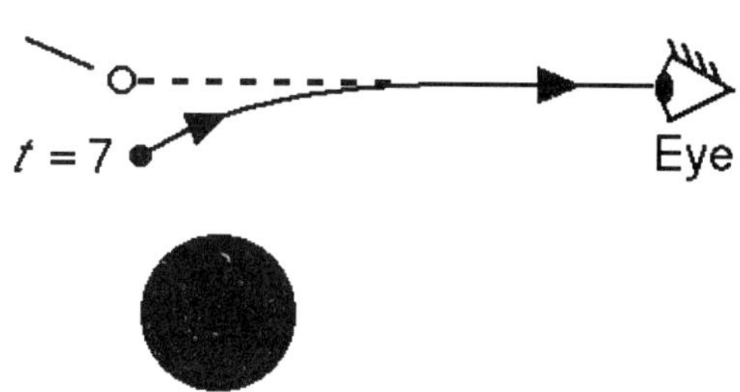

Figure 10 *Schwarzschild map of trajectory of light flash (solid curve) emitted at far-away time $t = 7$ from the plunging object (solid dot) whose trajectory is shown in Figure 9. The light is deflected as it moves outward, leading a remote observer to see the flash emission at a different location (dashed line and open dot).*

Visual appearance is distorted.

Visual appearance can be misleading near a black hole. This misinformation of visual appearance is not new in principle. When we set up the original latticework of rods and clocks that formed the reference frame of special relativity, we limited the observer to collecting data from the recording clocks (Chapter 1, Section 9). We expressly warned the observer about reporting events that he views by eye. Why warn him? Because the speed of light is finite. Light from a distant event can arrive at the observer's eye long after light from a nearer event that actually occurred later as recorded on the latticework of clocks. It is not easy to analyze events when their order is scrambled in the process of observation and recording.

Similar light-delay problems occur in viewing by eye objects in orbit around a black hole. Added to light delay is the visual misinformation about direction due to the deflection of light that results from the curvature of spacetime (Figure 10).

With knowledge of how light moves in the neighborhood of a black hole (Chapter 5), you may be able to reconstruct the Schwarzschild map from your visual observations. But such a reconstruction is quite different from seeing the Schwarzschild map directly.

Schwarzschild metric provides a universal language to fix the location of events.

In the following chapters we use bookkeeper coordinates r, ϕ, and t to describe and predict orbits of satellites and the trajectories of light flashes near a black hole. We use these coordinates to draw Schwarzschild maps of the trajectories. Behind the Schwarzschild map of any orbit stand observations made by free-float observers or shell observers, or predictions of their observations. The Schwarzschild metric is central in the translation of coordinates back and forth between direct observers (shell observers and free-float observers) and between each of these observers and the Schwarzschild bookkeeper.

Schwarzschild lattice

Using what we have learned about spacetime near a nonrotating, spherically symmetric massive body, we can in principle set up a **Schwarzschild lattice**, at rest with respect to the center of gravitational attraction, from which one can read directly the Schwarzschild coordinates r, ϕ, t of an event right down to the horizon. The value of the reduced circumference r is stamped on every spherical shell. Scatter over each shell a set of clocks that read far-away time (box, page 2-29). For a given plane in which a particle orbits, mark the angle ϕ on the shell, with zero angle in some chosen direction. The shell observer and the free-float observer near an event can then read the Schwarzschild coordinates r and ϕ of this event directly from the place on the shell at which the event occurs and the time t on the far-away clock mounted on the shell next to that event.

Construct Schwarzschild lattice from horizon outward to infinity.

This Schwarzschild lattice can stretch from near the horizon outward indefinitely in every direction (from an isolated body). Then the remote Schwarzschild bookkeeper receives, records, and manipulates Schwarzschild coordinates without the need for translation from shell or free-float coordinates. Everyone involved understands the Schwarzschild metric and its predictions, leading to a tidy, agreed-upon system that describes the location of all events. This collection of shells and clocks can then be called the "Schwarzschild observer."

I understand the idea of the three different coordinate systems described in this section, but I don't understand why we can't just use the Schwarzschild lattice alone. It provides time and space measures that each experimenter can agree to by looking at a nearby Schwarzschild clock, measuring tangential distances directly, and reckoning radial separations by subtracting the r-values stamped on each shell. Then we need no translating measurements from one coordinate system to another. The Schwarzschild lattice works fine all by itself. Get rid of all other coordinate systems!

Good point. No one can stop us from using Schwarzschild coordinates alone to design our experiments and predict results. And these predictions will describe what we observe. Then everything is totally consistent and convenient for experimenters scattered throughout the entire region outside the horizon of a black hole. There is a price for this convenience, however; are you willing to pay the following price? Our standard of time is based on the properties of particular atoms. Near a black hole (and near Earth!) an atomic clock "runs slow" when measured using the far-away Schwarzschild time coordinate. And the directly measured radial distance between shells is greater than the difference in r-values between these shells (while directly measured tangential distances are indeed correctly predicted by change in Schwarzschild coordinates). Does curved spacetime cause measuring rods to seem "rubbery," having different apparent lengths when oriented tangentially than when oriented radially? Does curved spacetime force "atomic time" to run at a different rate near a center of attraction than far from this center? Typically, such questions about "reality" are of no interest to people in the field. Whatever point of view leads to correct predictions is fine with them! And using the Schwarzschild lattice, with its seemingly rubbery measuring rods and time-changed atoms, leads to correct predictions. On the other hand, we may be more comfortable assuming that an atomic clock runs at its regular rate when observed at rest near a black hole. In this case we naturally adopt the shell frame for local observations, with its local measuring rods and atom-defined clock times. Each alternative reference frame has its own advantages and brings different perspectives to the structure of spacetime. In our opinion, life as a general relativist, however long, is more fun when you learn to jump mentally from frame to frame! For more, see Kip Thorne's *Black Holes and Time Warps,* Chapter 11, What Is Reality?

13 Summary

The Schwarzschild metric
In general, the metric provides a complete description of spacetime: the curvature of spacetime and the results of measurements carried out with rods and clocks. The metric for flat spacetime is the one that dominated our study of special relativity. However, special relativity cannot describe spacetime globally in the vicinity of a massive object. General relativity can do so, earning the name Theory of Gravitation.

The Schwarzschild metric describes spacetime exterior to the surface of any nonrotating, uncharged, spherically symmetric massive object. It describes spacetime everywhere around a nonrotating, uncharged black hole.

Several conventions make the Schwarzschild metric easy to understand and use:

1. Satellite motion in a plane. A light flash or test particle that moves through Schwarzschild geometry stays in a single spatial plane that passes through the center of the black hole. Describing motion on this plane requires only two space dimensions plus the time.

2. Polar coordinates. Motion with respect to a center is simply described using polar coordinates r and ϕ. For example, the metric for flat spacetime with two spatial dimensions goes from the Cartesian form

$$d\tau^2 = dt^2 - dx^2 - dy^2 \qquad \text{[34. flat spacetime]}$$

to the polar form

$$d\tau^2 = dt^2 - dr^2 - r^2 d\phi^2 \qquad \text{[9. flat spacetime]}$$

3. Mass in units of meters. We measure the mass M of a planet, star, or black hole in units of meters. Equation [5] makes the conversion from mass M_{kg} in kilograms to mass M in meters, using G, the gravitational constant of Newtonian mechanics and c, the speed of light:

$$M = \frac{G}{c^2} M_{\text{kg}} = \left(7.424 \times 10^{-28} \frac{\text{meter}}{\text{kilogram}}\right) M_{\text{kg}} \qquad [5]$$

In length units, the mass of Sun is 1.477 kilometers and the mass of Earth is 0.444 centimeters.

4. Radius as reduced circumference. The presence of the black hole renders impossible the direct measurement of the radial coordinate r of an object or satellite. Instead, define the radius as r = (circumference)/2π, where the circumference is measured around the great circle of a stationary spherical shell concentric to the black hole or center of attraction. As a reminder of this process, we often call r the *reduced circumference*.

5. Time t measured on far-away clocks. To avoid the effects of curvature on clocks, calculate the time, called *bookkeeper time* or *far-away time*, that would be measured on clocks located in flat spacetime far from the attracting body. Give far-away time the symbol t. Light flashes are used for comparison of clock rates and also for communication between a far-away clock and a clock in curved regions of spacetime.

Predictions from the Schwarzschild metric
With these simplifying conventions the Schwarzschild metric in its timelike form can be written

$$d\tau^2 = \left(1 - \frac{2M}{r}\right)dt^2 - \frac{dr^2}{\left(1 - \frac{2M}{r}\right)} - r^2 d\phi^2 \qquad [10]$$

This metric "measures" the separation of a pair of events that have a timelike relation and that occur near one another in spacetime. Various choices of these two events lead to predictions verified by experiment:

Prediction 1. Gravitational red shift. Let the two events be sequential ticks of a clock at rest on a spherical shell near a black hole. *At rest* means that the space separation between events is zero: $dr = d\phi = 0$. The proper time $d\tau$ (defined as the time between the events in a frame in which they occur at the same place) is just the time dt_{shell} read on the shell clock. Then the Schwarzschild metric tells us the relation between shell-time lapse and the lapse of far-away time:

$$dt_{shell} = \left(1 - \frac{2M}{r}\right)^{1/2} dt \qquad [19]$$

Instead of describing ticks on a clock, this equation can measure the period of a steady light wave emitted outward from a spherical shell at radius r. The equation predicts that the period dt measured by a remote observer is *greater* than the period dt_{shell} measured by the observer at the emitting clock. For visible light, longer period means redder light, so the general name for this effect is the *gravitational red shift*.

Prediction 2. Curvature of space. Let the two events occur at the ends of a measuring rod radially oriented with ends at two concentric spherical shells. And let these two events occur at the same far-away time. To analyze these two spacelike events, use the spacelike form of the Schwarzschild metric:

$$d\sigma^2 = -\left(1 - \frac{2M}{r}\right)dt^2 + \frac{dr^2}{\left(1 - \frac{2M}{r}\right)} + r^2 d\phi^2 \qquad [11]$$

For this example, $dt = d\phi = 0$, and the proper distance $d\sigma$ between them (defined as the separation between two events in a frame in which they occur at the same time) is just the radial separation measured by a shell observer:

$$dr_{\text{shell}} = \frac{dr}{\left(1 - \frac{2M}{r}\right)^{1/2}} \qquad [12]$$

The shell observer measures the distance dr_{shell} between shells to be greater than the difference dr between the reduced circumferences of the two shells.

Reference frames

General relativity allows use of any coordinate system whatsoever. We choose three coordinate systems convenient for our purposes: local free-float frames, local frames on spherical shells, and the global frame that employs Schwarzschild coordinates r, ϕ, t. Observers can take measurements directly in free-float frames and on spherical shells, but these measurements are local. In contrast, Schwarzschild coordinates describe events that can span all of spacetime near a massive body, but no one observer can make these measurements directly. Instead we speak of the *Schwarzschild bookkeeper* who records and analyzes events measured by others.

A shell observer and a passing free-float observer compare their local measurements using special relativity, including the Lorentz transformation. The shell observer and the Schwarzschild bookkeeper compare their measurements using equations [12] and [19]. The tangential distance $rd\phi$ is the same in both systems.

One can construct in imagination a *Schwarzschild lattice* of spherical shells, each stamped with the reduced circumference r, angle ϕ, and covered with clocks reading far-away time t. The Schwarzschild lattice can in principle start near the horizon and extend outward indefinitely (from an isolated body). The Schwarzschild coordinates of any event outside the horizon can then be read directly using this lattice. This collection of shells and clocks can collectively be called the "Schwarzschild observer."

Black holes just didn't "smell right"

During the 1920s and into the 1930s, the world's most renowned experts on general relativity were Albert Einstein and the British astrophysicist Arthur Eddington. Others understood relativity, but Einstein and Eddington set the intellectual tone of the subject. And, while a few others were willing to take black holes seriously, Einstein and Eddington were not. Black holes just didn't "smell right"; they were outrageously bizarre; they violated Einstein's and Eddington's intuitions about how our Universe ought to behave. . . . We are so accustomed to the idea of black holes today that it is hard not to ask, "How could Einstein be so dumb? How could he leave out the very thing, implosion, that makes black holes?" Such a reaction displays our ignorance of the mindset of nearly everybody in the 1920s and 1930s. . . . Nobody realized that a sufficiently compact object must implode, and that the implosion will produce a black hole.

—Kip Thorne

14 References

For a fascinating and readable account of the experimental verifications of Einstein's general relativity theory, see *Was Einstein Right? Putting General Relativity to the Test*, Second Edition, by Clifford M. Will, Basic Books/Perseus Group, New York, 1993.

For a graphic fictional description of the effects of tidal accelerations on the human body, see *Neutron Star* by Larry Niven, Del Ray Books, Ballantine Books, New York, 1968. In fact the dangers to the astronaut are much greater than those described in the story.

For a clear description of the physical and mathematical reasonings that underlie Einstein's field equations, see "On the 'Derivation' of Einstein's Field Equations" by S. Chandrasekhar, *American Journal of Physics*, Volume 40, pages 224–234 (1972).

Concerning possible derivations of the Schwarzschild metric, see "The Impossibility of a Simple Derivation of the Schwarzschild Metric" by Ronald P. Gruber, Richard H. Price, Stephen M. Matthews, William R. Cordwell, and Lawrence F. Wagner, *American Journal of Physics*, Volume 56, pages 265–269 (1988).

Initial quote from Einstein's general relativity paper, translated and reprinted as "The Foundation of the General Theory of Relativity," in *The Principle of Relativity* by H. A. Lorentz, H. Weyl, H. Minkowski, Dover Publications, New York, 1952, page 118.

Albert Einstein quote on page 2-4 from *Albert Einstein: Philosopher-Scientist*, edited by Paul Arthur Schilpp, Library of Living Philosophers, Evanston, IL, 1949, pages 65–67. (Second edition 1951.)

Figure 2 and caption, page 2-5, adapted from *Gravitation* by Charles W. Misner, Kip S. Thorne, and John Archibald Wheeler, W. H. Freeman and Company, New York, 1973, page 4. Several paragraphs in this chapter are adapted from the same source. Other material is adapted from *A Journey into Gravity and Spacetime* by John Archibald Wheeler, W. H. Freeman and Company, New York, 1990.

Kip S. Thorne quotes on page 2-7 and 2-42 from *Black Holes and Time Warps: Einstein's Outrageous Legacy*, W. W. Norton Company, New York, 1994, pages 111, 134, and 137. Here is the unedited quote from his page 111: "Thus, Einstein and Newton, with their very different viewpoints on the nature of space and time, give very different names to the agent that causes the crossing. Einstein calls it spacetime curvature; Newton calls it tidal gravity. But there is just one agent acting. Therefore, *spacetime curvature and tidal gravity must be precisely the same thing, expressed in different languages.*"

Woodcut of Copernicus, page 2-8, by Tobias Stimmer taken from "The Universe as Home for Man" by John Archibald Wheeler, in *The Nature of Scientific Discovery*, edited by Owen Gingerich, Smithsonian Institution Press, Washington, D.C., 1975. Caption from *Dictionary of Scientific Biography*, edited by Charles Gillispie, Scribner's, New York, 1971. The order of sentences has been rearranged.

References on the mass of the black hole in the center of our galaxy, page 2-16: "On the Nature of the Dark Mass at the Center of the Milky Way," R. Genzel, A. Eckart, T. Ott, and F. Eisenhauer, *Monthly Notices of the Royal Astronomical Society*, Volume 291, pages 219–234 (1997); "Stellar Proper Motions in the Central 0.1 pc of the Galaxy," A. Eckart and R. Genzel, *Monthly Notices of the Royal Astronomical Society*, Volume 284, pages 576–598 (1997); "High Proper-Motion Stars in the Vicinity of Sagittarius A*: Evidence for a Supermassive Black Hole at the Center of Our Galaxy," A. M. Ghez, B. L. Kelin, M. Morris, and E. E. Becklin, *The Astrophysical Journal*, Volume 509, pages 678–686, December 20, 1998.

Einstein comment to Schwarzschild, page 2-20, full reference in D. Howard and J. Stachel, *Einstein and the History of General Relativity*, Birkhauser, Boston, 1989, page 213.

Stephen Hawking quote on page 2-24 from "The Event Horizon" in *Black Holes*, edited by C. DeWitt and B. S. DeWitt, Gordon and Breach, New York, 1973. Reprinted in *Hawking on the Big Bang and Black Holes*, Advanced Series in Astrophysics and Cosmology, Volume 8, World Scientific, New York, 1993, page 34.

Figure 6, page 2-26, from *Black Holes: The Membrane Paradigm* by Kip S. Thorne, Richard H. Price, and Douglas A. MacDonald, Yale University Press, New Haven, 1986, page 16. This book also enriched our description of the shell frame.

John Wheeler quote, page 2-30, from *At Home in the Universe*, AIP Press, Woodbury, NY, 1994, page 132.

Alberto F. Viscarra suggested the Thinker and Bird interchange on page 2-38.

Quotation in exercise 8 of this chapter: John Archibald Wheeler with Kenneth Ford, *Geons, Black Holes, and Quantum Foam*, W. W. Norton, New York, 1998, page 300.

Chapter 2 Exercises

1. Proper Distance Between Spherical Shells

A black hole has mass $M = 5$ kilometers, a little more than three times that of our Sun. Two concentric spherical shells surround this black hole. The inner shell has r-coordinate (reduced circumference) r; the outer one has r-coordinate $r + dr$, where $dr = 1$ meter. What is the radial separation $d\sigma = dr_{shell}$ between these spherical shells as measured *directly* by an observer on one of these shells? Treat three cases of the reduced circumference r of the inner shell.

A. $r = 50$ kilometers

B. $r = 15$ kilometers

C. $r = 10.5$ kilometers

2. Grazing the Sun

Verify the statement in Section 4 (top of page 2-11) concerning two spherical shells around our Sun. The inner shell, of reduced circumference $r_1 =$ 695 980 kilometer, just grazes the surface of Sun. The outer shell is of reduced circumference one kilometer greater, namely $r_2 = 695\,981$ kilometers. Verify the prediction that the directly measured distance between these shells will be 2 millimeters more than 1 kilometer. (Outbursts and flares leap thousands of kilometers up from Sun's roiling surface, so this exercise is a bit unrealistic, even if we could build these shells!) *Hint:* Use the approximation

$$(1 + x)^n \approx 1 + nx \quad \text{when } |x| << 1 \text{ and } |nx| << 1$$

The exponent n can be a positive or negative integer or a positive or negative fraction.

3. Gravitational Red Shift

Consider a black hole with $M = 1.5$ kilometers, approximately equal to that of our Sun. An observer standing on a spherical shell of reduced circumference r shines a steady laser beam of wavelength 400 nanometers (4×10^{-7} meters: violet light) radially outward. This light is received by a remote observer at a radius very much greater than $2M$. What is the wavelength of the light received by this remote observer in each of the following cases? Note that red light has wavelength 700 nanometers and that, in conventional units,

$$\frac{\text{(wavelength)}}{\text{(period)}} = \frac{\lambda}{T_{sec}} = c$$

Treat three cases: The person shining the laser outward stands on a spherical shell of reduced circumference r with the value

A. $r = 20$ kilometers

B. $r = 5$ kilometers

C. $r = 3.01$ kilometers

D. *Guess:* Suppose the source is aimed in some other direction than the outward radial one, but the laser beam still arrives at a distant observer. Will this distant observer measure the same wavelength as computed in cases A, B, and C, or will the wavelength be different for a non-radial initial direction?

4. How Many Shells?

The President of the Black Hole Construction Company is waiting in your office when you arrive. He is waxing wroth. ("Let Roth wax him for a while."— Groucho Marx)

"You are bankrupting me!" he shouts. "We signed a contract that I would build spherical shells centered on Black Hole Alpha, the shells to be 1 meter apart extending down to the horizon. Now my staff relativist tells me that, starting at any radius whatever outside the horizon, I have to build an infinite number of these shells between that radius and the horizon. We do not have materials for that many!"

"Calm down a minute," you reply. "Black Hole Alpha has a horizon radius $r = 2M = 10$ kilometers = 10 000 meters. You agreed to build 1000 spherical shells starting at reduced circumference $r = 10\,001$ meters, then $r = 10\,002$ meters, then $r = 10\,003$ meters, and so forth, ending at $r = 11\,000$ meters. So what is the problem?"

"I don't know. Maybe we can figure it out if I describe our construction method. My worker robots hang 1-meter rods down vertically (radially) from each completed shell, measure them in place to be sure they are exactly 1 meter long, then weld to the ends of these rods the horizontal (tangential) beams of the next spherical shell of smaller radius."

"Ah, then your Black Hole Construction Company is indeed facing a large unnecessary expense," you conclude. "But I think I can help you."

A. Explain to the President of the Black Hole Construction Company how to alter his construction method in order to complete his obligation to build 1000 correctly spaced spherical shells. Be specific, but do not be a fussbudget.

B. Using the radius of the innermost shell in the relevant equation, make a first estimate of the directly measured separation between the innermost shell and the second shell, the one with the next-larger radius.

C. Using the radius of the second shell, the one just outside the innermost shell, make a second estimate of the directly measured separation between the innermost shell and the second shell.

D. *Optional.* If you are unhappy with the estimates of parts B and C, you may use calculus to make a correct calculation of the directly measured separation between the innermost shell and the one just outside it.

E. Was the contractor's staff relativist correct in predicting an infinite number of shells for this contract, using the method described in the fourth paragraph of this exercise?

5. A Dilute Black Hole

Most descriptions of black holes are so apocalyptic that one gets the impression that black holes are extremely dense objects. Of course, a black hole is not dense throughout, because all matter quickly plunges to the central crunch point. Still, one can speak of an artificial "average density," defined, say, by the total mass M divided by a spherical Euclidean volume of radius $r = 2M$. In terms of this definition, general relativity does not require that a black hole have a large average density. In this exercise you design a black hole with average density equal to that of the atmosphere you breathe on Earth, approximately 1 kilogram per cubic meter. Do all calculations to one-digit accuracy—we want an estimate!

A. From the Euclidean equation for the volume of a sphere

$$V = \frac{4}{3}\pi r^3$$

find an equation for the mass M of air contained in a sphere of radius r, in terms of the density ρ kilograms/meter^3. Use the conversion factor $G/c^2 = 7 \times 10^{-28}$ meter/kilogram (Section 6) to express this mass in meters. (The volume formula used here is for Euclidean geometry, and we are applying it to curved space geometry—so this exercise is only the first step in a more sophisticated analysis.)

B. Set $r = 2M$ for the Schwarzschild radius of the horizon of this black hole. What is the numerical value of $2M$ in meters? (*Hint:* Carry all units along to be sure you have not made a minor error somewhere.)

C. Compare your answer to the radius of our solar system. The mean radius of the orbit of the planet Pluto is approximately 6×10^{12} meters.

D. How many times the mass of our Sun is the mass of your designer black hole?

6. Astronaut Stretching According to Newton

As you plunge feet first radially inward toward the center of a black hole, you are not physically stress-free and comfortable! True, you detect no overall "force of gravity" accelerating you inward. But you do feel a tidal force pulling your feet and head apart and additional forces squeezing your middle inward from the sides like a high-quality corset. When do these tidal forces become uncomfortable? We cannot yet answer this question using general relativity, but Newton is available for consultation, so let's ask him. One-digit accuracy is plenty for numerical estimates in this exercise.

A. *Cleaning up the formula.* Let g_{conv} be the local acceleration of gravity in *conventional* units (meters per second squared). (Here and hereafter "conv" is always a subscript.) Set m times g_{conv} equal to the gravitational force in Newton's law of gravitation (equation [2], page 2-13). In Newton's law use a subscript M_{kg} to remind yourself that the mass of the center of attraction is in kilograms. Now cancel m (mass of your own body), divide through by c^2, convert units, and show that the resulting formula for the *local* acceleration of gravity is $g = M/r^2$. Here M is in meters (equation [5], page 2-14) and $g = g_{conv}/c^2$ is also in geometric units: meter/meter^2 = meter^{-1}. (*Note:* Exponent is *minus* one).

B. Convert g_{conv} at Earth's surface to geometric units to show that it has the approximate value $g_{Earth} = 10^{-16}$ meter^{-1}. Use this value in the remaining parts of this exercise.

What does "uncomfortable" mean? So that we all concur, let us say that you are uncomfortable when your head is pulled upward with half the usual force of gravity on Earth, your middle is in free-float and comfortable (except for being squeezed from the sides), and your feet are pulled downward with half Earth's usual force of gravity on them. In other words, the *difference* in gravitational acceleration between your head and feet as you fall is equal to the acceleration of gravity as it would be measured at Earth's surface: $dg = g_{Earth}$.

C. Take the derivative with respect to r of the local acceleration g found in part A to obtain an expression dg/dr in terms of M and r.

D. How massive a black hole do you want to fall into? Suppose M = 10 kilometers = 10 000 meters, or about seven times the mass of our Sun. Assume your head and feet are 2 meters apart. Find the radius r_{ouch}, in meters, at which you become uncomfortable according to our criterion. Compare this radius with that of Earth, namely 6.4×10^6 meters.

E. Will your discomfort increase or decrease or stay the same as you continue to fall inward toward the center from this radius?

F. Suppose you fall from rest at infinity. How fast are you going when you reach the radius of discomfort r_{ouch} according to Newton? Express this speed as a fraction of the speed of light.

G. Taking the velocity or part F to be constant from that radius to the center, find the corresponding (maximum) time in meters to travel from r_{ouch} to the center, according to Newton. This will be the maximum time lapse during which you will be—er—uncomfortable.

H. What is the maximum time of discomfort, according to Newton, expressed in seconds?

Note 1: If you carried the symbol M for the black hole mass through these equations, you found that it canceled out in expressions for the maximum time lapse of discomfort, parts G and H. In other words your discomfort time is the same for a black hole of *any* mass when you fall from rest at infinity—according to Newton. This equality of discomfort time for all M is also true for the general relativistic analysis.

Note 2: The wristwatch time lapse from any radius to the center according to general relativity is analyzed in Chapter 3 and in an exercise of that chapter. The "ouch time" is examined more thoroughly in Section 7 of Project B, Inside the Black Hole.

7. General Relativity over Chesapeake Bay

On November 22, 1975, a U.S. Navy P3C antisubmarine patrol plane flew back and forth for 15 hours at an altitude of 25 000 to 30 000 feet (7600 to 10 700 meters) over Chesapeake Bay in an experiment organized by Carroll Alley and collaborators. The plane carried atomic clocks that were compared by laser pulse with identical clocks on the ground. During the period of flight, the plane's clock gained 47.2 nanoseconds (47.2×10^{-9} seconds) compared with the ground clock. You have the tools to analyze this time difference. Take 9000 meters as an average altitude. Assume that the plane flew very slowly, just above stalling speed, so that time dilation due to relative speed can be neglected. (In fact, time stretching accounted for about 10 percent of the general relativity effect.) Call the clock on the surface of Earth the shell clock. Let t_{shell} be the time the airplane has been "on station" at 9000 meters altitude and let t be the corresponding far-away time. From equation [19],

$$t_{shell} = \left(1 - \frac{2M}{r}\right)^{1/2} t$$

where t is far-away time.

A. Take the derivative of the expression for t_{shell} with respect to r. (The plane's altitude h is much smaller than the radius r of Earth, so ignore h in places where it is added to r.) Use equations [12] and [19] to convert the resulting dr to dr_{shell} and the resulting t to t_{shell}. Show that the result gives the following relation between dt_{shell} and dr_{shell}:

$$dt_{shell} = \frac{M dr_{shell} t_{shell}}{r^2} \frac{1}{\left(1 - \frac{2M}{r}\right)^{1/2}}$$

B. Further approximations: Neglect any difference between the reduced circumference r of Earth and its radius $r = 6.4 \times 10^6$ meters as we may measure it using one or another method of Euclidean surveying. With this value of r and the mass of Earth in units of length, $M = 4.4 \times 10^{-3}$ meters (Section 6), show that the term $2M/r$ is completely negligible compared with unity so that the last factor in the equation above can be set equal to unity. Let $dr_{shell} = h$, the altitude of the airplane above the surface of Earth. The result is

$$dt_{shell} \approx \left(\frac{Mh}{r^2}\right) t_{shell}$$

C. Now substitute values for M and r into this approximate equation and let $h = 9000$ meters. The quantity in brackets is unitless, so we can express t_{shell} and dt_{shell} in any units we wish, as long as they are both in the same units. Let $t_{shell} = 15$ hours, compute dt_{shell}, the difference in readings between earthbound and airplane clocks, and convert the result to units of nanoseconds $= 10^{-9}$ seconds. Show that the difference in time between the Earth clock and the clock carried by the P3C patrol plane after 15 hours in the air should be approximately 50 nanoseconds, as observed.

Reference: Taylor and Wheeler, *Spacetime Physics*, page 133, which has further references.

8. Black Hole Area Never Decreases

Stephen Hawking discovered that the area of the horizon of a black hole never decreases, this area calculated using the Euclidean formula $A = 4\pi r^2$. Investigate the consequences of this discovery under alternative assumptions described in parts A and B that follow.

Note 1: The rule that the area of a black hole's horizon does not decrease is related in a fundamental way to the statistical law that the disorder (the so-called **entropy**) of an isolated physical system does not decrease. See Thorne, *Black Holes and Time Warps*, pages 422–426 and 445–446, and Wheeler, *A Journey into Gravity and Spacetime*, pages 218–222.

Note 2: The exception to Hawking's discovery is Hawking radiation, which ever so slowly milks energy and mass from the black hole and so decreases the area of its horizon—see the box on page 2-4.

Assume that two black holes coalesce. One of the initial black holes has mass M and the second has mass $2M$.

A. Assume, first, that the masses of the initial black holes simply add to give the mass of the resulting larger black hole. Then what is the r-value of the horizon of this combined black hole as a multiple of one of the initial masses M? What is the area of the resulting horizon?

B. Now make a different assumption about the final mass of the combined black hole. Listen to Wheeler and Ford (*Geons, Black Holes, and Quantum Foam*, page 300):

If two balls of putty collide and stick together, the mass of the new, larger ball is the sum of the masses of the balls that collide. Not so for black holes. If two spinless, uncharged black holes collide and coalesce—and if they get rid of as much energy as they possibly can in the form of gravitational waves as they combine—the square *of the mass of the new, heavier black hole is the sum of the* squares *of the combining masses. That means that a right triangle with sides scaled to measure the masses of two black holes has a hypotenuse that measures the mass of the single black hole they form when they join. Try to picture the incredible tumult of two black holes locked in each other's embrace, each swallowing the other, both churning space and time with gravitational radiation. Then marvel that the simple rule of Pythagoras imposes its order on this ultimate cosmic maelstrom.*

Following this more realistic scenario, find the r-value of the resulting horizon when black holes of masses M and $2M$ coalesce. What is the area of the horizon of the resulting black hole?

C. Do the results of both part A and part B follow Hawking's rule that the horizon's area of a black hole does not decrease?

9. Zeno's Paradox

Zeno of Elea (born about 495 B.C., died about 430 B.C.) developed several paradoxes of motion. One of these states that a body in motion can reach a given point only after having traversed half the distance to that point. But before traversing this half it must cover half of that half, and so on *ad infinitum*. Consequently the goal can never be reached.

A student raises a similar paradox about crossing the horizon. She refers us to the relation between dr_{shell} and dr given by equation [D] in the back of the book:

$$dr_{\text{shell}} = \frac{dr}{\left(1 - \frac{2M}{r}\right)^{1/2}} \qquad \text{[D]}$$

She then asserts, "As r approaches $2M$, the denominator of equation [D] goes to zero, so the distance between adjacent shells becomes infinite. Even at the speed of light, an object cannot travel an infinite distance in a finite time. Therefore nothing can travel across the horizon into the black hole." Analyze and resolve this paradox using the following argument or some other method.

As usual in relativity, the question is: Who measures what? In order to cross the horizon, the plunging object must pass every shell outside the horizon. Each shell observer measures the incremental distance dr_{shell} between his shell and the one below it. Then the observer on that next-lower shell measures the incremental distance between that shell and the one below *it*. By adding up these increments, we can establish a measure of the "summed shell distance" through which the object must move to reach the horizon as measured by the collection of shell observers.

Integrate equation [D] from some initial upper finite radius r_1 to a lower radius r_2, both outside the horizon. You can put the equation into a form to yield a result found in a table of integrals, as follows:

$$\begin{pmatrix}\text{Directly measured}\\ \text{distance from}\\ r_1 \text{ to } r_2\end{pmatrix}$$

$$= \left[r^{1/2}(r-2M)^{1/2} + 2M\ln\{r^{1/2} + (r-2M)^{1/2}\}\right]\Big|_{r_1}^{r_2}$$

Here ln is the natural log function and the expression is valid only for radii greater than that of the horizon. Enter your values for r_1 and r_2 and use the property of the natural log that $\ln B - \ln A = \ln(B/A)$.

Now let $r_2 \longrightarrow 2M$. Show that the resulting distance from r_1 to r_2 is finite as measured by the collection of collaborating shell observers. This is true even though the right side of [D] becomes infinite at $r = 2M$. As a result, the group of shell observers will agree that someone moving radially inward can traverse this distance in a finite time. [The material in Chapter 3 will allow you to calculate this "summed shell time" from r_1 to the horizon for an object falling from rest at infinity.]

Project A Global Positioning System

- *How does the Global Positioning System (GPS) work?*
- *How accurately can I locate myself with the GPS?*
- *How important is general relativity in its operation?*

Project A

Global Positioning System

There is no better illustration of the unpredictable payback of fundamental science than the story of Albert Einstein and the Global Positioning System [GPS] . . . the next time your plane approaches an airport in bad weather, and you just happen to be wondering "what good is basic science," think about Einstein and the GPS tracker in the cockpit, guiding you to a safe landing.

—Clifford Will

1 Operation of the GPS

Do you think that general relativity concerns only events far from common experience? Think again! Your life may be saved by a hand-held receiver that "listens" to overhead satellites, the system telling you where you are at any place on Earth. In this project you will show that this system would be useless without corrections provided by general relativity.

The Global Positioning System (GPS) includes 24 satellites, in circular orbits around Earth with orbital period of 12 hours, distributed in six orbital planes equally spaced in angle. Each satellite carries an operating atomic clock (along with several backup clocks) and emits timed signals that include a code telling its location. By analyzing signals from at least four of these satellites, a receiver on the surface of Earth with a built-in microprocessor can display the location of the receiver (latitude, longitude, and altitude). Consumer receivers are the size of a hand-held calculator, cost a hundred dollars or so, and provide a horizontal position accurate to approximately 25 meters. Originally the satellite signals contained a "jitter" introduced to make civilian receivers intentionally inaccurate. Military receivers could decode and eliminate this jitter. In the spring of the year 2000 this jitter was eliminated. GPS satellites are gradually revolutionizing driving, flying, hiking, exploring, rescuing, map making, and the study of geological motions of Earth's crust.

Airports use one GPS receiver at the control tower and one on the approaching airplane. The two receivers are close together, which cancels errors due to propagation of signals through the upper atmosphere.

Operation of the Global Positioning System

The goal of the Global Positioning System (GPS) is to determine your position on Earth in three dimensions: east-west, north-south, and vertical (longitude, latitude, and altitude). Signals from three overhead satellites provide this information. Each satellite sends a signal that codes where the satellite is and the time of emission of the signal. The receiver clock times the reception of each signal, then subtracts the emission time to determine the time lapse and hence how far the signal has traveled (at the speed of light). This is the distance the satellite was from you when it emitted the signal. In effect, three spheres are constructed from these distances, one sphere centered on each satellite. You are located at the single point at which the three spheres intersect.

Of course there is a wrinkle: The clock in your hand-held receiver is not nearly so accurate as the atomic clocks carried in the satellites. For this reason, the signal from a fourth overhead satellite is employed to check the accuracy of the clock in your hand-held receiver. This fourth signal enables the hand-held receiver to process GPS signals as though it contained an atomic clock.

Signals exchanged by atomic clocks at different altitudes are subject to general relativistic effects described using the Schwarzschild metric. Neglecting these effects would make the GPS useless. This project analyzes these effects.

As a result, measurement of the *relative* position of control tower and airplane is accurate to 1 or 2 meters. This configuration of receivers permits blind landing in any weather. Runway collisions can also be avoided by using this system to monitor positions of aircraft on the ground (a task impossible for the electromagnetic signals of radar).

The timing accuracy required by the GPS is so great that general relativistic effects are central to its performance. First, clocks run at different rates when they are at different distances from a center of gravitational attraction. Second, both satellite motion and Earth rotation must be taken into account; neither the moving satellite nor Earth's surface corresponds to the stationary spherical shell described in Chapter 2. In this project you will investigate these effects.

Your challenge in this project (and in all later projects) is to respond to the numbered queries. (Query 1 for this project appears on page A-4.) Typically, a query contains several related questions. Answer the queries in order, or as assigned to you, or skip to those that interest you the most.

2 Stationary Clocks

Earth rotates and is not perfectly spherical, so, strictly speaking, the Schwarzschild metric does not describe spacetime above Earth's surface. But Earth rotates slowly and the Schwarzschild metric is a good approximation for purposes of analyzing the Global Positioning System.

$$d\tau^2 = \left(1 - \frac{2M}{r}\right)dt^2 - \frac{dr^2}{1 - \frac{2M}{r}} - r^2 d\phi^2 \qquad [1]$$

Apply this equation twice, first to the orbiting satellite clock and second to a clock fixed at Earth's equator and rotating as Earth turns. Both the Earth clock and the satellite clock travel at constant radius around Earth's center.

So $dr = 0$ for each clock. Divide the Schwarzschild metric through by the square of the far-away time dt^2 to obtain, for either clock,

$$\left(\frac{d\tau}{dt}\right)^2 = \left(1 - \frac{2M}{r}\right) - r^2\left(\frac{d\phi}{dt}\right)^2 = \left(1 - \frac{2M}{r}\right) - v^2 \qquad [2]$$

Here $d\tau$ is the wristwatch time between ticks of either clock and $v = r\, d\phi/dt$ is the tangential velocity along the circular path of the same clock as measured by the bookkeeper using far-away time measurement. Write down equation [2] *first* for the satellite, using $r = r_{\text{satellite}}$, $v = v_{\text{satellite}}$, and $d\tau = dt_{\text{satellite}}$ between ticks of the satellite clock, *second* for the Earth clock, using $r = r_{\text{Earth}}$, $v = v_{\text{Earth}}$ and time $d\tau = dt_{\text{Earth}}$ between ticks of the Earth clock, all these for the same time lapse dt on the far-away clock. Divide corresponding sides of these two equations to obtain the squared ratio of time lapses recorded on the satellite and earth clocks:

$$\left(\frac{dt_{\text{satellite}}}{dt_{\text{Earth}}}\right)^2 = \frac{1 - \dfrac{2M}{r_{\text{satellite}}} - v_{\text{satellite}}^2}{1 - \dfrac{2M}{r_{\text{Earth}}} - v_{\text{Earth}}^2} \qquad [3]$$

The general relativistic effects we study are small. How small? Small compared to what? When *must* one use exact general relativistic expressions? When are approximations good enough? These questions are so central to the analysis that it is useful to begin with a rough estimate of the size of the expected effects, not worrying for now about the crudeness of this approximation.

Start by ignoring the motions of satellite clock and Earth surface clock. Ask instead what the difference in clock rates will be for *stationary* clocks at these two radii. Then equation [3] can be written

$$\frac{dt_{\text{satellite}}}{dt_{\text{Earth}}} \approx \frac{\left(1 - \dfrac{2M}{r_{\text{satellite}}}\right)^{1/2}}{\left(1 - \dfrac{2M}{r_{\text{Earth}}}\right)^{1/2}} = \left(1 - \frac{2M}{r_{\text{satellite}}}\right)^{1/2}\left(1 - \frac{2M}{r_{\text{Earth}}}\right)^{-1/2} \qquad [4]$$

You will show in Query 7 that the radius of a 12-hour circular orbit is about 26.6×10^6 meters from Earth's center. You will find values for the radius and mass of Earth among the constants inside the back cover.

We now make first use of an approximation that appears repeatedly in this project:

$$(1 + d)^n \approx 1 + nd \qquad \text{provided} \qquad |d| \ll 1 \qquad \text{and} \qquad |nd| \ll 1 \qquad [5]$$

Here the two vertical lines mean "absolute value."

Approximation [5] is accurate for any real (positive or negative, integer or fractional) value of the exponent n provided the absolute values of d and nd are both very much less than unity. Equation [5] is used so often in this book that it is rewritten for general reference as equation [E] on the last page of the book.

QUERY 1 **Formula: Clock rate difference due to height.** Apply approximation [5] to the two parenthetical expressions on the right of equation [4]. Multiply out the result to show that

$$\frac{dt_{\text{satellite}}}{dt_{\text{Earth}}} \approx 1 - \frac{M}{r_{\text{satellite}}} + \frac{M}{r_{\text{Earth}}} - \frac{M}{r_{\text{satellite}}}\frac{M}{r_{\text{Earth}}} \qquad [6.\ \text{for } v = 0]$$

QUERY 2 **Improved approximation.** What are the approximate values of M/r_{Earth} and $M/r_{\text{Satellite}}$? Make an argument that the last term on the right of [6] can be neglected in comparison with the other terms on the right, leading to the result for stationary satellite and Earth clocks:

$$\frac{dt_{\text{satellite}}}{dt_{\text{Earth}}} \approx 1 - \frac{M}{r_{\text{satellite}}} + \frac{M}{r_{\text{Earth}}} \qquad [7.\ \text{for } v = 0]$$

QUERY 3 **Numerical approximation.** Substitute numbers into equation [7] and find the numerical value of b in the following equation:

$$\frac{dt_{\text{satellite}}}{dt_{\text{Earth}}} \approx 1 + b \qquad [8.\ \text{for } v = 0]$$

The number represented by b in equation [8] is an estimate of the fractional difference in rates between stationary clocks at the position of the satellite and at Earth's surface. Is this difference negligible or important to the operation of the GPS? Suppose the timing of a satellite clock is off by 1 nanosecond (10^{-9} second). In 1 nanosecond a light signal (or a radio wave) propagates approximately 30 centimeters, or about one foot. So a difference of, say, hundreds of nanoseconds will create difficulties.

QUERY 4 **Synchronization discrepancy after one day.** There are 86,400 seconds in one day. To one significant figure, the satellite clocks and Earth clock go out of synchronism by about 50 000 nanoseconds per day due to their difference in altitude alone. Find the correct value to three-digit accuracy.

The satellite clock will "run fast" by something like 50 000 nanoseconds per day compared with the clock on Earth's surface due to position effects alone. Clearly general relativity is needed for correct operation of the Global Positioning Satellite System! On the other hand, the *fractional* difference between clock rates at the two locations (at least the fraction due to difference in radius) is small.

In addition to effects of position, we must include effects due to motion of satellite and Earth observer. In which direction will these effects influence the result? The satellite clock moves faster than the clock revolving with Earth's surface. But special relativity tells us that (in an imprecise summary) "moving clocks run slow." This prediction agrees with the negative sign of v^2 in equations [2] and [3]. Therefore we expect the effect of motion to *reduce* the amount by which the satellite clock runs fast compared to the Earth clock. In brief, when velocity effects are taken into account, we expect the satellite clock to run faster than the Earth clock by *less* than the estimated 50 000 nanoseconds per day. We will need to check our final result against this prediction.

3 Clock Velocities

Now we need to take into account the velocities of Earth and satellite clocks to apply the more complete equation [3] to our GPS analysis. What are the values of the clock velocities v_{Earth} and $v_{\text{satellite}}$ in this equation, and who measures these velocities? For the present we find the simplest measure of these velocities, using speeds calculated from Euclidean geometry and Newtonian mechanics. Newton uses a fictional "universal" time t, so Newtonian results will have to be checked later in a more careful analysis.

QUERY 5	**Speed of a clock on the equator.** Earth's center is in free float as Earth orbits Sun and rotates on its axis once per day (once per 86 400 seconds). With respect to Earth's center, what is the speed v_{Earth} of a clock at rest on Earth's surface at the equator? Use Newtonian "universal" time t. Express your answer as a fraction of the speed of light.

What is the value of the speed $v_{\text{satellite}}$ of the satellite? Newton tells us that the acceleration of a satellite in a circular orbit is directed toward the center and has the magnitude $v_{\text{conv}}{}^2/r$, where v_{conv} is measured in conventional units, such as meters per second. The satellite mass m multiplied by this acceleration must be equal to Newton's gravitational force exerted by Earth:

$$\frac{GM_{\text{kg}}m}{r_{\text{satellite}}^2} = \frac{mv_{\text{conv}}^2}{r_{\text{satellite}}} \qquad [9]$$

Equation [9] provides one relation between the velocity of the satellite and the radius of its circular orbit. A second relation connects satellite velocity and orbit radius to the period of one revolution. This period T is 12 hours for GPS satellites:

$$v_{\text{conv}} = \frac{2\pi r_{\text{satellite}}}{T_{\text{seconds}}} \qquad [10]$$

QUERY 6 **Geometric units.** In equations [9] and [10] convert the mass M to units of meters and convert satellite speed to a fraction of the speed of light. Leave T in units of seconds. Then eliminate the radius $r_{\text{satellite}}$ between these two equations to find an expression for $v_{\text{satellite}}$ in terms of M and T_{seconds} and numerical constants.

4 The Final Reckoning

QUERY 7 **Satellite radius and speed.** Find the numerical value of the speed $v_{\text{satellite}}$ (as fraction of the speed of light) for a satellite in a 12-hour circular orbit. Find the numerical value of the radius $r_{\text{satellite}}$ for this orbit—according to Newton and Euclid.

Now we have numerical values for all the terms in equation [3] and can approximate the difference in rates for satellite clocks and Earth clocks.

QUERY 8 **Formula: Clock rate difference.** Take the square root of both sides of equation [3]. Do not substitute numerical values yet. Rather, for both numerator and denominator in the resulting equation, use the approximation [5], as follows, In the numerator, set

$$d = -\frac{2M}{r_{\text{satellite}}} - v_{\text{satellite}}^2 \qquad [11]$$

In the denominator, use the same expression for d but with "Earth" as the subscripts. Carry out an analysis similar to that in Query 2 to preserve only the important terms. Show that the result is

$$\frac{dt_{\text{satellite}}}{dt_{\text{Earth}}} \approx 1 - \frac{M}{r_{\text{satellite}}} - \frac{v_{\text{satellite}}^2}{2} + \frac{M}{r_{\text{Earth}}} + \frac{v_{\text{Earth}}^2}{2} \qquad [12]$$

QUERY 9 **Numerical clock rate difference.** Substitute values for the various quantities in equation [12]. Result: To *two* significant figures, the satellite clock appears to run faster than the Earth clock by approximately 39 000 nanoseconds per day. Give your answer to *three* significant figures.

Section 2 described the difference in clock rates due only to difference in altitude. We predicted at the end of Section 2 that the full general relativistic treatment would lead to a *smaller* difference in clock rates than the altitude effect alone. Your result for Query 9 verifies this prediction. In the following section we examine some of the other approximations made in the analysis.

A practical aside: When Carroll O. Alley was consulting with those who originally designed the Global Positioning System, he had a hard time convincing them not to apply *twice* the correction given in equation [12]: first to account for the difference in clock rates at the different altitudes and second to allow for the blue shift in frequency for the signal sent downward from satellite to Earth. There is only one correction; moreover there is no way to distinguish what is the "cause" of this correction. Hear what Clifford Will has to say on the subject, as he describes the difference in rates between one clock on a tower and a second clock on the ground:

> *A question that is often asked is, Do the intrinsic rates of the emitter and receiver or of the clock change, or is it the light signal that changes frequency during its flight? The answer is that it doesn't matter. Both descriptions are physically equivalent. Put differently, there is no operational way to distinguish between the two descriptions. Suppose that we tried to check whether the emitter and the receiver agreed in their rates by bringing the emitter down from the tower and setting it beside the receiver. We would find that indeed they agree. Similarly, if we were to transport the receiver to the top of the tower and set it beside the emitter, we would find that they also agree. But to get a gravitational red shift, we must separate the clocks in height; therefore, we must connect them by a signal that traverses the distance between them. But this makes it impossible to determine unambiguously whether the shift is due to the clocks or to the signal. The observable phenomenon is unambiguous: the received signal is blue shifted. To ask for more is to ask questions without observational meaning. This is a key aspect of relativity, indeed of much of modern physics: we focus only on observable, operationally defined quantities, and avoid unanswerable questions.*

5 Justifying the Approximations

We calculated the speed of a satellite in circular orbit and the speed of the clock on Earth's surface using Euclidean geometry and Newtonian mechanics with its "universal time." Now, the numerator in each expression for speed, namely $rd\phi$, is the same for Euclidean geometry as for Schwarzschild geometry because of the way we defined r in Schwarzschild spacetime. However, the time dt in the denominator of the speed is not the same for Newton as for Schwarzschild. In particular, the derivation of equation [3] assumes that the speeds in that equation are to be calculated using changes in far-away time dt. Think of a spherical shell constructed at the radius of the satellite orbit and another "shell" that is the surface of Earth. Then our task boils down to estimating the difference between far-away time dt and shell time dt_{shell} in each case, which can be done using our equation [C] in Selected Formulas at the end of this book.

$$dt_{\text{shell}} = \left(1 - \frac{2M}{r}\right)^{1/2} dt \qquad [13]$$

QUERY 10 **General-relativistic times *vs.* Newton's universal time.** For comparison, equate Schwarzschild far-away time with Newton's universal time and see what difference there is between this time and "shell time" either at the radius of the satellite orbit or the surface of Earth. Use equation [13] and the approximation equation [5] to set up an approximate relation between two measures of velocity in each case:

$$\frac{rd\phi}{dt} \approx \frac{rd\phi}{dt_{\text{shell}}}(1-q) \qquad [14]$$

where q is a small number. Find an algebraic expression for q. Then find numerical values of q both for Earth's surface and at the orbital radius of the satellite. Use these results to estimate the difference that changed velocity values will make in the numerical result of Query 9. Is this difference significant?

Two Notes

Note 1: The approximate analysis in this project also assumed that the radius $r_{\text{satellite}}$ of the circular orbit of the satellite is correctly computed using Newtonian mechanics. The Schwarzschild analysis of circular orbits is carried out in Chapter 4. When you have completed that chapter, you will be able to show that this approximate analysis is sufficiently accurate for our purposes.

Note 2: Our analysis assumed the speed v_{Earth} of the Earth clock to be that of the speed of the equator. One might expect that this speed-dependent correction would take on different values at different latitudes north or south of the equator, going to zero at the poles where there is no motion of the Earth clock due to rotation of Earth. In practice there is no latitude effect because Earth is not spherical; it bulges a bit at the equator due to its rotation. The smaller radius at the poles increases the M/r_{Earth} term in equation [12] by the same amount that the velocity term decreases. The outcome is that our calculation for the equator applies to all latitudes.

6 Summary

A junior traveler, making her first trip by train from the United States into Mexico, sees the town of Zacatecas outside her window and reassures herself by the marginal note in the guidebook that this ancient silver-mining town is 1848 kilometers from San Diego, California, and 1506 kilometers from New Orleans, Louisiana. On a surface, two distances thus suffice to fix location. But in space it is three. Find those three distances, to each of three nearest satellites of the Global Positioning System, by finding the time taken by light or radio pulse to come from each satellite to us. Simple enough! Or simple as soon as we correct, as we must and as we have, for the clock rates at each end of the signal path. (1) General relativity predicts that both the relative altitudes and the relative speeds of satellite and

Earth clocks affect their relative rates. (2) The clock in the hand-held receiver on Earth is far less accurate than the atomic clock in each satellite, so the signal from a fourth satellite is employed to correct the Earth clock. With these corrections, we can use the Global Positioning System to locate ourselves anywhere on Earth with an uncertainty of only a few meters.

7 References and Acknowledgments

Initial quote from "A Tale of Einstein, Congress and Safe Landings" by Clifford Will, an online posting by the Editorial Services, Washington University, St. Louis.

Carroll O. Alley provided much of the information used in this project.

Carroll O. Alley, "Proper Time Experiments in Gravitational Fields With Atomic Clocks, Aircraft, and Laser Light Pulses," in *Quantum Optics, Experimental Gravity, and Measurement Theory*, edited by Pierre Meystre and Marlan O. Scully, Plenum Publishing, New York, 1983, pages 421–424.

Neil Ashby, "Relativistic Effects in the Global Positioning System," in *Gravitation and Relativity at the Turn of the Millennium*, Proceedings of the GR-15 conference held at IUCAA, Pune, India, December 16–21, 1997, Naresh Didhich and Jayant Narlikar, editors, Inter-University Centre for Astronomy and Astrophysics, Pune, India, 1998, pages 231–258.

Quote about clocks on the ground and on a tower: Clifford M. Will, *Was Einstein Right? Putting General Relativity to the Test*, Second Edition, Basic Books/Perseus Group, New York, 1993, pages 49–50.

Chapter 3 Plunging

- *Am I comfortable as I fall toward a black hole?*
- *How fast am I going when I reach the horizon? Who says?*
- *How long do I live once I fall into a black hole?*
- *What happens to the mass of a black hole when a stone falls into it?*

CHAPTER 3

Plunging

If you will not take the answer too seriously, and consider it only as a kind of joke, then I can explain [general relativity] as follows. It was formerly believed that if all material things disappeared out of the universe, time and space would be left. According to the relativity theory, however, time and space disappear together with the things.

— Albert Einstein

1 The Principle of Extremal Aging

"Go straight!" spacetime shouts at the stone.
The stone's wristwatch verifies that its path is straight.

All the exotic talk about curved spacetime geometry near stars and black holes leaves us unprepared for a revelation about motion right at home: Schwarzschild geometry correctly describes the motions of baseballs and stones near the surface of Earth. Even more surprising: Analyzing trajectories of near-Earth objects using Schwarzschild geometry prepares us to go back and describe trajectories around stars, white dwarfs, neutron stars, and black holes.

Thrown stone follows straight path in free-float frame.

Throw a stone and let it fall back to Earth. The stone follows a parabolic path in space, the solid curve in the diagram to the left in Figure 1. At the beginning and end of this path, fix two events in space and time: Event 1, initial launch; Event 2, final impact. Why does the stone follow the particular path in space between Event 1 and Event 2, shown as a solid line in Figure 1? Why not hurry faster along a higher parabolic path, the upper dashed line in Figure 1, to get back in time for the appointed impact? Or move slower along a lower parabolic path, the lower dashed line? Why not some entirely different trajectory between these two events? What command does spacetime give to the stone, telling it how to move?

Spacetime shouts, "Go straight!" The free stone obeys. What does "straight" mean? Straight with respect to what? We know the answer: The path of the stone is straight in a free-float frame. Ride in a free-float frame that rises and falls vertically in concert with the stone, as shown in the right diagram of Figure 1. With respect to the free-float frame, the stone moves on a straight path during the entire trip between launch (Event 1) and impact (Event 2).

"Straight" means straight *worldline*.

Not only must the trajectory of the stone be straight in an inertial frame, but the stone must also move with constant speed as measured in that frame. Figure 2 shows a plot of the position of the stone (horizontal axis)

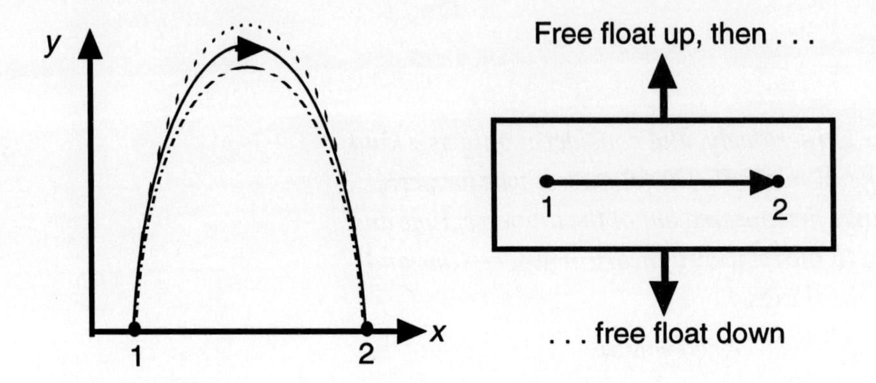

Figure 1 *Parabolic path of a stone (solid line, left diagram) connecting launch (Event 1) and impact (Event 2). Dashed lines show alternative spatial paths between these two events, alternatives that the stone does not take.* **(Why not?)** *On the right is a free-float frame that rises and falls with the stone. With respect to this free-float frame, the stone follows a straight path. Plotting its motion as a function of time yields a straight worldline (Figure 2).*

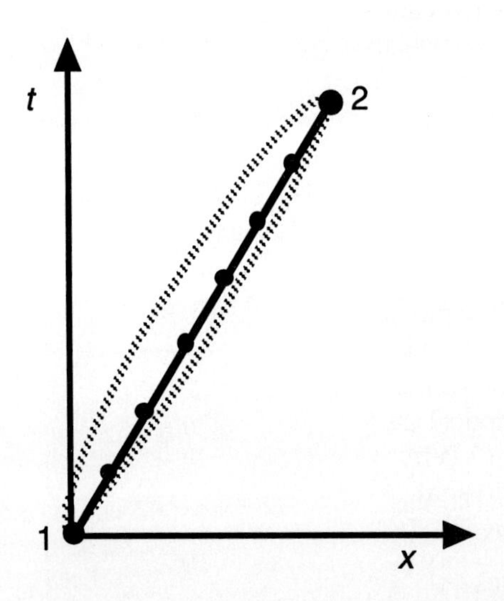

Figure 2 *Spacetime diagram of the stone's worldline in the free-float frame that rises and falls with the stone (right diagram of Figure 1). This worldline is straight between launch (Event 1) and impact (Event 2). Intermediate clock ticks are shown as event points along the worldline. Curved dashed lines between events 1 and 2 represent alternative worldlines of smaller aging, alternative worldlines that the stone does not take.* **(I said, Why not?!)**

as time passes (vertical axis). The line traced out by the motion of the stone as it changes spatial location as a function of time is called a **worldline.** Constant velocity results in a *straight worldline*. Nature's command to the stone in its general form is *"Follow a straight worldline in a local inertial frame."* No description could be simpler.

Stone carries a wristwatch.

"Follow a straight worldline!" is the command by which spacetime grips mass, telling it how to move. The stone carries a wristwatch. During the trip the stone's wristwatch ticks off the time lapse between events of launch and impact. Between Event 1 and Event 2 in Figure 2, the wristwatch ticks off intermediate events along the worldline of the particle in the spacetime diagram of the free-float frame (event points on the straight worldline of Figure 2).

More About the Black Hole

The term "black hole" was adopted in 1967 (by John Wheeler), but the concept is old. As early as 1783, John Michell argued that light must "be attracted in the same manner as all other bodies" and therefore, if the attracting center is sufficiently massive and sufficiently compact, "all light emitted from such a body would be made to return toward it." Pierre-Simon Laplace came to the same conclusion in 1795, apparently independently, and went on to reason that "it is therefore possible that the greatest luminous bodies in the universe are on this very account invisible."

Michell and Laplace used Isaac Newton's "action at a distance" theory of gravitation in analyzing escape of light from or its capture by an already existing compact object. (See the box "Newton Predicts the Horizon of a Black Hole?" on page 2-22.) But is such a static compact object possible? In 1939, J. Robert Oppenheimer and Hartland Snyder published the first detailed treatment of gravitational collapse within the framework of Einstein's theory of gravitation. Their paper predicts the central features of nonspinning black holes described in this book.

Ongoing theoretical study has shown that the black hole is the result of natural physical processes. A nonsymmetric collapsing system is not necessarily blown apart by its instabilities but can quickly—in seconds!—radiate away its turbulence as gravitational waves and settle down into a stable structure. In its final form a black hole has three properties and three properties only: mass, charge, and angular momentum. No other property remains of anything that combined to form the black hole, from pins to palaces. This absence of all detail beyond these three properties has led to the saying (also by Wheeler) "The black hole has no hair."

An uncharged nonspinning black hole is completely described by the Schwarzschild metric (the generalization of equation [1] to three spatial dimensions) derived in 1915 by Karl Schwarzschild from Einstein's equations for general relativity. The energy of a nonspinning black hole is not available for use outside its horizon. For this reason, a nonspinning black hole is called a "dead black hole."

In contrast to the spinlessness of a dead black hole, the typical black hole, like the typical star, has a spin, sometimes a great spin. The energy stored in this spin, moreover, is available for doing work: for driving jets of matter and for propelling a spaceship. In consequence, the spinning black hole deserves and receives the name "live black hole." It has an angular momentum of its own.

A spinning black hole—or any spinning mass, it turns out—drags around with it spacetime in its vicinity. This "frame-dragging effect" is unquestionably measurable, even near our spinning Earth, by techniques now under development. One technique employs a small gyroscope (a quartz ball 4 centimeters in diameter housed in Gravity Probe B, soon to be launched); the other uses a large "gyroscope" (a satellite orbiting Earth). In both cases the axis of the spinning gyroscope is dragged around by a fraction of a second of arc per year. To measure this tiny precession against background effects is the challenge. It will be exciting to see for the first time this new general relativistic effect. Theory predicts that near a rapidly spinning black hole, frame-dragging effects can be large, even inexorable, dragging along nearby spaceships no matter how strong their rockets.

The metric for an uncharged spinning black hole was derived by Roy P. Kerr in 1963. (See Project F, The Spinning Black Hole.) In 1965 Ezra Theodore Newman and others solved the Einstein equations for the spacetime geometry around a *charged* spinning black hole. Subsequent theorems have proved that around a steady-state black hole of specified mass, charge, and angular momentum, Kerr-Newman geometry is the *only* solution to Einstein's field equations.

Natural motion has maximum wristwatch time.

How is the straight worldline different from all other possible worldlines that connect Event 1 and Event 2 (dashed lines in Figure 2)? We know the answer to that question too, from the Principle of Extremal Aging: The actual worldline has the longest wristwatch time of all possible worldlines between these two events. The free stone progresses uniformly from one event to the other, without jerks, jolts, or accelerations, thereby recording the longest possible time on its wristwatch between these two events. In contrast, a frantic traveler starting at the same Event 1 races at near-light speed to Moon, then streaks back in time for obligatory Event 2. The frantic traveler's wristwatch reads less elapsed time between Events 1 and 2 than does the wristwatch of the relaxed stone. The essential lesson of the Twin Paradox (Section 4 of Chapter 1) is that the *natural* motion between two events has *maximum* wristwatch time.

No frantic trip as far as Moon is necessary to demonstrate the basic principle: *any* deviation whatsoever from the straight worldline, no matter how small, leads to a shorter elapsed wristwatch time. The stone's wristwatch, accurate beyond all human timepieces, detects this difference and traces out the worldline of maximum wristwatch time. Wonder of wonders, the stone sniffs out and follows the worldline of maximum proper time without any wristwatch at all! How? Simply by going straight in local spacetime.

"Aging" measures total elapsed wristwatch time.

We use the word **aging** to describe the total elapsed proper time—the elapsed wristwatch time—along any worldline a particle takes from some initial event to another final event. Then the actual worldline the stone takes through spacetime is the worldline of maximal aging. Spacetime's command to the stone can be rephrased: "Follow the worldline of maximal aging!" From this simple command flows every description of motion in the remainder of this book. Amen.

Natural motion implies extremal aging.

Well, not quite "Amen." As described in Section 4 of Chapter 1, it is possible that the stone will follow a worldline not of *maximal* aging but of *minimal* aging. A noun that covers both cases is *extremum.* The corresponding adjective is *extremal*. The technical term for such a worldline is **geodesic**. To cover this unusual case, from now on we shift from maximum and maximal to the noun *extremum* and the adjective *extremal*. The *Principle of Extremal Aging* summarizes the result that a free particle follows a geodesic, a worldline of extremal aging. This fussy detail is presented for completeness. Tuck it away; all the worldlines we examine in this book, all our geodesics, result in *maximal* aging.

> ***Principle of Extremal Aging** (repeated): The path that a free particle takes between two events in spacetime is the path for which the time lapse between these events, recorded on its wristwatch, is an extremum.*

Einstein: There is no "gravitational force"!

Figures 1 and 2 witness that, for slow speed and weak gravitational interaction, Newton's mechanics correctly describes the contrast between a straight worldline in spacetime and a curved path in space. So what is new about relativity? On the theory side, Einstein says that you can do away entirely with Newton's gravitational force, substituting instead the idea of geodesic: A free test particle moves along a worldline *straight* in *spacetime* as described with respect to every *local* free-float frame through which it passes. But the result may be a path *curved* in *space* as described by *global* Schwarzschild coordinates.

You claim that a worldline straight in every local free-float frame can nevertheless be curved in space, observed using the global Schwarzschild coordinates. You predict curved satellite orbits around a star. But your whole idea is obviously false; there is no way that tiny STRAIGHT motions can be added up to give overall CURVED motion!

Straight or curved? The description depends on the reference frame and on what kind of graph you draw. Figure 1 shows, in the left diagram, the curved path in space traced by a projectile observed with respect to Earth's surface and, in the right diagram, the straight path in space of the same projectile observed in a free-float frame. The projectile moves also with constant velocity in the free-float frame, a fact witnessed by the worldline of constant slope in Figure 2—the straight worldline in spacetime. The motion so described is as straight as it can possibly be—a geodesic. Yet for the Earth observer the path in space is curved.

Moon moves straight in local spacetime.

Figure 1 describes motion confined to a local region of spacetime, where we can switch back and forth between a frame at rest with respect to Earth and a free-float frame in which spacetime is effectively flat for the flight time of the stone. In contrast, no free-float frame spans the entire orbit of Moon around Earth. Yet here too Moon moves straight in the spacetime of the local free-float frame. It follows a *geodesic* in spacetime, while its *trajectory in space* is curved.

General relativity stitches together "quilt squares" of local free-float frames.

General relativity stitches together the quilt squares of local free-float frames into a full quilt that covers wide regions of spacetime. Einstein predicts basically the same orbits as Newton does for motion near Earth and Sun. But even here Einstein corrects small discrepancies, while predicting motions different from Newton around compact stellar objects (and for the Universe as a whole!). In all known cases in which the two theories conflict, experiment verifies the predictions of general relativity. (See Projects A, C, D, and E.)

2 Rebel Stone and Obedient Computer

Try all possible worldlines between initial and final events.
The free stone chooses the worldline of extremal aging.

Suppose that the stone rebels; let it disobey the command issued by spacetime to follow the worldline of extremal aging. Or, more realistically, think of an external experimenter grasping the stone and forcing it to move along a worldline that it would not freely follow. This rebellion, this deviation from the natural, is partial: the stone is present at the two obligatory ceremonial events of launch and impact. However, the stone does not keep its appointments with intermediate events along the standard, the natural, the actual worldline. Perhaps it moves slower than normal between adjacent points on parts of the spatial path and faster than normal on other parts of the path. Or perhaps it wanders off the spatial path entirely, taking some other trajectory. Nevertheless, its wristwatch continues conscientiously to tick off wristwatch time—accumulated aging—along this alternative worldline. In due course the stone arrives at the event of impact. The stone's penalty for its errant behavior? A mild punishment! At the event of impact the stone's wristwatch will read less time than it would if it had obeyed the command of spacetime (or more time if the natural worldline happens to be one of minimal aging). The errant stone's aging for this worldline will not be extremal among all possible worldlines between the events of launch and impact.

Pick the actual worldline: the one with extremal wristwatch time.

The disobedient stone shows us how to predict—simply, accurately, powerfully—the worldline of any test object moving freely in any region of spacetime with known metric, no matter whether the spacetime region is curved or flat. The recipe could hardly be simpler: "Behave like a large number of rebellious stones!" Each rebellious stone follows a different worldline from initial event to final event. Compute the aging along each alternative worldline—the sum of incremental wristwatch times between each pair of adjacent events along the candidate worldline. Among all these candidates, select the worldline with extremal aging. The extremal-aging worldline is the one taken by the real stone, the stone moving freely between fixed initial and final events.

Your theory is fundamental and interesting—and useless! How can we predict the motion of the stone if we need to know from the beginning the "fixed" final event on the worldline—the place and time of impact? The location of that final event is just what the laws of motion are supposed to TELL us: Given the launch point and the initial velocity, where will the projectile impact? Usually we don't even care WHEN it reaches that point. For such an analysis, your prescription is useless.

No, not useless. Think of trapshooting (or skeet shooting), a sport in which we fire buckshot pellets at a ceramic target ("clay pigeon") launched by a spring. We know the trajectory of the clay pigeon in advance, or we can predict this trajectory. Hitting the clay pigeon requires taking account of both location and time of impact between shot and clay pigeon. The tight packet of shotgun pellets must cross the trajectory of the clay pigeon WHEN the clay pigeon is at that particular point in space. In brief, fix both the space and time location of a final impact *event*. The initial launch event is the firing of the shotgun. Think of a computer program that selects spacetime events of launch and impact, tries out various alternative worldlines between the two events, selects the worldline of extremal aging, and specifies for us the aiming direction (for a given muzzle velocity) to achieve a hit in terms of the specified events of launch and impact. This procedure yields the same result as the more common analysis that starts from initial conditions and predicts subsequent motion. However, one can also do it your way: The following two sections use the Principle of Extremal Aging to derive the expression for energy in curved Schwarzschild geometry. These results help to carry out the more conventional analysis ("predict subsequent motion from data on initial position and velocity").

3 Energy in Curved Schwarzschild Geometry

Extremal aging finds energy as a constant of the motion.

This section reveals a new expression for the energy of a particle falling radially into a black hole. This expression grows naturally out of our conviction that the Principle of Extremal Aging can be used, along with the Schwarzschild metric, to find quantities that remain constant during the motion of an object. In Section 5 of Chapter 1 we derived the expression for energy of a particle in flat spacetime. Our present derivation is an extension of that analysis and applies to orbiting as well as plunging particles. In Chapter 4, the Principle of Extremal Aging leads to a second constant of the motion, angular momentum. These two constants of motion help us to predict motions ranging from the sweep of a comet to the bending of light by Sun and from the precession of the elliptic orbit of the planet Mercury to the long "teetering on the verge" that can precede a satellite's plunge into a black hole.

Here is the Schwarzschild metric in its timelike form (page 2-19):

$$d\tau^2 = \left(1 - \frac{2M}{r}\right)dt^2 - \frac{dr^2}{1 - \frac{2M}{r}} - r^2 d\phi^2 \qquad [1]$$

Think of a stone plunging radially ($d\phi = 0$) toward the center of attraction, as shown in Figure 3. The stone emits three flashes rapidly, one after the other. These three events of flash emission bracket two adjacent segments of its trajectory. These segments, A and B in the figure, need not be the same length.

The following analysis examines the wristwatch time and far-away time separations among these events. For simplicity, replace the differential notations $d\tau$ and dt with symbols τ and t, with the understanding that these time separations are small. Also, we will be interested only in the parts of the Schwarzschild metric that contain t. With these simplifications, equation [1] becomes

$$\tau^2 = \left(1 - \frac{2M}{r}\right)t^2 + \text{(terms without } t\text{)} \qquad [2]$$

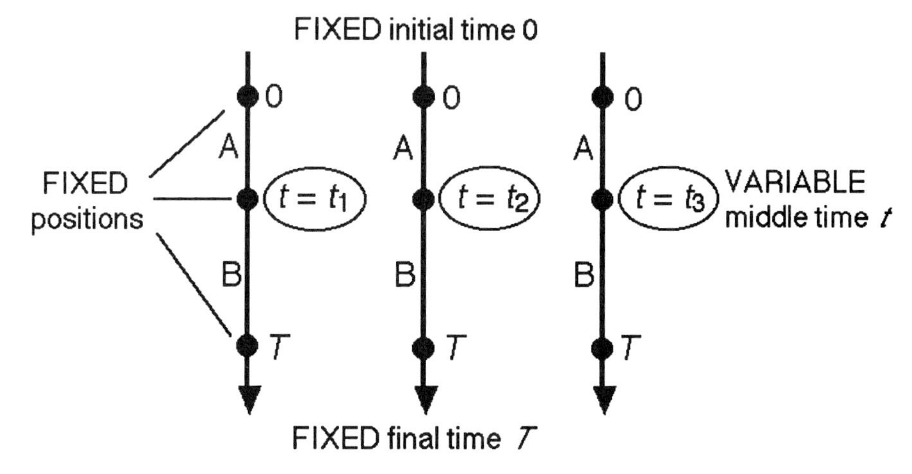

Figure 3 *Three possible times t_1, t_2, t_3 for the intermediate event as a stone carrying a wristwatch plunges radially inward toward the center of attraction. The stone emits three flashes. All flashes are fixed in position and the first and last are also fixed in time. We ask: At what time t will the stone pass through the intermediate dot? Answer this question by demanding that the total wristwatch time from first to last events be an extremum. From this requirement comes an expression for the energy as a constant of the motion.*

Principle of Extremal Aging helps us time flashes during plunge.

Fix all three events in space, and fix the first and last events in far-away time. Call these first and last times 0 and T, respectively. How can we find the time t at which the stone will pass the intermediate point and emit the intermediate flash (Figure 3)? Answer: Choose the intermediate time t that makes the total wristwatch time τ from first to last events an extremum. Use the Principle of Extremal Aging to derive an expression for the energy of the stone, an energy that remains the same as the stone descends.

The elapsed time for segment A is t. Write down an expression for the wristwatch time τ_A for this segment. Let r_A be an appropriate average value of the radius over segment A. We are going to take a derivative with respect to the intermediate time t, so ignore all terms in the metric that do not contain time. Then the wristwatch time τ_A for the first segment is

$$\tau_A = \left[\left(1 - \frac{2M}{r_A}\right)t^2 + (\text{terms without t})\right]^{1/2} \qquad [3]$$

To prepare for the derivative that leads to extremal aging, take the derivative of τ_A with respect to t:

$$\frac{d\tau_A}{dt} = \frac{\left(1 - \frac{2M}{r_A}\right)t}{\left[\left(1 - \frac{2M}{r_A}\right)t^2 + (\text{terms without } t)\right]^{1/2}} = \left(1 - \frac{2M}{r_A}\right)\frac{t}{\tau_A} \qquad [4]$$

The elapsed time for segment B is $T - t$. Let r_B be a suitable average value of the radius over segment B. The wristwatch time τ_B for this segment is

$$\tau_B = \left[\left(1 - \frac{2M}{r_B}\right)(T-t)^2 + (\text{terms without } t)\right]^{1/2} \qquad [5]$$

Again, to prepare for the derivative that leads to extremal aging, take the derivative of τ_B with respect to intermediate time t:

$$\frac{d\tau_B}{dt} = \frac{-\left(1 - \frac{2M}{r_B}\right)(T-t)}{\left[\left(1 - \frac{2M}{r_B}\right)(T-t)^2 + \left(\begin{matrix}\text{terms}\\ \text{without } t\end{matrix}\right)\right]^{1/2}} = -\left(1 - \frac{2M}{r_B}\right)\frac{(T-t)}{\tau_B} \qquad [6]$$

Now add the two wristwatch times to obtain a total wristwatch time between the first and last events:

$$\tau = \tau_A + \tau_B \qquad [7]$$

The Principle of Extremal Aging says that the natural motion yields an extremum for the total wristwatch time τ. To find this extremum, take the derivative of [7] with respect to t, substitute from equations [4] and [6], then set the result equal to zero:

$$\frac{d\tau}{dt} = \frac{d\tau_A}{dt} + \frac{d\tau_B}{dt} = \left(1 - \frac{2M}{r_A}\right)\frac{t}{\tau_A} - \left(1 - \frac{2M}{r_B}\right)\frac{(T-t)}{\tau_B} = 0 \qquad [8]$$

From the last equality in equation [8],

$$\left(1 - \frac{2M}{r_A}\right)\frac{t}{\tau_A} = \left(1 - \frac{2M}{r_B}\right)\frac{(T-t)}{\tau_B} \qquad [9]$$

Set $t = t_A$ and $T - t = t_B$, so equation [9] becomes

$$\left(1 - \frac{2M}{r_A}\right)\frac{t_A}{\tau_A} = \left(1 - \frac{2M}{r_B}\right)\frac{t_B}{\tau_B} \qquad [10]$$

The expression on the left side of equation [10]depends only on the parameters of the first segment A; the expression on the right side depends only on the parameters of the second segment B. This equation displays a quantity that is constant from one segment of the path to another. Expressed in words, it says,

$$\begin{pmatrix}\text{effective value of}\\ (1-2M/r)\\ \text{along 1st segment}\end{pmatrix}\frac{\begin{pmatrix}\text{advance of}\\ \text{far-away time}\\ \text{along 1st segment}\end{pmatrix}}{\begin{pmatrix}\text{advance of}\\ \text{wristwatch time}\\ \text{along 1st segment}\end{pmatrix}} = \begin{pmatrix}\text{effective value of}\\ (1-2M/r)\\ \text{along 2nd segment}\end{pmatrix}\frac{\begin{pmatrix}\text{advance of}\\ \text{far-away time}\\ \text{along 2nd segment}\end{pmatrix}}{\begin{pmatrix}\text{advance of}\\ \text{wristwatch time}\\ \text{along 2nd segment}\end{pmatrix}} \qquad [11]$$

The value of either side of this equation must be independent of which segment we choose to look at. We have found a constant of the motion, the same for all segments. Return to the differential notation to identify this constant of the motion as the energy, which has the same form for *any* segment of the path of the plunging particle:

Energy in Schwarzschild geometry

$$\frac{E}{m} = \left(1 - \frac{2M}{r}\right)\frac{dt}{d\tau} \qquad [12]$$

Energy takes special relativity form far from black hole.

Identification with energy E follows by noting that for large r (far from the center of attraction where spacetime must be flat), the expression reduces to that for energy in special relativity, $E/m = dt/d\tau$ (equation [25] on page 1-11).

Rather than focus on E alone, we usually look at the dimensionless ratio E/m. Why? For two reasons: (1) We recognize that particles of different mass m follow the same worldline through spacetime. What counts for motion is neither the mass of the plunging particle by itself nor its energy by itself but only the ratio of the two, the energy per unit mass. (2) The ratio E/m has no units. This encourages us to express E and m in the same unit, a unit that we may choose according to convenience and the experiment being described. Both numerator and denominator in E/m may be expressed in kilograms or the mass of the proton or million electron-volts or the mass of Sun, and so on.

Same energy expression is also correct for non-radial motion of the stone.

Figure 3 implies that the stone is plunging radially. Yet if you look back over this derivation, you will see that it is equally valid for segments of nonradial motion, in which the angle ϕ changes. Therefore equation [12] is the *general* expression for the energy of a stone moving around a spherically symmetric, nonrotating center of gravitational attraction.

4 Energy Measured at Infinity

A new way to measure total energy

Energy in general relativity is not divisible.

Isaac Newton (1642–1727) developed the laws of orbital motion and Leonard Euler (1707–1783) perfected the ideas of energy and energy conservation. In describing the energy of a satellite orbiting a black hole or plunging into it, anyone who has studied Newton's account of gravity and Euler's account of gravitational energy will have to gulp. Their nonrelativistic viewpoint distinguished two forms of energy. One form, "kinetic," depended on speed, not distance from the center of attraction. The other form, "potential," depended on distance from the center of attraction, not speed. Einstein's account of motion in curved spacetime recognizes no such distinction. This new outlook presents us with a new unity. Energy associated with speed, energy associated with location, and energy associated with mass are stirred together inextricably into a greater and simpler whole.

"Energy measured by an observer at infinity"

How can the "unified energy" of an orbiting satellite be observed, assigned a numerical value? In principle it can be measured through its gravitational effects on a remote test particle. The rest energy of the combined star-plus-satellite system—their total mass M_{total}—is measured using a beacon in a circular orbit so remote that Newtonian attraction supplies an accurate tool for measuring mass, as described in the box on page 3-11. From this total subtract the rest energy of the star—its mass M_{star} The difference is the "energy of the satellite measured by an observer at infinity" (provided this energy is small compared with the star's mass):

$$E = M_{total} - M_{star} \qquad [13]$$

What is the value of this energy? It must remain constant as the particle orbits or plunges. Equation [12] gives us such a constant. Evidently equation [12] provides the general expression for the energy-to-mass ratio for a particle orbiting near or plunging into a spherically symmetric nonspinning gravitating mass.

Our derivation of equation [13] says nothing about where the inwardly plunging particle starts its trajectory. In particular, this equation says that energy is a constant, even when the particle does not have enough energy to reach infinite separation from the center of attraction. An example is a clock bolted rigidly to the local shell (Sample Problem 1, page 3-12). According to equation [16] or [13], when the stationary clock is added to the system, the mass of the system is increased by the energy E of this clock: $M_{total} = M_{Sun} + E$.

Measuring the Energy *E* of a Satellite

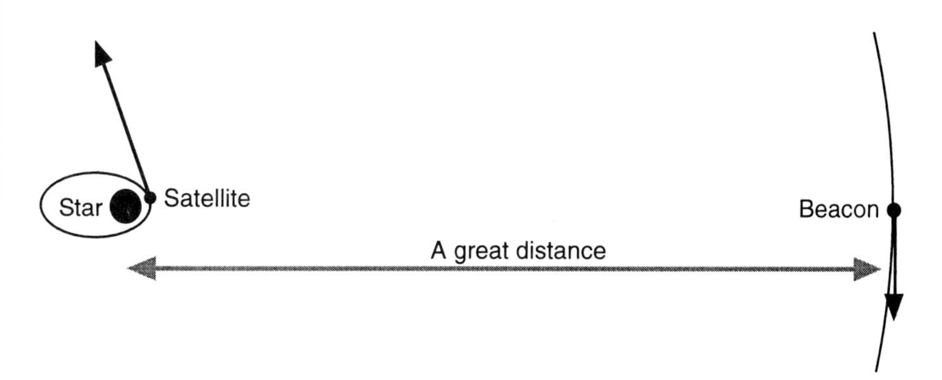

Figure 4 *Measuring the energy observed at infinity, E, of a small satellite in orbit around a star. A beacon is put into a distant circular orbit around the star-satellite system. From the acceleration of the beacon on its path, derive the mass M_{total} of the star-satellite system using Newtonian mechanics. Subtract the mass M_{star} of star from this total to obtain the "energy observed at infinity E" of the satellite alone. This unified, single energy replaces multiple energies—the potential energy of Newtonian mechanics plus the kinetic energy and rest energy of special relativity.*

Of all the ways to capture the essence of the total particle energy E for a small satellite, none is more compelling than this: Perform a "weighing" of center of attraction plus satellite M_{total} using Newton's mechanics. Far outside the center of attraction, far outside the orbit of that satellite, place a test particle, a mini satellite, a beacon in circular orbit around the star-satellite system and so distant that Newtonian mechanics may be used (Figure 4).

Measure the time for the beacon to go once around the circle at this distance. Measure the circumference of the circular orbit of this far-away beacon. From circumference and time compute the speed v of the beacon in its orbit. From the circumference compute the r-coordinate. From speed and radius, reckon the inward acceleration v^2/r implied by a circular orbit. That acceleration times the particle's mass must equal the gravitational force applied by the center of attraction plus inner satellite—in Newton's view. In conventional units we write

$$\begin{pmatrix}\text{force on}\\ \text{beacon}\end{pmatrix} = m_{\text{beacon}}\begin{pmatrix}\text{inward}\\ \text{acceleration}\\ \text{of beacon}\end{pmatrix} \qquad [14]$$

$$\frac{GM_{\text{total}}m_{\text{beacon}}}{r^2} = m_{\text{beacon}}\frac{v^2}{r}$$

In the second of these equations, divide out the common factor m_{beacon}. Solve the resulting equation for the only unknown (all quantities in convential units):

$$M_{\text{total}} = \begin{pmatrix}\text{mass of the}\\ \text{satellite + star}\\ \text{system}\end{pmatrix} = \frac{v^2 r}{G} \qquad [15]$$

From this total mass subtract now the mass, M_{star} of the star (or whatever that center of attraction may be), measured by the same method before the satellite was added to the system. The difference gives the energy of the small satellite as sensed by a far-away observer, the beacon. In geometric units:

$$E = M_{\text{total}} - M_{\text{star}} \qquad [16]$$

To speak of satellite energy as a constant of the motion is to say that the quantity E—the satellite's energy measured at infinity—always keeps the same value during free flight of the satellite.

SAMPLE PROBLEM 1 Energy of a Clock Bolted to a Spherical Shell

A shell clock of mass m is bolted to a spherical shell at r-coordinate r_0. What is its energy E that we measure using the orbit of a remote beacon?

SOLUTION

The clock's energy is computed from the ratio $dt/d\tau$ (equation [12]):

$$\frac{E}{m} = \left(1 - \frac{2M}{r}\right)\frac{dt}{d\tau} \qquad [12]$$

Choose two events to be two sequential ticks of this shell clock. Then the proper time between these ticks is just the time read on the clock. In brief, $d\tau = dt_{shell}$. Equation [C] in Selected Formulas at the back of the book gives us the relation between dt_{shell} of a clock at rest on a shell and dt of a far-away clock:

$$dt = \left(1 - \frac{2M}{r}\right)^{-1/2} dt_{shell} \qquad [17]$$

Combine this equation with [12] and with $d\tau = dt_{shell}$ to obtain, for an object such as a shell clock,

$$\frac{E}{m} = \left(1 - \frac{2M}{r_0}\right)^{1/2} \qquad [18.\ \text{at rest at } r_0]$$

Note the square root in this equation. This expression is correct for an object at rest on the shell at any reduced circumference r_0 outside the horizon. In the limiting case of a particle at rest at large r_0, the rest energy is equal to mass m, as special relativity requires. When the clock sits at a radius r_0 near a black hole, the ratio E/m is less than unity, so energy is less than the mass of the clock (evidence of the negative "energy of gravitational binding"). This total energy goes to zero at the horizon, $r_0 \rightarrow 2M$.

Suppose the bolt on the clock works loose and the clock falls. The value of E remains the same during the drop, as the clock crosses the horizon, and as the clock reaches the crunch at the center of the black hole. During this fall, the remote observer, using the technique described in the box on page 3-11, detects no change in the combined mass of the system: black hole plus clock.

Now the mounting bolt works loose so that the clock falls into the black hole. The energy of the total system does not change. The remote beacon cannot tell the difference. After the plunge, only the black hole remains. The mass of the black hole simply increases by E.

5 Falling from Rest at Infinity

Drift slowly inward, then plunge toward the center.

Special case: Stone starts from rest at $r = \infty$.

The fact that E/m is constant for a free particle yields great simplification in describing the motion of a radially plunging particle. As an example, think of a stone originally at rest a very great distance from a spherically symmetric nonspinning black hole. Over the eons this stone moves gradually toward the center of attraction, finally plunging radially inward to oblivion. Formally we say that the stone starts at rest at an infinite distance from the black hole in the infinite past. The Newtonian analysis of this plunge appeared in the box on page 2-22.

Stone has constant energy during plunge.

According to the law of constant E/m, the stone has constant energy during its entire trip, whether it is far from the black hole or close to it. But when at rest far from the black hole, this stone has energy identical to its mass, or $E/m = 1$. From equation [12], therefore, for every radius r on the inward plunge we have

$$\frac{E}{m} = \left(1 - \frac{2M}{r}\right)\frac{dt}{d\tau} = 1 \qquad [19.\ \text{from rest at } r = \infty]$$

Do all particles moving radially inward have energy $E/m = 1$?

Far from it! This value applies only to a particle (a stone, a key, a coffee cup) of mass m that starts from zero velocity at a great distance from the black hole. An entirely different case is a stone already moving when it is far away, a stone that has an inward velocity far away. Call this velocity v_{far}. Then its energy in that remote region of flat spacetime has the greater value $E/m = 1/(1 - v_{far}^2)^{1/2}$, says special relativity. Later on in its plunge, the particle will keep the same value of this energy, even though it leaves flat spacetime and enters the curved Schwarzschild geometry near the black hole. Sample Problem 3, page 3-25, analyzes this case.

In a third example, a stone of mass m is released initially at rest from a shell of radius r_0. Sample Problem 1 on page 3-12 shows that the value of its energy per unit mass is, from equation [18], $E/m = (1 - 2M/r_0)^{1/2}$. This value is smaller than that of a stone falling from rest at infinity. In this case too the stone maintains its value of E/m as it falls to smaller values of r.

From the energy equation [19] and the Schwarzschild metric [1], we can find an expression for dr/dt, the rate of change of the r-coordinate with far-away time t for a stone starting from rest at a very great distance. To obtain this derivative, square terms on either side of the right-hand equality in equation [19], multiply through by $d\tau^2$, and equate it to the Schwarzschild-metric expression for $d\tau^2$ (equation [1]) in the case of radial fall ($d\phi = 0$):

$$\left(1 - \frac{2M}{r}\right)^2 dt^2 = d\tau^2 = \left(1 - \frac{2M}{r}\right)dt^2 - \frac{dr^2}{\left(1 - \frac{2M}{r}\right)} \qquad [20]$$

Divide through by dt^2, solve for dr/dt, and take the square root to obtain

Bookkeeper measure of radial velocity

$$\frac{dr}{dt} = -\left(1 - \frac{2M}{r}\right)\left(\frac{2M}{r}\right)^{1/2} \qquad [21. \text{ from rest at } r = \infty]$$

We take the minus square root because the expression describes a *decreasing* radius as the object falls toward the black hole outside the horizon. Equation [21] gives the bookkeeper velocity that describes the plunge as the rate of change of reduced circumference r with time t measured on far-away clocks.

Why is this analysis so COMPLICATED? How can a stone carry out all these calculations as it drops freely toward a center of attraction? A stone is brainless, yet in order to follow equation [21] it must be better at quick computation than we are. Do you seriously believe that spacetime—or anything else—is issuing such complicated directions to the poor stone and that the stone is actually **following** *these instructions?*

The stone does not "follow equation [21]"; the stone does not care about the Schwarzschild metric or shell time or far-away time or the reduced circumference r. The stone can be totally brainless because it obeys the simplest instruction imaginable: "Go straight!" The fine print in the directive rephrases the command more precisely at the cost of slightly greater length: "Follow a straight worldline in the local free-float frame." This command is all the stone hears. Obeying this command is all the stone does.

We, however, are not satisfied with the local instruction given to the stone. We ask more global questions: "What is the entire path followed by the stone?" "What is its speed at various points on that path?" The local free-float frame does not suffice to answer these questions because spacetime curvature causes differences in motion from place to place that limit the size of the spacetime domain of a single inertial frame. Global questions require a global reference frame. Now we need the heavy machinery of spherical shells and the Schwarzschild metric and the complicated equation [21]. Ask a more complicated question, get a more complicated answer! Whose fault is that? The stone's fault? Nature's fault? No! It is our fault. If we were satisfied with local instructions, we could be as serene and unthinking as the stone.

Bookkeeper prediction: Stone slows down as it approaches the horizon!

Equation [21] for free fall from rest at an infinite radius has some surprising consequences. As the particle approaches the horizon, as $r \longrightarrow 2M$, the curvature factor $(1 - 2M/r)$ goes to zero, so the bookkeeper's velocity dr/dt also goes to zero. The Schwarzschild bookkeeper, keeping track of the reduced circumference r as a function of far-away time t, reckons that the freely falling particle slows down as it approaches the event horizon. As it gets closer and closer to the event horizon at $r = 2M$, the velocity dr/dt goes to zero. When tracked in these coordinates, the particle itself reaches the event horizon only after infinite far-away time t.

Impossible! The particle is propelled ever inward by what we would describe in Newtonian language as a fierce gravitational force. How can you possibly ask us to believe that this force results in the particle slowing down? Will someone clinging to the shell just outside the event horizon at $r = 2M$ observe the particle to decrease speed and settle gently—over an infinite time!—to rest at the horizon? The whole idea is simply insane!

What seems insane resolves itself into a more believable result when we follow this questioner's lead and ask who observes this zero speed at the event horizon. The answer is, Nobody! Not the shell observer; not a passing free-float observer. Nobody near the black hole observes *directly* a velocity whose magnitude is given by equation [21].

The r-coordinate, remember, is obtained by dividing the circumference around a great circle of the spherical shell by 2π. And far-away time t is recorded by the Schwarzschild bookkeeper remote from the black hole. As a result, equation [21] is merely the result of a calculation, it is a "bookkeeper's velocity." The bookkeeper tracks changes dr in the r-coordinate and divides each such change by the corresponding computed change dt in far-away time. The ratio dr/dt, sure enough, approaches zero as the particle approaches the event horizon at $r = 2M$.

I do not care what one or another observer measures or writes in a notebook. I am interested in REALITY! Stop beating around the bush; does the in-falling stone REALLY come to rest at the horizon or not?

Already in special relativity we learned to concentrate on predicting the result of an experiment. We were forced to acknowledge, for example, that "the time between two events" and "the velocity of a particle" are not invariants; typically they do not have equal values as measured by different inertial observers. In this sense "the real time between two events" and "the real velocity of a particle" have no unique meaning. Similarly, here in general relativity "the velocity at the horizon" must refer to the records of some reference frame. The phrase "real velocity" has no unique meaning. According to the remote bookkeeper, the in-falling object comes to rest at the horizon. Next we will find that for the shell observer the falling object passes across the horizon with the speed of light. What a contrast!

The shell observer—the observer standing on a fixed spherical shell—measures a velocity of the free-falling particle quite different from that given by equation [21]. Yet what he observes is easily derived from that equation. The shell observer does not measure dr. Rather, he lets down his tape measure between radially separated clocks bolted to his shell. Thus he measures directly the proper distance $d\sigma = dr_{shell}$ between their locations. From equation [D] in Selected Formulas at the back of the book, the two radial distances dr and $d\sigma$ are related by the expression

$$d\sigma = dr_{shell} = \left(1 - \frac{2M}{r}\right)^{-1/2} dr \qquad [22]$$

Moreover, shell observer clocks tick off not far-away time dt but shell time dt_{shell}. This wristwatch time, $d\tau$, is defined as the time recorded in a frame in which two events occur at the same place; in the shell frame the shell clock does not change spatial coordinates between ticks. The relation between shell time dt_{shell} and far-away time dt is given by equation [C] in Selected Formulas at the back of the book:

$$dt_{shell} = \left(1 - \frac{2M}{r}\right)^{1/2} dt \qquad [23]$$

Shell observer measures a different speed of in-falling stone.

The last two equations and equation [21] tell us that the radial velocity of the free-fall observer as measured by the shell observer has the value

$$\frac{dr_{shell}}{dt_{shell}} = -\left(\frac{2M}{r}\right)^{1/2} \qquad [24. \text{ from rest at } r = \infty]$$

This expression for radial velocity is the same as the result of the Newtonian analysis (box on page 2-22). However, that Newtonian expression failed to distinguish between shell coordinates and bookkeeper coordinates. Equation [24] makes clear that the expression refers to shell coordinates and shell measurements. (It is all right to leave the reduced circumference r in the right-hand side of this equation, since each spherical shell is stamped with its individual radius. By looking at this stamp, *any* observer can determine which shell he is talking about.)

Shell observer clocks stone at speed of light at horizon.

At the reduced circumference $r = 8M$, or four times the Schwarzschild radius ($2M$) at the horizon, the particle falling from rest at infinity is moving inward at half the speed of light, as witnessed by shell observers. As the particle crosses the event horizon at $r = 2M$, nearby shell observers record it as moving at the speed of light. [Shells—and shell observers—cannot exist inside the horizon (see the following section) or even *at* the horizon, where the spherical shell experiences infinite stresses. The prediction of equation [24] at the horizon must be taken as a limiting case, an extrapolation.]

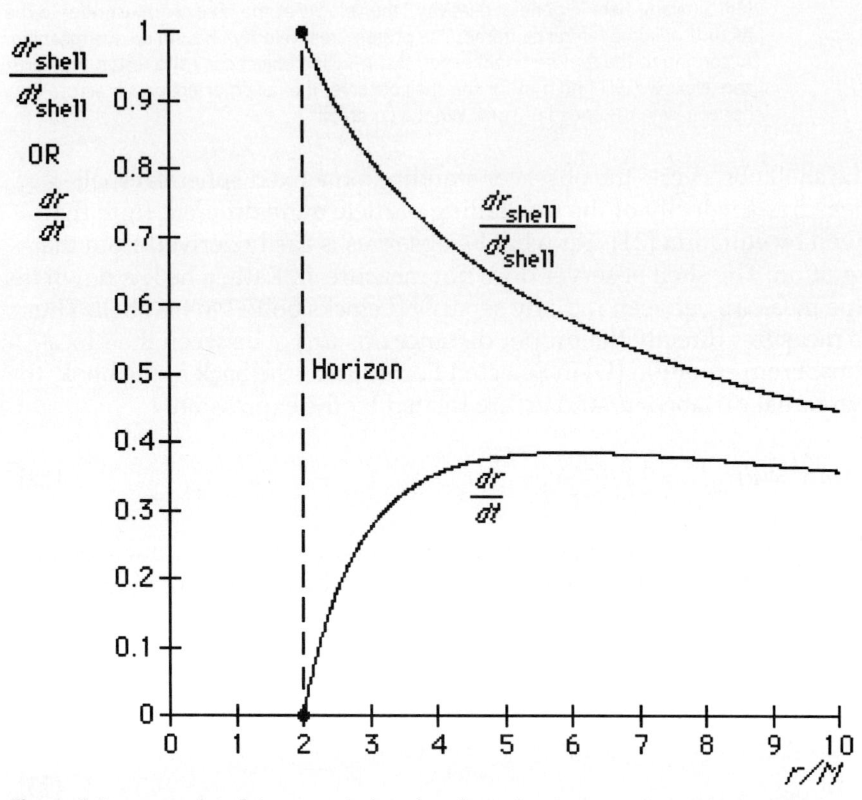

Figure 5 *Computer plot of the two velocity values for a plunging stone (treated here as positive). A stone falling radially from rest at infinity has speed dr_{shell}/dt_{shell} as measured by observers on shells through which the stone plunges and speed dr/dt as derived from the records of the Schwarzschild bookkeeper. At the horizon, the shell speed rises to the speed of light (equation [24]), while the bookkeeper speed drops to zero (equation [21]).*

Radical difference between "shell speed" and "bookkeeper speed" of stone at horizon

What a contrast between these nearby measurements and the bookkeeper speed *dr/dt*, a speed that goes to zero at the event horizon! (See Figure 5.) There the in-falling stone moves with the speed of light as recorded directly by one observer (equation [24]); it moves with zero speed as reckoned by another observer (equation [21]). Nothing demonstrates more dramatically how far we have come from the phenomena that take place in flat spacetime as described by special relativity!

How can you use the time transformation [23] to describe an in-falling stone plunging from one spherical shell to another shell of different reduced circumference? Let a firecracker explode at each shell as the stone passes. Clocks on these different shells "run at different rates" according to that very equation [23]. How can you possibly combine readings from these two different-rate clocks to meter the shell time the stone takes between the two flash emissions?

For once we are caught by the sloppy way physicists do calculus! In writing the Schwarzschild metric, we use differentials *dr*, *dt*, and so forth to remind ourselves that event pairs so separated must be near enough to one another in spacetime to justify using a single value of *r*. In many cases this criterion can be met for events separated by many meters of distance and time. When we measure velocity, however, we employ calculus in the conventional sense of a limiting process, with *dr* and *dt* and dr_{shell} and dt_{shell} all tending to zero. Then the firecrackers go off right next to each other; the two clocks are—in the limit—on the same shell, so they run at the same rate at this limit.

In summary, for a radially plunging object that starts from rest at infinity, the principle of constancy of energy tells us that the inward speed measured by shell observers increases steadily for smaller values of r, rising to the speed of light at the horizon. In contrast, the inward speed of the object drops to zero at the horizon when reckoned from the accounts of the Schwarzschild bookkeeper.

6 Energy Measured by Shell Observer

The shell observer uses special relativity to analyze motion locally.

The farther toward the center a stone falls, the faster it goes as observed by a sequence of local shell observers (equation [24]). But a shell observer sees a faster-moving stone as having more energy; the faster the stone moves past him, the more energy he can in principle extract from it by slowing it down to rest. In this sense a shell observer at smaller radius attributes more energy to the faster-moving stone as it passes him. How can this increased shell energy at smaller radii be reconciled with the concept of energy as a constant of the motion (equation [19])? To answer this question, start from equation [24]:

$$v_{\text{shell}} = \frac{dr_{\text{shell}}}{dt_{\text{shell}}} = -\left(\frac{2M}{r}\right)^{1/2} \qquad \text{[25. from rest at } r = \infty]$$

"Shell energy" is computed using special relativity.

The shell observer uses the results of special relativity to calculate the energy of the in-falling stone. He applies equation [24], page 1-10, to this equation [25] and obtains

$$\frac{E_{\text{shell}}}{m} = \frac{1}{(1 - v_{\text{shell}}^2)^{1/2}} = \frac{1}{\left(1 - \frac{2M}{r}\right)^{1/2}} \qquad \text{[26. from rest at } r = \infty]$$

Shell energy is a LOCAL quantity.

This shell energy per unit mass E_{shell}/m is a local quantity, measured by the shell observer from the speed of the passing stone and its mass m. The shell observer needs to know nothing about "energy measured at infinity" or "energy as a constant of the motion." He simply observes a stone come whizzing past and reckons its local energy per unit mass E_{shell}/m using the formulas of special relativity. In special relativity, E_{shell} represents the total energy, which includes the rest energy m plus the kinetic energy. As we have said, no such division between rest energy and kinetic energy is possible for the energy measured at infinity E of general relativity.

There is, however, a relation between E_{shell} and energy E, the constant of the motion. A stone at rest at infinity has energy measured at infinity $E = m$. Therefore we can write equation [26] as

$$\frac{E_{\text{shell}}}{E} = \frac{1}{\left(1 - \frac{2M}{r}\right)^{1/2}} \qquad \text{[27. \textit{anything} falling into black hole]}$$

In this equation, E is energy measured at infinity, a constant of the motion, and E_{shell} is the local energy measured by a shell observer. Equation [27] is general, it turns out. In the exercises of Chapter 5 you show that the same equation even applies to a flash of light fired outward from a fixed shell.

E_{shell} is a local quantity; E is a constant of the motion.

In brief, E_{shell} is a local quantity measured by the shell observer, a quantity that he analyzes using the laws of special relativity. But E_{shell} is not a constant of the motion; it is different for shell observers at different radii. Equation [27] connects this local energy to the energy measured at infinity E, which *is* a constant of the motion.

The farther toward the center a stone falls, the faster it goes as measured by shell observers (equation [24]). Therefore the more energy E_{shell} the stone has, as measured by the local shell observer. But general relativity says that there is no gravitational force on the stone, that the stone "moves freely along a locally straight worldline, a geodesic." If there is no gravitational force on the stone, where does the descending stone GET its INCREASING energy?

The story is told of a proper Bostonian lady who was asked where she and her friends "buy your awful hats." "*Buy* our hats?" she exclaimed, "*Buy* our hats? We don't *buy* our hats; we *have* our hats!" Similarly, the descending stone does not *get* its energy from anywhere. It *has* its total energy E, the energy measured at infinity, which remains constant as the stone descends in free float. Equation [24] results from this constant energy. The beacon satellite orbiting the system of black hole plus stone (Figure 4, page 3-11) detects no increase in total energy of that system as the stone descends.

Whoa! Is it not true that shell observers at smaller and smaller r-values measure larger and larger energy E_{shell} for the passing stone?

Yes. One must make a distinction between E and E_{shell}, which are connected by equation [27]. E is a constant of the motion ruled by general relativity, an energy that cannot be divided into kinetic, potential, and rest energy. In contrast, E_{shell} is an energy that the shell observer reckons locally according to the rules of special relativity, dividing it into kinetic plus rest energy (mass) if he wishes. True, as r decreases, different shell observers measure larger values for E_{shell}. However, each shell observer is stuck with a local analysis. He cannot expand his view to embrace the entire spacetime surrounding the black hole without adopting the energy measured at infinity E, which is a constant of the motion and cannot be divided up into kinetic, potential, and rest components.

Come on! What's wrong with the basic results of Newtonian mechanics? As the stone descends, kinetic energy goes up as gravitational potential energy goes down—leaving total energy of the stone unchanged. Why cannot the same words fit the general relativistic description?

Newton's analysis does not apply to any of the reference frames we use. Each shell observer experiences only local events, each watches the stone streak past and disappear. Moreover, the shell observer near the horizon measures the stone to move with a speed approaching that of light, therefore with a kinetic energy (and a total energy) that increases without limit as observed in his frame. What can be conserved in these circumstances? The astronaut in free fall next to the stone also uses special relativity over a limited region of spacetime. For her the stone has zero kinetic energy at every event of the plunge, even at the horizon. No conversion of potential energy to kinetic energy for her! Only the Schwarzschild bookkeeper has the big picture. But for the bookkeeper the stone *decreases* its speed as it comes close to the horizon, the speed going to zero at the horizon (equation [21]), and therefore any so-called kinetic energy also goes to zero there. Even the bookkeeper cannot divide up the stone's energy measured at infinity into kinetic energy plus gravitational potential energy. General relativity, it must be said again, treats energy E as a unified whole, with no separation possible between gravitational potential energy, kinetic energy, and rest energy (mass).

7 Over the Edge: Entering the Black Hole

No jerk. No jolt. A hidden doom.

No feature of the black hole excites more curiosity than the Schwarzschild horizon at $r = 2M$. It is the point of no return beyond which no traveler can find the way back—or even send signals—to the outside world. What is it like to fall into a black hole? No one from Earth has yet experienced it. Moreover, future explorers who do so will not be able to return to tell about it or transmit messages to us about their experience—or so we believe! In spite of the impossibility of receiving a final report, there exists a well-developed and increasingly well-verified body of theory that makes clear predictions about our experience as we approach and cross the horizon of a black hole. Here are some of those predictions.

We are not sucked into black hole.

We are not "sucked into" a black hole. Unless we get quite close to it, a black hole will no more grab us than Sun grabs Earth. If our Sun should suddenly collapse into a black hole, Earth and the other planets would continue on their present courses undisturbed (even though perpetual night would prevail!). The Schwarzschild solution would continue then to describe Earth's path through spacetime, just as it does now. The exercises of Chapter 4 show that for orbits that stay at radii greater than about $300M$, Newtonian mechanics predicts the motion to a good approximation. We will find that when we drop to a radius $6M$ or less, no stable circular orbit is possible; for a smaller radius we spiral inward and cross the horizon. Even then we can always escape, given sufficient rocket power. Only when we descend as close as or nearer than the Schwarzschild radius $r = 2M$ are we irrevocably "sucked in," our fate sealed.

We cannot tell directly when we are crossing the horizon.

No special event occurs as we fall through the Schwarzschild horizon. Even when we cross into a black hole at the event horizon $r = 2M$, we experience no shudder, jolt, or jar. True, the tidal forces are ever increasing as we fall inward, and this increase continues smoothly at the horizon. But we are not suddenly torn apart at $r = 2M$. True also, the curvature factor $(1 - 2M/r)$ in the Schwarzschild metric goes to zero at this radius. But the resulting zero in the time term of the metric and the infinity in the radial term turn out to be singularities of the bookkeeper coordinates r and t, not singularities in spacetime geometry. They do not lead to discontinuities in our experience as we pass through this radius. There are other coordinate systems whose metrics do not have a discontinuity at the Schwarzschild radius. (See Project B, Inside the Black Hole, Section 4, page B-12.)

There are no shell frames inside the horizon.

Inside the horizon there are no shell frames. Outside the horizon of the black hole we have erected, in imagination, a set of nested spherical shells. We say "in imagination" because no known material is strong enough to stand up under the "pull of gravity," which increases without limit as one approaches the horizon from outside. Locally such a stationary shell can be replaced by a rocket ship with rockets blasting to keep it stationary with respect to the black hole. Inside the horizon, however, nothing can remain at rest. No stationary shell. No stationary rocket ship, however ferocious the blast of its engines. The material composing the original star, no matter how strong, was unable to resist the collapse that formed the black hole. The same irresistible collapse forbids any stationary structure or object inside the horizon.

Packages can be sent inward but not outward.

"Outsiders" can send packages to "insiders." Different free-float frames still move with relative speeds inside the black hole. For example, one traveler may drop from rest just outside the horizon. Another unpowered spaceship may have fallen in from rest at a great distance. A third may be hurled inward. In fact, the hurled-inward ship—if launched soon enough—can be used to carry packages and information from outside to the drop-from-just-outside traveler inside. (Exercises at the end of this chapter concern these three different cases.) Light and radio waves can carry messages inward as well. We who have fallen inside the horizon can still see the stars, though with changed directions, colors, and intensities. (See Sections 9 and 11 of Chapter 5 and Section 9 of Project B.) Packages and communications sent inward across the horizon? Yes. Outward? No!

Inside there is an interchange of the character of the *t*-coordinate and *r*-coordinate. For an *r*-coordinate less than the Schwarzschild radius $2M$, the curvature factor $(1 - 2M/r)$ in the Schwarzschild metric becomes negative. In consequence, the signs reverse between the radial part and the time part of this metric, making the dt^2 term negative and the dr^2 term positive. Space and time themselves do not interchange roles. Coordinates do: The *t*-coordinate changes in character from a timelike coordinate to a spacelike coordinate. Similarly, the *r*-coordinate changes in character from a spacelike coordinate to a timelike one.

Timelike *r*-coordinate inside horizon means motion toward center is inevitable!

What does it mean to say that inside the Schwarzschild radius the *r*-coordinate "changes in character from a spacelike coordinate to a timelike one"? It means that our free-float frame moves to ever-smaller r with all the inevitability that we ordinarily associate with the passage of time. The explorer in his jet-powered spaceship prior to arrival at $r = 2M$ always has the option to turn on his jets and change his motion from decreasing r (infall) to increasing r (escape). Quite the contrary is the situation once he has allowed himself to fall inside $r = 2M$. Then the further decrease of r represents the passage of time. No command the traveler can give to his jet engine will turn back time. That unseen power of the world that drags everyone forward willy-nilly from age twenty to forty and from forty to eighty also drags the rocket in from the coordinate $r = 2M$ to the later value of the "time" coordinate $r = 0$. No human act of will, no engine, no rocket, no force can make time stand still. As surely as cells die, as surely as the traveler's watch ticks away the unforgiving minutes, with equal certainty r drops from $2M$ to 0 with never a halt along the way. Section 5 of Project B demonstrates these results in detail.

Wait a minute! How can far-away time t have any meaning at all inside the horizon? In order to verify the far-away time of an event, one must communicate with the far-away observer. But for someone inside the horizon such communication is impossible; the signal from inside never passes outward through the horizon—so you say! So isn't far-away time meaningless for the inside plunger?

You are essentially right; far-away time is pretty useless inside the horizon. One can conceive of a complicated scheme in which the far-away observer emits timing pulses that are received inside the horizon and used to time local events there. However, it is much easier for the plunger simply to read her wristwatch as she analyzes events in her vicinity, as in the following paragraph. This "plunger time" is incorporated into a new version of the metric described in Project B, Inside the Horizon, page B-13.

Wristwatch time from horizon to crunch point

Time to arrive at the crunch ($r = 0$). How long do we live once we pass through the Schwarzschild horizon? Sample Problem 2 on page 3-22 gives a derivation of the horizon to crunch wristwatch time for an object falling radially from rest at a great distance. Expressed in meters and seconds,

$$\tau[\text{meters}] = \frac{4}{3}M \qquad \text{[28. horizon to crunch]}$$

$$\tau[\text{seconds}] = 6.57 \times 10^{-6}\frac{M}{M_{\text{Sun}}} \qquad \text{[29. horizon to crunch]}$$

Here M_{Sun} is the mass of our Sun. Let a stone fall from rest at infinite distance. The proper time given by equations [28] and [29] is the time recorded on the wristwatch of this stone between its crossing the horizon and its arrival at the center of the black hole.

Maximum horizon to crunch wristwatch time

The *maximum* horizon to crunch wristwatch time occurs for a free-fall traveler dropping from rest just outside the horizon. This maximum wristwatch time τ_{max} is given by the following expression (derivation in the exercises):

$$\tau_{\text{max}}[\text{meters}] = \pi M \qquad \text{[30. horizon to crunch]}$$

or

$$\tau_{\text{max}}[\text{seconds}] = 15.5 \times 10^{-6}\frac{M}{M_{\text{Sun}}} \qquad \text{[31. horizon to crunch]}$$

A maximum of 15 microseconds does not seem long to live when we fall into a black hole of mass equal to that of our Sun. However the horizon to crunch time increases in direct proportion to the mass of the black hole. The calculated maximum survival time increases to almost a minute for the plunge into the object believed to reside at the center of our galaxy, a black hole of mass 2.6 million Sun masses. To survive a maximum of 40 more years (1.26×10^9 seconds) after crossing the horizon, find a black hole with a mass of roughly 10^{14} Sun masses (combined mass of some thousands of galaxies)!

SAMPLE PROBLEM 2 Wristwatch Time from Horizon to Crunch

Starting from rest at a great distance, you plunge straight toward a black hole of mass M. You set your watch to noon as you determine—by one means or another—that you are crossing the horizon. How much more time does the wristwatch—and you—have until the instant of crunch?

SOLUTION

This question concerns wristwatch time, tau-time τ, proper time, and the correlation between this time and the r-coordinate. Equation [19] gives us $dt/d\tau$, the correlation between time at infinity and spaceship proper time:

$$\left(1-\frac{2M}{r}\right)\frac{dt}{d\tau} = 1 \qquad [19]$$

Equation [21] gives us dr/dt, the correlation between radial r-coordinate and far-away time:

$$\frac{dr}{dt} = -\left(1-\frac{2M}{r}\right)\left(\frac{2M}{r}\right)^{1/2} \qquad [21]$$

We use equations [21] and [19] to find the correlation between changes in r-coordinate and wristwatch time:

$$\frac{dr}{d\tau} = \frac{dr}{dt}\frac{dt}{d\tau} = -\left(\frac{2M}{r}\right)^{1/2} \qquad [32]$$

This equation gives the change $d\tau$ in proper time racked up during the shrinkage dr in the r-coordinate:

$$d\tau = -\frac{r^{1/2}dr}{(2M)^{1/2}} \qquad [33]$$

Integrate both sides of this expression from $r = 2M$ to $r = 0$:

$$\tau = -\int_{2M}^{0}\frac{r^{1/2}dr}{(2M)^{1/2}} = -\frac{1}{(2M)^{1/2}}\frac{2}{3}r^{3/2}\Big|_{2M}^{0}$$

$$= \frac{2(2M)^{3/2}}{3(2M)^{1/2}} = \frac{4}{3}M \qquad [34]$$

This result, the same as equation [28], is in meters of light-travel time. To write the result in seconds, divide by the speed of light:

$$\tau_{\text{seconds}} = \frac{\frac{4}{3}M[\text{meters}]}{c[\text{meters/second}]} = \frac{4}{9}\times 10^{-8}M \qquad [35]$$

where M is still measured in meters. This result is conveniently expressed in terms of the mass of our Sun, which has the value M_{Sun} = 1477 meters. Make the substitution

$$M = \frac{M}{M_{\text{Sun}}}M_{\text{Sun}} = \frac{M}{M_{\text{Sun}}}(1477 \text{ meters}) \qquad [36]$$

which leads to equation [29]:

$$\tau[\text{seconds}] = 6.57\times 10^{-6}\frac{M}{M_{\text{Sun}}} \qquad [29]$$

Make this daring investigation of an already existing black hole? No. We and our exploration team want to be still more daring, to follow a black hole as it forms. We go to a multiple-galaxy system so crowded that it teeters on the edge of gravitational collapse. The day after our arrival at the outskirts, it starts the actual collapse, at first slowly, then more and more rapidly. Soon a mighty cataract thunders toward the center from all sides, a cataract of objects and radiation, a cataract of momentum-energy. The matter of the galaxies and with it our group of enterprising explorers pass smoothly across the horizon at the Schwarzschild radius $r = 2M$.

Life goes on inside the horizon.

From that moment onward we lose all possibility of signaling to the outer world. However, radio messages from that outside world, light from the familiar stars, and packages fired after us at high speed continue to reach us. Moreover, communications among us explorers take place now as they did before we crossed the horizon. We express our findings to each other in the familiar categories of space and time. With our laptop computers we turn out an exciting journal of our measurements and conclusions. (Our motto: Publish *and* perish.)

Nothing rivets our attention more than the tide-producing forces that pull heads up and feet down with ever-increasing tension. Before many years have passed, we can predict, this differential pull will have reached the point where we can no longer survive. Moreover, we can foretell still further ahead and with absolute certainty an instant of total crunch. In that crunch are swallowed up not only the stars beneath us, not only we explorers, but space itself. An instant comes after which there is no "after."

After crunch there is no "after."

Figure 6 shows the worldlines of the plunger who starts at rest from infinity as timed by her own wristwatch and also as timed by the clocks of the Schwarzschild bookkeeper.

I am still bothered by the idea of a "material" particle traveling across the event horizon as a particle. The shell observer sees it moving at the speed of light, but it takes light to travel at light speed. Does the particle becomes a flash of light at the horizon?

No. As you free-fall across the horizon, you feel nothing. You certainly do not turn into a flash of light! The idea of a shell observer at the horizon makes sense only as a limiting case. The shell observer just outside the horizon measures the in-falling particle to move at less than the speed of light. AT THE HORIZON no shell is possible, because the "local acceleration of gravity" increases without limit (see exercise 9 at the end of this chapter). So no shell observer can be stationed at the horizon to verify that the in-falling particle moves at light speed. AT and INSIDE the horizon, the only dependable measurements are made by free-float observers.

Go back to bookkeeper coordinates. A serious objection still remains: If all objects as they fall into the black hole come to rest at the horizon as reckoned by the bookkeeper, then shouldn't the black hole be eternally surrounded by all the junk that has ever fallen into it, including the original star that collapsed to form the black hole in the first place? Our Russian colleagues originally called the black hole a "frozen star." I call it a "frozen junk pile"!

It is true that, measured in terms of reduced circumference r and far-away time t, all radially in-falling objects coast to rest at the horizon. But along with this effect is another: the gravitational red shift of light from these objects. As each object approaches the horizon, its emitted light is shifted farther and farther into the red as observed far from the black hole. You can calculate how rapidly this downshift occurs in far-away time (exercise 9 in Chapter 5). Very quickly, light from the object becomes invisible to the eye; the object turns black—thus becoming part of the black hole as far as the remote observer is concerned. Stationary black junk is still black!

8 Summary

Central results of this chapter derive from (1) the Schwarzschild metric and (2) the Principle of Extremal Aging. A stone falling radially toward a center of gravitational attraction passes three goal posts, each at a fixed r-coordinate, and at each passage emits a flash. Fix the time of the first and the third flashes, but try different times for the second flash. Find the emission time of the second flash such that the total wristwatch time from first to third flashes is maximum. The result is an expression for a constant of the motion, which we identify as the total energy of the stone:

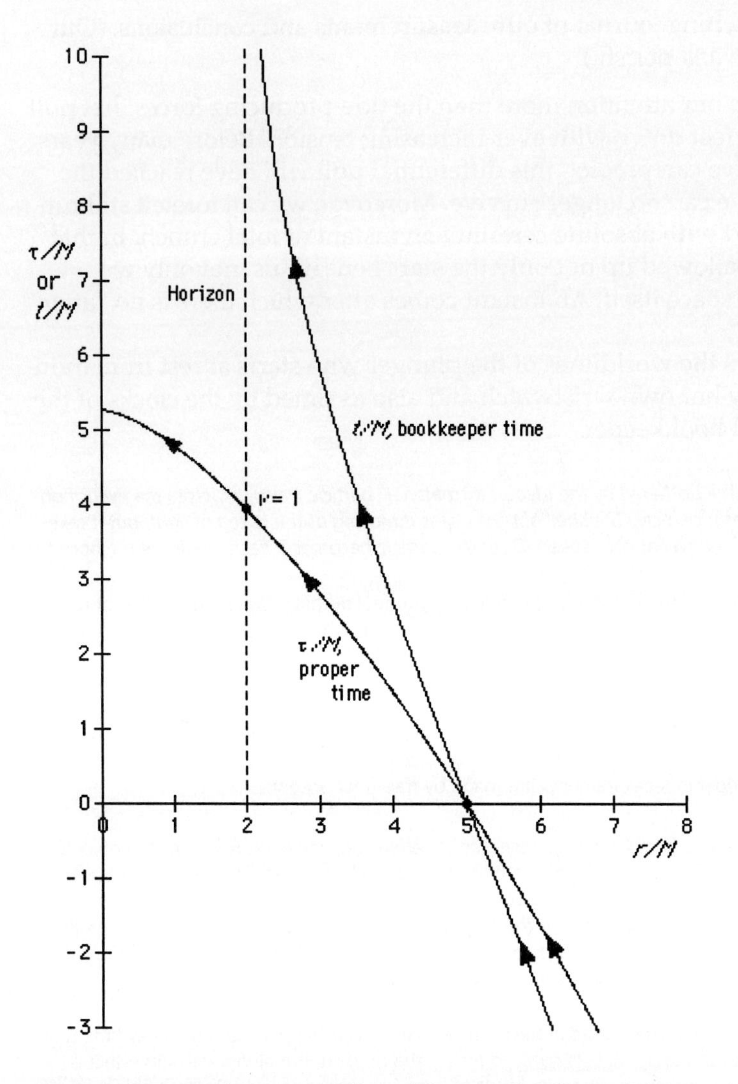

Figure 6 *Computer plot: Two worldlines of the SAME free-float radial plunger who starts at rest at a large radius—plotted using (1) Schwarzschild bookkeeper time t and (2) proper time τ read on the wristwatch of the in-faller. Use of coordinates r/M, t/M, and τ/M makes the curves independent of M and thus valid for all Schwarzschild black holes. By prearrangement, the plunger wristwatch and bookkeeper clocks are set to zero as the plunger passes r = 5M. Derivation of the bookkeeper plot comes from equation [21]. Derivation of the plunger plot is in Section 2 of Project B, Inside the Black Hole.*

$$\frac{E}{m} = \left(1 - \frac{2M}{r}\right)\frac{dt}{d\tau} \qquad \text{[12. free particle]}$$

At large radius *r*, this expression reduces to that for the total energy of special relativity, as it must. In general relativity the total energy *E* is a seamless whole and cannot be divided into kinetic, potential, and rest energy (mass). Its magnitude is measured by the distant gravitational

SAMPLE PROBLEM 3 Hurling a Stone into a Black Hole

According to equation [24], a stone falling into a black hole from zero initial velocity at infinity moves with the speed of light as it crosses the event horizon as measured by nearby shell observers. Can we make this "final observed speed" *greater* than the speed of light by *hurling* the stone inward from a great distance, rather than letting it start from rest?

SOLUTION

Start with equation [12]:

$$\frac{E}{m} = \left(1 - \frac{2M}{r}\right)\frac{dt}{d\tau} = \text{constant} \qquad [12]$$

We evaluated this constant to obtain equation [19] by demanding that at remote distances ($2M/r << 1$) the stone be at rest ($dr/dt = 0$). Here we deal with the more general case: The stone has an initial velocity inward along the radial direction $dr/dt = -v_{far}$, where the subscript "far" refers to conditions at a remote distance from the black hole. For large r, spacetime is flat; both dr and dt have the usual special-relativity meaning—and therefore so does the velocity dr/dt. In flat spacetime there is no need to distinguish among different measures of velocity. Then the value of the constant is equal to the special relativity value of $E/m = \gamma_{far}$, defined as follows:

$$\frac{E}{m} = \left.\frac{dt}{d\tau}\right|_{\text{large } r} = \gamma_{far} \equiv \frac{1}{(1 - v_{far}^2)^{1/2}} \qquad [37]$$

Substitute this result into equation [12]:

$$\frac{E}{m} = \left(1 - \frac{2M}{r}\right)\frac{dt}{d\tau} = \gamma_{far} \qquad [38]$$

Multiply through by $d\tau$ and substitute the Schwarzschild metric [20] for $d\tau$, as we did in deriving equation [21]:

$$\left(1 - \frac{2M}{r}\right)dt = \gamma_{far}d\tau$$

$$= \gamma_{far}\left(\left(1 - \frac{2M}{r}\right)dt^2 - \frac{dr^2}{\left(1 - \frac{2M}{r}\right)}\right)^{1/2} \qquad [39]$$

Divide through by dt, solve for dr/dt, and take the negative square root corresponding to a decreasing radius:

$$\frac{dr}{dt} = -\left(1 - \frac{2M}{r}\right)\left[1 - \frac{1}{\gamma_{far}^2}\left(1 - \frac{2M}{r}\right)\right]^{1/2} \qquad [40]$$

From equations [22] and [23], we know that

$$\frac{dr_{shell}}{dt_{shell}} = \left(1 - \frac{2M}{r}\right)^{-1}\frac{dr}{dt} \qquad [41]$$

Hence:

$$\frac{dr_{shell}}{dt_{shell}} = -\left[1 - \frac{1}{\gamma_{far}^2}\left(1 - \frac{2M}{r}\right)\right]^{1/2} \qquad [42]$$

What happens to the two velocities dr/dt and $(dr/dt)_{shell}$ as the stone approaches the Schwarzschild radius at $r = 2M$? The bookkeeper velocity dr/dt (equation [40]) goes to zero as before. The "shell velocity" dr_{shell}/dt_{shell} (equation [42]) takes on the value unity—the velocity of light—at $r = 2M$, also as before. Hurling the stone inward with any possible velocity v_{far}—and hence with any gamma factor γ_{far}—does not increase its velocity as it passes across the horizon. The speed of light remains the fastest directly observable speed, even in general relativity!

effects of stone plus black hole, thus earning for E the name **energy measured at infinity**.

For a stone falling from rest starting an infinite distance away, this energy has the value

$$\frac{E}{m} = \left(1 - \frac{2M}{r}\right)\frac{dt}{d\tau} = 1 \qquad [19. \text{ free fall from rest at } r = \infty]$$

whereas the total energy of a stone hurled in from a great distance with initial speed v_{far} is given by the expression (Sample Problem 3)

$$\frac{E}{m} = \left(1 - \frac{2M}{r}\right)\frac{dt}{d\tau} = \gamma_{far} = \frac{1}{(1 - v_{far}^2)^{1/2}} \qquad \text{[37, 38. flung in from } r = \infty]$$

The total energy of an object held at rest on a shell of radius r_o is given by the expression (Sample Problem 1)

$$\frac{E}{m} = \left(1 - \frac{2M}{r_o}\right)^{1/2} \qquad \text{[18. from rest at } r_o]$$

We study further the case of a stone released from rest at infinity. Using the Schwarzschild metric, we derive the bookkeeper velocity *dr/dt* as a function of radius as it falls inward—

$$\frac{dr}{dt} = -\left(1 - \frac{2M}{r}\right)\left(\frac{2M}{r}\right)^{1/2} \qquad \text{[21. from rest at } r = \infty]$$

—and the velocity of the stone as measured by an observer on a shell of radius r:

$$\frac{dr_{shell}}{dt_{shell}} = -\left(\frac{2M}{r}\right)^{1/2} \qquad \text{[24. from rest at } r = \infty]$$

The shell observer stationed at radius r uses this velocity and expressions of *special relativity* to reckon the local energy of the passing stone:

$$E_{shell} = \frac{E}{\left(1 - \frac{2M}{r}\right)^{1/2}} \qquad [27]$$

Shell energy is *not* a constant of the motion but represents energy measured directly by the shell observer. This equation is a special case of the relation between local energy E_{shell} and the total energy E measured at infinity. This equation applies to flashes of light, stones hurled inward from infinity, and all other forms of energy. In the case of a stone, E_{shell} represents the total energy computed from special relativity, which includes the rest energy m plus the kinetic energy.

Returning again to the stone falling from rest at a great distance, we can calculate the (proper) time τ recorded on its wristwatch between passing the horizon and arriving at the crunch point at the center of the black hole:

$$\tau[\text{meters}] = \frac{4}{3}M[\text{meters}] \qquad \text{[28. horizon to crunch]}$$

$$\tau[\text{seconds}] = 6.57 \times 10^{-6}\frac{M}{M_{Sun}} \qquad \text{[29. horizon to crunch]}$$

Maximum time from horizon to crunch is given by equations [30] and [31].

9 References and Acknowledgments

Quote on page 3-1 by Einstein to a group of reporters in 1921 who asked him for a short explanation of relativity. Quoted in Ronald W. Clark, *Einstein, The Life and Times*, Avon Books, New York, 1971 (paperback edition), page 469.

This chapter owes a large intellectual debt in ideas, figures, and text to *Gravitation* by Charles W. Misner, Kip S. Thorne, and John Archibald Wheeler, W. H. Freeman and Company, San Francisco (now New York), 1973. At least one paragraph is quoted from *A Journey into Gravity and Spacetime* by John Archibald Wheeler, W. H. Freeman and Company, New York, 1990. In addition, our treatment was helped by reference to "Nonrotating and Slowly Rotating Holes" by Douglas A. Macdonald, Richard H. Price, Wai-Mo Suen, and Kip S. Thorne in the book *Black Holes: The Membrane Paradigm*, edited by Kip S. Thorne, Richard H. Price, and Douglas A. Macdonald, Yale University Press, New Haven, 1986.

References for the box "More about the Black Hole." This box is excerpted in part from John Archibald Wheeler, "The Lesson of the Black Hole," *Proceedings of the American Philosophical Society*, Volume 125, Number 1, pages 25–37 (February 1981); J. Michell, *Philosophical Transactions of the Royal Society*, London, Volume 74, pages 35–37 (1784), cited and discussed in S. Schaffer, "John Michell and Black Holes," *Journal for the History of Astronomy*, Volume 10, pages 42–43 (1979); P.-S. Laplace, *Exposition du système du monde*, Volume 2 (Cercle-Social, Paris, 1795), modern English translation in S. W. Hawking and G. F. R. Ellis, *The Large Scale Structure of Space-Time*, Cambridge University Press, Cambridge, U.K., 1973, pages 365–368; J. R. Oppenheimer and H. Snyder, *Physical Review*, Volume 56, pages 455–459 (1939) (published the day World War II began), quoted in Stuart L. Shapiro and Saul A. Teukolsky, *Black Holes, White Dwarfs, and Neutron Stars: The Physics of Compact Objects*, John Wiley and Sons, New York, 1983, page 338; R. P. Kerr, *Physical Review Letters*, Volume 11, pages 237–238 (1963); E. T. Newman, E. Couch, K. Chinnapared, A. Exton, A. Prakash, and R. Torrence, *Journal of Mathematical Physics*, Volume 6, pages 918–919 (1965); S. W. Hawking "Black Hole Explosions?" *Nature*, Volume 248, pages 30–31 (1 March 1974); See also *Black Holes: Selected Reprints*, edited by Steven Detweiler, American Association of Physics Teachers, New York, December 1982, which includes reprints of papers by John Michell, Karl Schwarzschild, S. Chandrasekhar, J. Robert Oppenheimer, and H. Snyder, Roy P. Kerr, S. W. Hawking, etc.

Porus Lakdawala suggested the exercise on light speed for the fastest possible stone. Rae Yip suggested the exercise on Newton approximation of the energy of a plunging stone.

Chapter 3 Exercises

1. Plunging from Rest at Infinity

Black Hole Alpha has a mass $M = 5$ kilometers and a horizon at $r = 2M = 10$ kilometers. A stone starting from rest far away falls radially into Black Hole Alpha.

A. At what velocity does a shell observer at $r = 35$ kilometers measure the stone to be going as the stone passes him? (Answer to nearest digit is –0.5. Supply three-digit accuracy.)

What is the bookkeeper velocity dr/dt of the stone as it passes r = 35 kilometers? (Answer to nearest digit is –0.4. Supply three-digit accuracy.)

B. At what velocity does a shell observer at $r = 25$ kilometers measure the stone to be going as it passes him? (Answer to nearest digit is –0.6. Supply three-digit accuracy.)

What is the bookkeeper velocity dr/dt of the stone as it passes $r = 25$ kilometers? (Answer to nearest digit is –0.4. Supply three-digit accuracy.)

C. Qualitatively, what do the formulas in the text lead you to *expect* about the relative shell speeds (greater or smaller) at the two radii? the relative values of the shell and bookkeeper speeds (greater or smaller) at each radius?

D. In the limit as r—> 2M, what is the shell speed of the stone? What is the bookkeeper speed of the stone?

2. Maximum Bookkeeper Speed

A stone is released from rest far from a black hole of mass M. The stone drops radially inward. Bookkeeper records show that the stone's inward speed initially increases but declines toward zero as the stone approaches the horizon. Bookkeeper speed must therefore reach a maximum at some intermediate radius r. Find this radius for maximum bookkeeper speed. Check your answer using Figure 5. *Optional, probably hard:* Find the radius of maximum bookkeeper speed for the more general case of a stone *hurled* into the black hole (Sample Problem 3). Verify that your result reduces to the dropped-from-rest expression when the initial speed is zero.

3. Hitting a Neutron Star

A typical neutron star has a mass equal to approximately 1.4 times the mass of Sun (magnitude well-known observationally) and a radius of roughly 10 kilometers (magnitude not well-known). A stone falls from rest at a great distance onto the surface of a nonrotating neutron star with these values of radius and mass.

A. If this neutron star were a black hole, what would be the r-value of its horizon? What fraction is this of the radius of the neutron star?

B. With what speed does the stone hit the surface of the neutron star as measured by someone standing (!) on the surface?

C. With what speed does the stone hit the surface according to the far-away bookkeeper?

D. With what kinetic energy per unit mass does the stone hit the surface according to the surface observer?

E. What is the energy per unit mass of the stone as it hits the surface according to the far-away observer? (Gotcha!)

F. With what speed and kinetic energy per unit mass does the stone hit the surface according to Newton? Compare with your results of parts B through D.

4. Shell Energy of a Stone Hurled Inward

Find a relation between the total energy E of a stone hurled radially inward toward a black hole from a great distance and the energy E_{shell} of the stone measured by a shell observer when the stone later passes that observer. Let v_{far} be the speed of the stone at infinite distance and $\gamma_{far} = (1 - v_{far}^2)^{-1/2}$ be the stretch factor for that speed.

Start from equation [42] for a stone hurled into a black hole from a great distance:

$$\frac{dr_{\text{shell}}}{dt_{\text{shell}}} = -\left[1 - \frac{1}{\gamma_{\text{far}}^2}\left(1 - \frac{2M}{r}\right)\right]^{1/2} \qquad [42. \text{ hurled}]$$

Now assume that the special relativity expression for energy applies to observations made by the shell observer. Use equation(s) from special relativity, along with equation [42], to find an expression for the energy of the falling stone E_{shell} as measured by shell observers. Verify that for this stone the shell energy and the total energy E satisfy the general equation [27].

5. Light Speed for the Fastest Possible Stone

A stone is hurled radially inward toward a black hole from a great distance, with initial stretch factor γ_{far}. Sample Problem 3 describes the shell and bookkeeper speeds of this stone after it falls inward to reduced circumference r. Now hurl the stone inward with greater and greater initial energy so that the remote stretch factor γ_{far} approaches infinity. In other words, in the limiting case let the stone take on the properties of a flash of light.

A. Show that the resulting velocities of light reckoned by the bookkeeper (dr/dt) and measured directly by the shell observer ($dr_{\text{shell}}/dt_{\text{shell}}$) have the values, respectively,

$$\frac{dr}{dt} = -\left(1 - \frac{2M}{r}\right) \qquad [43. \text{ light}]$$

$$\frac{dr_{\text{shell}}}{dt_{\text{shell}}} = -1 \qquad [44. \text{ light}]$$

The second of these results seems reasonable enough. But the first expression looks strange! Investigate further.

B. What is the "bookkeeper radial velocity of light" very far from the black hole ($r \longrightarrow \infty$)?

C. Does the "bookkeeper radial velocity of light" increase or decrease in magnitude as one nears the black hole?

D. What is this "bookkeeper radial velocity of light" in the limiting case of approach to the horizon ($r \longrightarrow 2M$)?

Indeed, equation [43] is shown on page 5-3 to be the bookkeeper radial speed of light.

6. Energy Conversion Using a Black Hole

Advanced civilizations may use black holes as energy sources. Most useful will be "live" black holes, those that have angular momentum. See Project F, Spinning Black Hole. Unfortunately the nonrotating black holes we study in this chapter are "dead": no energy can be extracted from them (except for Hawking radiation—see the box on page 2-4). However, it is possible to use a dead black hole to help convert mass into energy, as you can verify in this exercise.

A bag of garbage of mass m at rest at a dumping point far from a black hole has energy measured at infinity equal to its mass: $E/m = 1$. In contrast, a mass m at rest on a shell at reduced circumference r_o has a smaller energy, given by equation [18] derived in Sample Problem 1, page 3-12:

$$\frac{E}{m} = \left(1 - \frac{2M}{r_o}\right)^{1/2} \qquad [45. \text{ object at rest}]$$

For r_o near the event horizon, $r_o \longrightarrow 2M$, this energy approaches zero. Is there some way, by tossing garbage toward a black hole and stopping it near the horizon, that we can convert the rest energy (mass!) of the garbage into a useful form of energy? And what are the various energy-conversion relations on the way to this result? Investigate these questions using the following outline or some other method.

A. Suppose we mount on a shell at reduced circumference r_o a machine that slows to rest the bag of incoming garbage of mass m and converts its local kinetic energy entirely into light. How much kinetic energy is available for the machine to convert to light?

B. The machine now directs the resulting flash of light radially outward. Apply equation [27] to the light to calculate the energy of this light as it arrives at the power station back at the original remote dumping radius.

C. Take the limit as the r-coordinate of the converter approaches that of the event horizon. Show that in this limit *all* of the rest energy of the garbage at the remote dumping point returns to the remote point as light energy.

D. Now the garbage that has been stopped at the conversion machine is released into the black hole. Construct a convincing argument that, in the limiting case that the conversion machine is at the horizon, the mass of the black hole does not change as a result of the garbage dumping described in this exercise. Thus the total conversion of garbage mass into light energy does not violate the law of conservation of energy of an isolated system (garbage plus black hole).

E. Uh oh! If you have carried out exercise 5 on radial light speed, state how long it will take for the energy from the horizon to become available at the original far-away dumping point. What other difficulties are there with locating the conversion machine at the horizon?

Discussion: Suppose the neighborhood of a black hole is strewn with garbage. We can tidy up the vicinity by dumping the garbage into the black hole. This act reduces disorder in the surroundings of the black hole. Powerful principles of thermodynamics and statistical mechanics say that the disorder of an isolated system (garbage plus black hole) will not spontaneously decrease. Therefore the disorder of the black hole must increase when we dump garbage into it. Jacob Bekenstein and Stephen Hawking have quantified this argument to define a measure of the disorder. See Kip S. Thorne, *Black Holes and Time Warps,* pages 422–448.

7. Dropping in on a Black Hole

A clock of mass m near a black hole is bolted to the shell at r-coordinate r_o outside the event horizon. The bolt works loose and the clock drops radially toward the center.

A. What is the bookkeeper velocity dr/dt and the shell velocity dr_{shell}/dt_{shell} of the clock as it falls through a spherical shell of smaller radial coordinate r? With what velocity does it cross the horizon of the black hole?

Find these answers:

$$\frac{dr}{dt} = -\left(1-\frac{2M}{r}\right)\left[\frac{\frac{2M}{r}-\frac{2M}{r_o}}{1-\frac{2M}{r_o}}\right]^{1/2} \qquad [46]$$

$$\frac{dr_{shell}}{dt_{shell}} = -\left[\frac{\frac{2M}{r}-\frac{2M}{r_o}}{1-\frac{2M}{r_o}}\right]^{1/2} \qquad [47]$$

Note that at the horizon, at $r = 2M$, the shell speed is always that of light, no matter from which shell the clock has dropped.

Now this really is too much! Let the object be released from a shell of reduced circumference r_o 1 meter greater than that of the horizon at $r = 2M$. Do you mean to tell me that this object is accelerated to the speed of light in 1 meter of distance? Do you mean to say also that the bookkeeper velocity dr/dt goes to zero during the same time lapse? You really have "gone over the edge" yourself!

Motion described in Schwarzschild coordinates certainly has peculiarities near the horizon! Look at the relation between shell radial distance dr_{shell} and increment dr in reduced circumference (equation [D] in Selected Formulas at the end of this book):

$$dr_{shell} = \frac{dr}{\left(1-\frac{2M}{r}\right)^{1/2}} \qquad [D]$$

You ask if the object dropped from a reduced circumference r_o one meter greater than that of the horizon accelerates to the speed of light in one meter. But this equation says that as we move close to the horizon, the radial distance measured by shell observers stretches out more and more compared with the change in r-coordinate. The last meter of r-coordinate before the horizon stretches away to a great distance in shell dimension. Plenty of shell distance for acceleration to the speed of light!

Near the horizon the shell coordinates are practically useless. *At* the horizon they go berserk. Where can we turn for some sanity? As always, return to the free-float frame. When events are clocked on the wristwatch carried by the in-falling object, the shells labeled with reduced circumference r pass by in stately succession. As observed from this free-float frame, nothing weird happens at the horizon, as demonstrated in exercise 8.

B. A robot worker of the Black Hole Construction Company is repairing the shell at $r = 40$ kilometers around Black Hole Alpha, which has a mass $M = 5$ kilometers and a horizon at $r = 2M = 10$ kilometers. The worker drops from rest his wrench (British: *spanner*).

At what velocity does a shell observer at $r = 25$ kilometers measure the wrench to be going as it passes him? *Should* this velocity have a greater or lesser magnitude than that of the shell velocity at $r = 25$ kilometers of the wrench dropped from infinity (part B of Exercise 1)? Does it turn out as you predict?

C. What is the bookkeeper velocity dr/dt of the wrench as it passes $r = 25$ kilometers? *Should* the magnitude of this velocity be more than or less than the bookkeeper speed at $r = 25$ kilometers of the stone dropped from infinity (part B of Exercise 1)? Does it turn out as you predict?

D. What are the shell-observer and bookkeeper velocities as the wrench approaches the horizon at $r = 2M$?

E. Returning to Part A, let the shell observer on a shell of radius r use special relativity to set $E_{shell} = m/(1 - v_{shell}^2)^{1/2}$. Substitute for v_{shell} from the result of Part A and use the result of Sample Problem 1 to eliminate r_0 from the resulting equation. Show that your result agrees with equation [27] assuming $r < r_0$.

8. Timetable to the Center

An astronaut drops from rest off a shell of radius r_0. How long a time elapses, as measured on her wristwatch, between letting go and arriving at the center of the black hole? If she jumps off the shell just outside the horizon, what is her horizon-to-crunch time (the maximum possible horizon-to-crunch time, see equations [30] and [31] on page 3-21).

Several hints: The first goal is to find $dr/d\tau$, the rate of change of r-coordinate with wristwatch time τ, in terms of r and r_0. Then form an integral whose variable of integration is r/r_0. The limits of integration are from $r/r_0 = 1$ (the release point) to $r/r_0 = 0$ (the center of the black hole). The integral is

$$\tau = -\frac{r_0^{3/2}}{(2M)^{1/2}} \int_1^0 \frac{(r/r_0)^{1/2} d(r/r_0)}{(1 - r/r_0)^{1/2}} \qquad [48]$$

Solve this integral using tricks, nothing but tricks: Simplify by making the substitution $r/r_0 = \cos^2 \psi$. (The "angle" ψ is not measured anywhere; it is simply a variable of integration.) Then $(1 - r/r_0)^{1/2} = \sin \psi$ and $d(r/r_0) = -2 \cos \psi \sin \psi \, d\psi$. The limits of integration are from $\psi = 0$ to $\psi = \pi/2$. With these substitutions, the integral for proper time becomes

$$\tau = 2\frac{r_0^{3/2}}{(2M)^{1/2}} \int_0^{\pi/2} \cos^2 \psi \, d\psi = 2\frac{r_0^{3/2}}{(2M)^{1/2}} \left[\frac{\psi}{2} + \frac{\sin 2\psi}{4}\right]_0^{\pi/2} \qquad [49]$$

The answer follows immediately. Its units are meters of light-travel time. Now convert this result to seconds and examine the special case of release from just outside the horizon.

9. Gravitational Acceleration on the Spherical Shell

A robot worker on the shell at $r = r_0$ drops a tool from rest. What initial acceleration will the robot measure for the tool? Answer this question using the following outline or some other method. (Note that part A is identical to part A of Exercise 6 in Chapter 2.)

A. Express Newtonian acceleration in geometric units. According to Newton, what is the radial acceleration at a distance r_0 from a spherically symmetric center of attraction, in conventional units? Express the result as $g_{conv} = d^2r/dt^2$, where "conv" means "in conventional units." Now set dt(meters) $= c\, dt$(seconds) and show that the Newtonian prediction, expressed in geometric units is

$$g = \frac{d^2 r}{dt^2} = \frac{M}{r_0^2} \qquad [50. \text{ Newton}]$$

(ignoring the fact that acceleration is in the negative radial direction). In these units, what is the value of g_E, the acceleration of gravity at the surface of Earth? Let g_{Econv} be the acceleration of gravity in conventional units. Show that g_E in geometric units has the units meter^{-1} and the approximate value

$$g_E = \frac{g_{conv}}{c^2} = 10^{-16}\ \text{meter}^{-1} \qquad \text{[51. Newton]}$$

B. What is the corresponding prediction of general relativity? First we need to decide which dr and dt we are talking about. The statement of the exercise specifies that it is the shell worker whose measurements we are to predict. Therefore we want $d^2r_{shell}/dt^2_{shell}$. Start with the result of Exercise 7:

$$\frac{dr_{shell}}{dt_{shell}} = -\left[\frac{\frac{2M}{r} - \frac{2M}{r_o}}{1 - \frac{2M}{r_o}}\right]^{1/2} \qquad [52]$$

Take the derivative of this expression with respect to dt_{shell}, remembering that r_o is fixed, a constant. On the right side of the result you will have a factor dr/dt_{shell}, where dr is the change in reduced circumference r, not shell coordinate. Use equation [D] in Selected Formulas at the end of this text to eliminate dr from your derivative. Then substitute for dr_{shell}/dt_{shell} from equation [52]. Evaluate the result at $r = r_o$ to obtain the simple expression (again ignoring the minus sign)

$$g_{shell} = \left.\frac{d^2r_{shell}}{dt^2_{shell}}\right|_{\text{(from rest)}} = \frac{M}{r_o^2}\left(1 - \frac{2M}{r_o}\right)^{-1/2} \qquad [53]$$

What are the limiting cases (1) as r_o approaches $2M$ and (2) as r_o becomes very large (but *not* infinite)?

C. The robot worker stands on a shell of radius $r_o = 4M$ near a black hole of mass 5000 meters. How many "gee"—that is, how many times the value of g_E at Earth's surface—is the initial acceleration of his dropped tool? What is the Newtonian prediction? (A fighter pilot risks blacking out when she makes her plane turn or rise at an acceleration of $7g_E$ or more.)

D. What is the acceleration of gravity at the surface of the typical neutron star described in Exercise 3?

E. *Optional.* You want to hover, rockets blasting, just above the horizon of a black hole. Call your reduced circumference $r_o = 2M + dr$, where $dr << 2M$. Find an approximate expression for g_{shell} under these circumstances. Now, in order to stay conscious, you want g_{shell} in your spaceship to be $7g_E$. If $dr = 1$ kilometer, what is the mass M of the black hole you should choose for this maneuver? Express your answer as a multiple of the mass of our Sun.

10. Horizon Alarm for Your Spaceship

This is a thought question. The Space Agency is anxious about the fate of their rocket ship (and you) and requests that you carry an automatic alarm designed to warn you when you are in danger of an irretrievable plunge into a black hole. The alarm should warn you when you are approaching the horizon. As a first cut at the design, assume that you are dropping from rest at a great distance and that the device has a register that allows you to enter the known value of the mass M of the black hole. Beyond this single setting, the device must depend only on readings and experiments made internal to the spaceship and cannot require information from outside. On what principle or principles can such a device operate? Can you present a back-of-the-envelope design? How far can you go toward setting the parameters for the device, that is, specifying its numerical sensitivity for a black hole of known mass M? *Optional:* Add the following features: (a) A second alarm that warns when tidal forces are approaching values that may injure you. (b) If an alarm goes off, you will want to activate rockets to bring you to rest on a stationary shell at your current radius r. A third alarm should warn you when your radial position r is small enough so that the gravitational field experienced on that shell is greater than, say $3g_{Earth}$.

11. Newton Approximates Plunging Energy

Show that the general relativistic expressions for energy of a plunging particle reduce (sort of!) to the Newtonian result for small velocities and small values of $2M/r$. Use the following outline or your own method.

A. Set up the Newtonian expression for the total energy of a particle in free fall around a center of gravitational attraction. Convert to geometric units and show that the result can be written

$$\frac{E}{m} = \frac{1}{2}v^2 - \frac{M}{r} \qquad \text{[54. Newton]}$$

B. Now consider the general-relativistic energy of a particle, equation [12]:

$$\frac{E}{m} = \left(1 - \frac{2M}{r}\right)\frac{dt}{d\tau} \qquad [55]$$

Set

$$\frac{dt}{d\tau} = \frac{dt}{dt_{\text{shell}}}\frac{dt_{\text{shell}}}{d\tau} \qquad [56]$$

and use the approximation (for $|d| << 1$ and $|nd| << 1$)

$$(1 + d)^n \approx 1 + nd \qquad [57]$$

several times to show that, approximately,

$$\frac{E}{m} \approx 1 + \frac{1}{2}v_{\text{shell}}^2 - \frac{M}{r} \qquad \text{[58. Newton?]}$$

C. How do equations [54] and [58] differ from one another? How do you account for the difference(s)?

Project B Inside the Black Hole

- *When I fall into a black hole, do I move faster than light?*
- *Why can't I send a message to my friends outside the horizon?*
- *Is my death quick and painless?*
- *What is the last thing I see?*

Project B

Inside the Black Hole

Inside the Horizon of a Nonspinning Black Hole

Alice had not a moment to think about stopping herself before she found herself falling down what seemed to be a very deep well. Either the well was very deep, or she fell very slowly, for she had plenty of time as she went down to look about her, and to wonder what was going to happen next. First she tried to look down and make out what she was coming to, but it was too dark to see anything; then she looked at the sides. . . . For, you see, so many out-of-the-way things had happened lately that Alice had begun to think that very few things indeed were really impossible.

—Lewis Carroll, *Alice in Wonderland*

Note: Sections 8–10 in this project make use of material in Chapters 4 and 5.

1 Interview of a Diving Candidate

So you are applying to be a member of the black hole diving research team.

Yes.

Have you had experience diving into black holes?

This question is a joke, right?

Why do you want to be part of this project, since your research results cannot be reported back to us outside the horizon?

Our diving team went through the Astronautics School together. We have studied black holes intensively all our professional lives and are deeply curious to discover for ourselves whether for not predictions for conditions inside the black hole are correct. We all feel strongly that we want to cap our careers with this trip.

Tell me, why do black-hole divers use a free-float frame inside the horizon?

Inside the horizon stationary spherical shells are not possible, even in principle. Free-float frames are the simplest and most universal.

Then how will you measure your radial position r without a spherical shell?

We derive the value of r from the distance to test particles on either side of us that are also diving radially inward.

What clocks will you use in your experiments?

Our carry-along wristwatches.

What do you know about the trajectory?

In effect we start from rest at a great distance from the black hole, then drop radially inward.

When do you cross the horizon?

As measured on whose clock?

You *are* savvy. When do you cross the horizon as read on your wristwatch?

Actually zeroing our wristwatches is arbitrary. The Astronautics Commission has a fancy scheme for synchronizing all clocks, including our wristwatches, mostly for convenience in scheduling. Want more details?

Not now. Have you been briefed on the plans to resupply you?

There is no need to send us supply packages. Either these packages would have to be hurled after us at great speed or dropped from a shell near the horizon before we leave so that we catch up with them on our descent. It is simpler and cheaper to carry all supplies with us.

Is there anything at all we can do for you after you cross the horizon?

Sure. We would appreciate receiving radio and television bulletins of the latest news and reports of scientific developments outside the horizon.

And will your outward radio transmissions change frequency during their transit to us?

Another joke, I see.

Of course. All such predictions, including those essential to your health and safety, depend on the validity of the Schwarzschild metric inside the horizon.

I am betting my life on the correctness of these predictions. The selected black hole is an old one, so we presume that the sea of gravitational waves trapped inside the horizon at its formation will have died away. Only we divers will find out if these calculations are correct.

Speaking of your life, we predict your death 20 years on your wristwatch after crossing the horizon.

Yes. Unlike you on the outside, we will know when to expect it. Gives a feeling of assurance. Actually, we all feel privileged to be part of this expedition. You know very well that 27 percent of qualified Galaxy Fleet personnel volunteered for this mission.

Does the—ah—end seem mercifully quick to you?

Yes, I have determined the terminal stretch-compress interval. Apparently it will be over faster than pain signals can move from extremities to the brain. Many of you outside the horizon would welcome assurance of such a quick end.

What will you do for relaxation during the trip?

I am lunar champion in zero-g football and a grand master chess player. Also, my fiancé has already been accepted as part of the team. We will be married before launch.

What kind of science are these people talking about? Obviously nothing more than science fiction! No one who crosses the event horizon of a black hole can report observations to the scientific community outside the horizon. So measurements made inside the horizon lack the essential feature of reportability. Science requires that the wider community examine observations, compare them with predictions, and discuss published analyses. Therefore all observations carried out inside the horizon—and conclusions drawn from them—remain pure speculation. Speculation is not science!

Yours is one sensible view of science. On the other hand, nothing stops us from forming an in-falling community of investigators that moves together in free float across the horizon. For a sufficiently massive black hole, we in such an in-falling community have decades of life ahead of us, as recorded on our wristwatches. We receive signals and possibly packages of supplies from friends outside the horizon. We view the ever-changing pattern of stars in the heavens overhead. We carry out our investigations, communicate among ourselves, discuss our observations, publish our own journals, and reach consensus about the correctness of predictions based on the Schwarzschild metric. "Inside science" has all essential features of "outside science": scientists predict, verify, discuss, dispute, and concur. The fact that our results are not available to some other group of scientists—those who stay outside—may make our "inside science" parochial, but it is not necessarily invalid.

Ha! What kind of science can it be when all records of your investigations are destroyed—crunched to nothingness—at the center of the black hole?

Such is the prediction! But is that so different from the eventual fate of the publications of human science? The records and journals of "outside" civilizations will no doubt decay or be destroyed over the millennia that lie ahead. They will surely disappear if the Cosmos eventually recontracts upon itself to a final crunch, as some predict.

In summary, we recognize that the horizon separates two communities of investigators with a one-way membrane. Outsiders cannot verify predictions about our life inside the horizon of the black hole. Outsiders must leave it to us insiders to substantiate or disprove predictions with all the rigor of our lively in-falling research community.

QUERY 1 **Mass of the "20-year black hole."** Our chosen black hole has a mass such that it takes 20 years of wristwatch time for a diver who falls from rest starting from an infinite distance to pass from the horizon to the center.

A. Find the approximate mass of the "20-year black hole" (i) in meters, (ii) as a multiple of the mass of our Sun, and (iii) in years. (Sample Problem 2, page 3-22.)

B. On average, a galaxy consists of approximately 10^{11} stars similar to our Sun. The "20-year black hole" has the mass of approximately how many galaxies?

C. What is the value of $r = 2M$ at the horizon for the "20-year hole" in light-years?

D. **Discussion questions:** How will you know when you are crossing the horizon? How can the value of the Schwarzschild radius $r = 2M$ in light-years possibly be greater than the wristwatch time of 20 years that it takes to fall this distance?

2 The Rain Frame

We dive to the center of a black hole riding in an unpowered spaceship that moves radially inward. To simplify further, we ride on a spaceship launched in a particular way, namely starting from rest at a great distance from the black hole. We call such a free-float frame a **rain frame**, because on Earth rain also starts from rest at a great height (even though near Earth the braking effect of air keeps rain from falling freely). The rain frame starts from rest at so great a height—a radius so remote from the center—that spacetime there is flat. We use the term **raindrop** or **diver** to label a local inertial frame (an unpowered spaceship) that falls radially inward from rest at infinite radius.

Every raindrop is a local frame, limited to a region of spacetime in which relative tidal accelerations of test particles are too small to be detected with equipment of given sensitivity (Section 8 of Chapter 1). Many different local rain frames exist, each with its own unique radial line of motion and time for passing a given radius.

There are other kinds of local free-float frames besides rain frames. An unpowered spaceship that moves slower than a raindrop at a given r is one released from rest at a radius r_0 outside the horizon. We call this free-float frame a **drip frame**. An unpowered vessel that moves faster than a raindrop at a given r is one hurled inward from a great distance with ini-

Eggbeater Spacetime?

Being torn apart at the center of a black hole is bad enough. But according to some calculations, you will not even make it to the center alive: Your atoms will be scrambled by violent, chaotic tidal forces some distance from the center—especially if you fall into a young black hole.

The first theory of the time development of a black hole by J. Robert Oppenheimer and Hartland Snyder (1939) assumed that the collapsing structure was spherically symmetric. The result is a black hole that settles immediately into a placid final state. A diver approaching the singularity at the center of the Oppenheimer black hole is stretched with steadily increasing force along the radial direction and compressed steadily and increasingly along the tangential direction.

A real collapsing structure is almost never spherically symmetric. Theory shows that when a black hole forms, the asymmetries exterior to the horizon are quickly—in a few seconds of far-away time!—radiated away in the form of gravitational waves. However, a fraction of the waves is captured inside the horizon of the black hole. This sea of trapped gravitational waves evolves inside the black hole and influences the structure of the singularity inside it.

So what happens? We cannot personally verify and report about the truth of any theory that describes conditions inside the horizon of a black hole. But this deficiency on our part does not keep us from making predictions! According to Einstein's field equations, there is more than one kind of possible singularity. Vladimir Belinsky, Isaac Markovich Khalatnikov, Evgeny Mikhailovich Lifshitz, and independently Charles Misner found a singularity that is very strange. According to their theory, as a plunging observer approaches the center point, spacetime oscillates chaotically, squeezing and stretching the poor traveler in random directions like an electric mixer (eggbeater) or the old-fashioned taffy-pulling machine sometimes seen at carnivals and candy stores.

These oscillations increase in both amplitude and frequency as the astronaut approaches the center of the black hole. Charles Misner called this a *mixmaster singularity,* referring to a now-discontinued brand of electric mixer. Any physical object, no matter what stresses it can endure, is necessarily utterly destroyed by a mixmaster singularity. There is theoretical basis for predicting that mixmaster oscillations die away with time, so an astronaut who waits to dive for a long time after the black hole is formed may not encounter them.

In addition to the mixmaster singularity, Einstein's field equations also predict a second possible type of singularity that exists only inside a charged or spinning black hole (see Project F). This second type has been called by Eric Poisson and Werner Israel a *mass-inflation singularity.* Investigators agree that the mass-inflation singularity is a milder, or softer, type of singularity; physical objects that approach it are not stretched to infinite length. However, at this point we reach the kind of disagreement that typifies ongoing research. Some believe that it is theoretically possible to traverse the mass-inflation singularity peacefully and even speculate that under some circumstances the astronaut might use a spinning or charged black hole as a portal for hyperspace travel. Others report results that at a mass-inflation singularity different parts of the astronaut's body, while not stretched directly, are given relative velocities equal to the speed of light—not a comfortable prospect! A theory of quantum gravity may be required to resolve these disagreements; see references at the end of this project.

The black hole we are studying in this project is not charged and does not spin, so we do not expect our descending astronauts to encounter a mass-inflation singularity. We also ignore possible eggbeater (mixmaster) oscillations of spacetime and assume that as our astronaut colony approaches the center the "spacetime weather" is clear and calm.

tial velocity v_{far}. We call this third kind of free-float frame a **hail frame,** so named because on Earth hailstones falls faster than raindrops—or at least they hurt more when they hit you! Drip frames and hail frames are useful when we try to resupply divers after they cross the horizon. Taken together, rain frames (dropped from initial rest at infinity), drip frames (dropped from rest at various radii), and hail frames (hurled radially inward from infinity at various velocities) cover all possible radially moving free-float frames. In this project we concentrate on the rain frame; some of our conclusions will be true for all free-float frames.

Equation [24], page 3-15, gives the velocity of the raindrop as clocked by the shell observer past whom it falls:

$$\frac{dr_{shell}}{dt_{shell}} = -\left(\frac{2M}{r}\right)^{1/2} \qquad [1]$$

Equation [32], page 3-22, develops a similar equation that relates the change dr in the Schwarzschild r-coordinate to the proper time lapse dt_{rain} (called $d\tau$ in that sample problem):

$$\frac{dr}{dt_{rain}} = -\left(\frac{2M}{r}\right)^{1/2} \qquad [2]$$

Note that equation [2] is expressed in mixed coordinates: t_{rain} and dt_{rain} measure the time read on the wristwatch of the rain observer, while r and dr are measured in terms of Schwarzschild r-coordinate. The r-coordinate is measured by means described in the box on page B-8.

Equation [2] is plotted in Figure 1. Notice that at the horizon ($r = 2M$), the magnitude of this quantity, the radial speed, is equal to unity, the conventional speed of light. Inside the horizon ($r < 2M$), the quantity on the right has a magnitude greater than unity. At the center ($r = 0$), this quantity becomes infinite. Does this expression really represent the speed of the infalling local rain frame? Inside the horizon, does a raindrop really move "faster than light"? These questions are debated in Section 3.

QUERY 2 **Rain frame time between given radii**

A. Show that the elapsed rain time from falling past radius r_1 to zipping past a smaller radius r_2 is given by the expression

$$t_{2\,rain} - t_{1\,rain} = \frac{1}{3}\left(\frac{2}{M}\right)^{1/2}(r_1^{3/2} - r_2^{3/2}) \qquad [3]$$

Rewrite this equation so that the variables are t_{rain}/M and r/M and so that M occurs nowhere else in the equation. A typical curve is plotted in Figure 4 for a raindrop that passes $r = 5M$ at wristwatch time zero ($t_{1\,rain} = 0$). (These radii are measured by the method shown in the box on page B-8.)

B. What happens to the value of the time lapse in equation [3] when the initial radius r_1 becomes infinite? How do you account for this result?

3 Faster than Light Inside the Horizon? Debate!

True or false: Inside the horizon the rain observer falls faster than the speed of light. Take both sides, arguing as fiercely as possible for and against this proposition. Follow the outline here for each side or develop your own independent cases.

YES! FASTER THAN LIGHT

YES argument 1: Formula for the speed proves it.

Figure 1 uses equation [2] to plot raindrop speed dr/dt_{rain} as a function of r/M for the local rain frame. *Continued at the bottom of page B-9.*

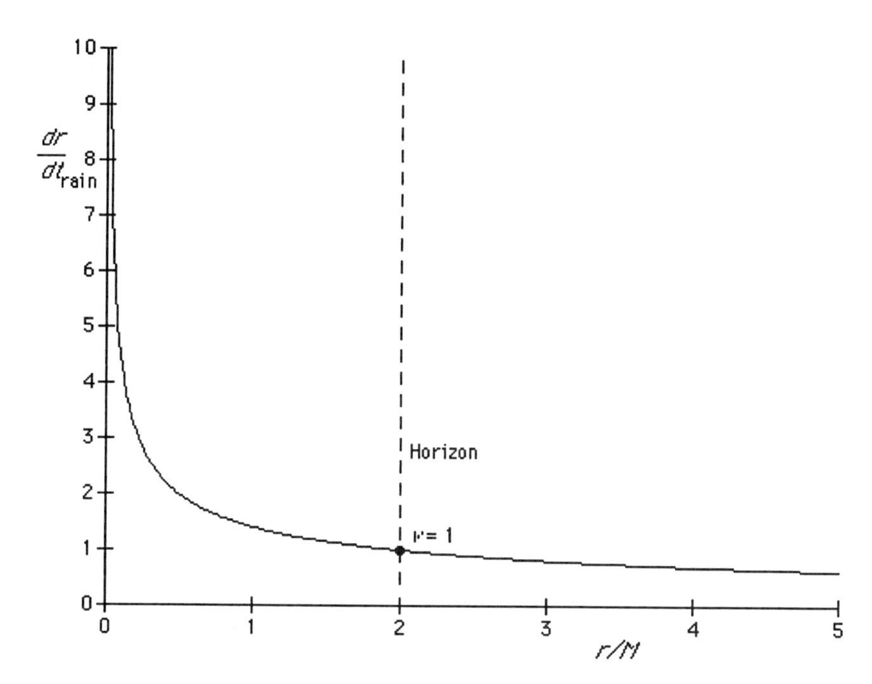

Figure 1 *Computer plot: The value of the mixed-coordinate expression dr/dt_{rain}, the "speed of the local rain frame," plotted as a function of r/M. The value of this speed reaches unity at the horizon and increases without limit as the frame reaches the center at r = 0.*

QUERY 3 **Shell separation as measured in the local rain frame.** A local rain frame observer falls radially inward but has not yet reached the horizon. On the fly she measures the distance dr_{rain} between adjacent shells. According to *special* relativity, she measures the distance between adjacent shells to be *shorter* than the shell measurement, by the factor $(1 - v^2)^{1/2}$, where v is the relative speed (equation [1]) between in-faller and local shell. According to *general* relativity, the measured distance between adjacent shells is *longer* than the bookkeeper distance dr, according to equation [D] in Selected Formulas at the back of the book. Show that these two factors cancel, so that $(dr_{rain}/dr_{shell})(dr_{shell}/dr) = dr_{rain}/dr = 1$ and we have the surprising result:

$$dr_{rain} = dr \qquad \text{[4. meterstick measurement]}$$

In other words, the in-falling rain observer measures the separation between adjacent shells to be equal to the difference dr in their Schwarzschild r-coordinate. From now on we can use dr instead of dr_{rain}.

Note 1: Relation [4] is true only for the *rain* observer, not observers falling slower ("drip frame") or faster ("hail frame") than the rain observer at a given radius. *Note 2:* Equation [4] is correct only for separation between events that are simultaneous in the rain frame, as is true for measurement of length by any observer.

Measuring the Value of *r* Inside the Horizon

Outside the horizon, *r*-values are determined by measuring the circumference of each spherical shell and dividing this circumference by 2π (Section 4 of Chapter 2). Hence we call *r* the reduced circumference. But no stationary concentric spherical shell exists inside the horizon. *Question:* How can the rain observer possibly determine the reduced circumference *r* of her location? *Answer:* By measuring a *piece* of circumference.

Two raindrops fall radially side by side past shell AA′ outside the horizon, as shown schematically in the Schwarzschild map of Figure 2. One raindrop falls along the straight radial path ABCO and its companion along the nearby straight radial and converging path A′B′C′O. Draw a circular segment connecting AA′ and similar circular segments connecting BB′ and CC′. Now the angle AOA′ at the center is the same as the angle BOB′. Hence each of the circular segments AA′ and BB′ represents the same fraction of the entire circumference of the circle of corresponding radius.In equation form,

$$\frac{\begin{pmatrix}\text{Length of}\\ \text{circular segment AA'}\end{pmatrix}}{\begin{pmatrix}\text{Circumference of}\\ \text{spherical shell}\\ \text{passing through A}\end{pmatrix}} = \frac{\begin{pmatrix}\text{Length of}\\ \text{circular segment BB'}\end{pmatrix}}{\begin{pmatrix}\text{Circumference of}\\ \text{spherical shell}\\ \text{passing through B}\end{pmatrix}} \qquad [5]$$

Call r_A the reduced circumference at point A and r_B the reduced circumference at point B. Then the denominators of the two sides of the equation become $2\pi r_A$ and $2\pi r_B$, respectively, and we can cancel the 2π from both sides. Now, if the angle at the center is small, the length of the circular segment is approximately equal to the straight-line distance between A and A′. Call this distance AA′. And call BB′ the corresponding straight-line distance between B and B′. Then equation [5] becomes, for equal angles,

$$\frac{AA}{r_A} \approx \frac{BB}{r_B} \qquad [6]$$

Suppose a diver reads the labeled radius r_A of a spherical shell as she passes it and at that instant measures the distance AA′ between the two raindrops. Then at a later time she measures the raindrop separation to be BB′. From this measurement she deduces the radius r_B at that later time to be

$$r_B \approx \frac{BB}{AA} r_A \qquad [7]$$

Both points A and B lie outside the horizon. We now assume that the same formula [7] holds for any radius, even for a radius r_C less than that of the horizon. In other words, the infalling observer can measure her radial position r_C at point C inside the horizon using the equation

$$r_C \approx \frac{CC'}{AA'} r_A \qquad [8]$$

How will the infalling raindrop observer measure separations AA′ or CC′? One possibility employs a simple ruler, as shown in Figure 3. The ruler will, of course, be crushed by tidal forces near the singularity. An alternative method, radar ranging, allows measurement of *r* even closer to the crunch.

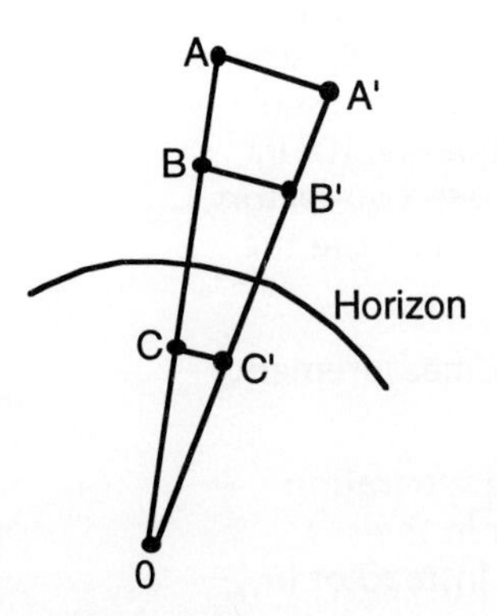

Figure 2 *Schwarzschild map of paths of two in-falling raindrops. By measuring their separations, such as AA′ and CC′, the in-falling observer can deduce her radius, even when she has passed inside the horizon.*

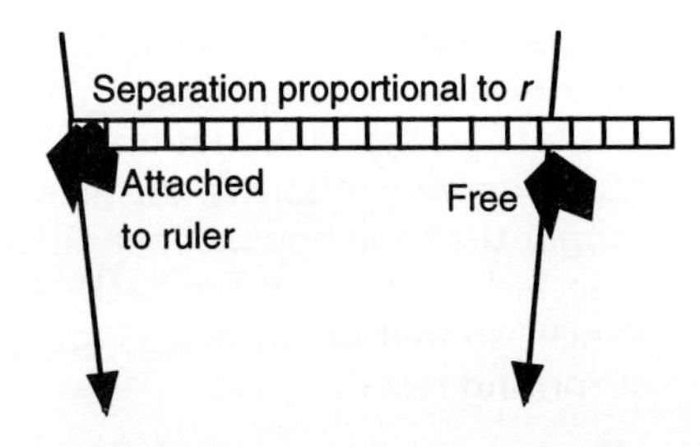

Figure 3 *Device for measuring the reduced circumference r of an observer falling freely inward along a radius. Two objects fall inward with matched speeds along straight radially converging paths. A ruler is attached to one; determine the distance between objects by locating the second object along the ruler. Details: Make an initial ruler reading as the device passes a spherical shell of known reduced circumference outside the horizon. Then assume that the magnitude of a later separation is proportional to the reduced circumference r.*

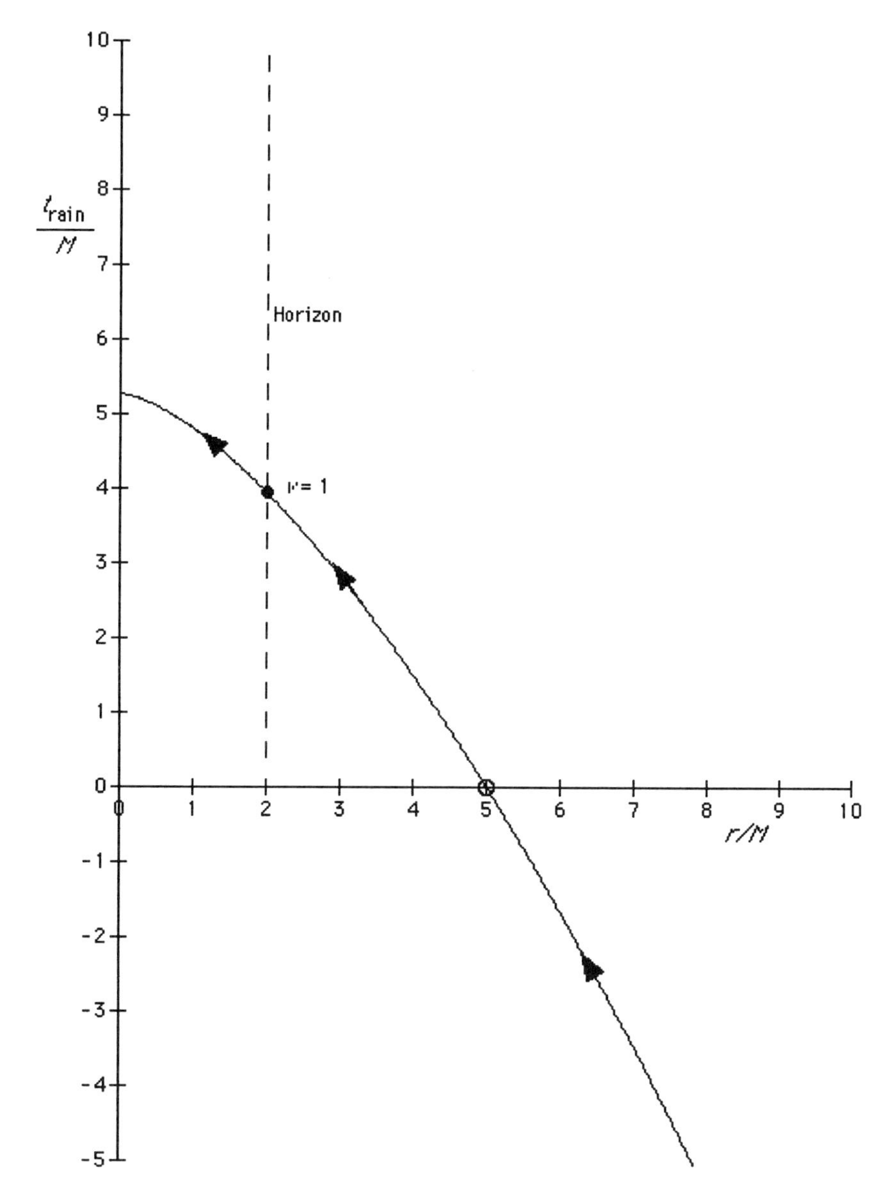

Figure 4 *Computer plot: Worldline of r versus* t_{rain} *for a rain observer (one who falls from rest at infinity). The zero of rain time is set arbitrarily at the event of coincidence between the in-falling observer and the spherical shell at r = 5M (open circle on the horizontal axis). For other choices of zero time the curve can be moved bodily up or down on the graph without change of shape.*

$$\frac{dr}{dt_{\text{rain}}} = -\left(\frac{2M}{r}\right)^{1/2} \qquad [2]$$

Inside the horizon ($r < 2M$), this raindrop speed certainly takes on a magnitude greater than unity, greater than the speed of light. This argument proves that we have finally broken the light barrier—that inside the horizon the rain observer moves faster than light.

YES argument 2:

QUERY 4 **Wristwatch time to the center proves it.** Compare the radius of the horizon to the wristwatch time for the rain observer to fall from horizon to center. Show that the ratio of the two represents an average speed greater than unity, greater than the speed of light. Recall Query 1, part D.

YES argument 3:

QUERY 5 **INSIDE the horizon, the raindrop must move faster than it does AT the horizon.**

A. AT the horizon, as a limiting case, the shell observer measures the in-falling rain observer to move inward with what speed dr_{shell}/dt_{shell}?

B. The in-falling observer measures the last passing spherical shell at the horizon (as a limiting case) to be moving past at what speed?

C. Equation [2] agrees with the result of part B. And thereafter the in-faller surely increases her speed. Therefore inside the horizon . . .

NO! SLOWER THAN LIGHT

NO argument 1: Local rain frame observer measures light to move at $v = 1$.

The local rain frame is a free-float frame. As measured in such a frame, light moves with its conventional speed: 1 meter of distance per meter of time. We expect that the rain observer measures inward-moving light to pass her with the conventional light-speed unity even after she has passed inside the horizon. Since light passes the rain frame, it is ridiculous to say that the local rain frame moves "faster than light."

The gift of fantasy

When I examine myself and my methods of thought, I come to the conclusion that the gift of fantasy has meant more to me than my talent for absorbing positive knowledge.

—Albert Einstein

NO argument 2:

QUERY 6 **The "proper velocity" of light is infinite.** The expression dr/dt_{rain} of equation [2] does not clock an ordinary frame speed. Its numerator, sure enough, is the change dr in Schwarzschild r-coordinate. Its denominator, however, is the proper time (wristwatch time) of the in-falling raindrop. This ratio has a counterpart in special relativity of flat spacetime. Explore the special relativity case as follows:

A. Forget black holes and general relativity and return to flat spacetime. Think of a trip from Earth to Alpha Centauri, 4 light-years from Earth, in an unpowered spaceship moving at speed dx/dt as observed in the Earth frame. Now change the measure of time from Earth-frame time dt to proper time $d\tau$ as recorded on the wristwatch carried by the spaceship. Then change the measure of speed from dx/dt to $dx/d\tau$. Call $dx/d\tau$ the **proper speed**, since the spaceship covers the proper distance between Earth and Alpha Centauri (the distance measured in the Earth frame, in which both are at rest) in the proper time for the trip (the time measured in the spaceship frame, in which events of departure from Earth and arrival at Alpha Centauri occur at the same place). Find an expression for proper speed $dx/d\tau$ in terms of Earth-frame speed dx/dt.

B. Now let the spaceship repeat the trip from Earth to Alpha Centauri at greater and greater Earth-frame speeds dx/dt. For what Earth-frame speed dx/dt does the proper speed $dx/d\tau$ become unity?

C. As the Earth-frame speed dx/dt of the spaceship increases further and approaches the speed of light, what value is approached by the proper speed $dx/d\tau$?

D. As a limiting case of high speed, what is the "proper speed of light" in flat spacetime?

E. Now return to the speed of the in-falling local rain frame given by equation [2]. Here, too, the time increment dt_{rain} is equal to the proper time $d\tau$ read on the wristwatch of the in-falling raindrop. Display a conclusion that proves that even inside the horizon the proper speed dr/dt_{rain} of the raindrop is less than the limiting-case "proper speed of light"? Is there any location inside the horizon at which the two have the same value? Is there more than one such location inside the horizon?

Discussion: One way to resolve the debate is to acknowledge the radical difference between the circumstances in which velocity is measured inside the horizon and the usual situations in which we define velocity. The differences in these circumstances are so great that our usual ideas of velocity have simply been transcended. However, even inside the horizon at least one fundamental observation may remain true, as described in the following query.

QUERY 7 **Debaters agree?** What do the YES! advocates and the NO! advocates in the debate have to say about the following statement: No LOCAL observer directly measures any object to move faster than light. Do they both agree that it is true, or do they disagree? Is this statement true for an observer who drops from rest at an intermediate radius? Is it true for an observer who starts toward the black hole from a great distance with an initial inward velocity?

4 Metric for the Rain Frame

Up until now we have examined only events in the immediate neighborhood of a single in-falling rain observer. In this near-surround, one can construct a local free-float frame consisting of a limited latticework of clocks. Now we want to describe the motion of light from distant stars and the motion of resupply packages hurled radially inward for our use. These trajectories span large regions of spacetime. To describe them we need a global reference frame.

For the region outside the black hole horizon, the global reference frame was originally provided by the bookkeeper coordinates r, ϕ, and t used in the Schwarzschild metric. Inside the horizon the coordinates r and ϕ can still be measured, but far-away time t is no longer useful, because reports of events can no longer reach the distant bookkeeper. Instead we used time t_{rain} measured on the wristwatch of a single in-falling rain observer. In Query 3 we showed that outside the horizon $dr = dr_{rain}$ for a pair of events simultaneous in the rain frame (true *only* for plungers who start at rest at infinity), so dr is a useful measure of radial increment for the rain observer. We assume that this relation also is correct inside the horizon.

The box on page B-13 derives the metric in the coordinates r, ϕ, and t_{rain}. This metric is valid both outside and inside the horizon. As usual, the metric allows us to answer all possible scientific questions about (nonquantum) features of spacetime near the Schwarzschild black hole and about the motion of particles and light flashes in its vicinity. With the metric in this form, the analysis can proceed seamlessly from outside to inside the horizon and (almost!) all the way to $r = 0$.

QUERY 8 **Flat space in the rain frame?** Set $dt_{rain} = 0$ and show that the space part of the metric [15] is what one would expect from Euclidean geometry. That is, the space part is *locally flat.* Is this result a surprise? In the Schwarzschild description, dr_{shell} is greater than dr outside the horizon (equation [D] in the box). Why is there no similar stretch in the dr^2 term in the metric for the global rain frame? (*Hint:* Recall the results of Query 3.)

Metric for the Rain Frame

We want a metric in the coordinates r, ϕ, and t_{rain}. We make this transition in two jumps for events outside the horizon: from bookkeeper coordinates to shell coordinates, then from shell coordinates to rain coordinates. Assume that the resulting metric is valid inside the horizon as well as outside. The transition from bookkeeper coordinates to shell coordinates is given by equations [C] and [D] in Selected Formulas at the end of this book:

$$dr_{shell} = \frac{dr}{\left(1 - \frac{2M}{r}\right)^{1/2}} \qquad \text{[D]}$$

$$dt_{shell} = \left(1 - \frac{2M}{r}\right)^{1/2} dt \qquad \text{[C]}$$

Now, to go from shell coordinates to rain coordinates, use the Lorentz transformation of special relativity. Choose the "rocket" coordinates to be those of the rain frame and the "laboratory" coordinates to be those of the shell frame. The Lorentz transformation for differentials (page 103 of *Spacetime Physics*) is expressed for motion along the x-axis, which in this case lies along the radially inward direction:

$$dt_{rain} = -v_{rel}\gamma dr_{shell} + \gamma dt_{shell} \qquad \text{[9]}$$

Substitute equations [C] and [D] into the Lorentz transformation equation [9] to obtain:

$$dt_{rain} = -\frac{v_{rel}\gamma dr}{(1 - 2M/r)^{1/2}} + \gamma(1 - 2M/r)^{1/2} dt \qquad \text{[10]}$$

Solve for dt:

$$dt = \frac{dt_{rain}}{\gamma(1 - 2M/r)^{1/2}} + \frac{v_{rel}dr}{(1 - 2M/r)} \qquad \text{[11]}$$

Substitute v_{rel} from equation [1] on page B-5 into the expression for the stretch factor γ:

$$v_{rel} = -\left(\frac{2M}{r}\right)^{1/2} \qquad \text{[12]}$$

$$\gamma \equiv \frac{1}{(1 - v_{rel}^2)^{1/2}} = \frac{1}{\left(1 - \frac{2M}{r}\right)^{1/2}} \qquad \text{[13]}$$

Substitute equations [12] and [13] into [11] to obtain

$$dt = dt_{rain} - \frac{(2M/r)^{1/2} dr}{1 - \frac{2M}{r}} \qquad \text{[14]}$$

The Schwarzschild metric is equation [A] in the Selected Formulas at the end of this book:

$$d\tau^2 = \left(1 - \frac{2M}{r}\right)dt^2 - \frac{dr^2}{1 - \frac{2M}{r}} - r^2 d\phi^2 \qquad \text{[A]}$$

Substitute expression [14] into the Schwarzschild metric and collect terms to obtain the global rain metric in r, ϕ, t_{rain}:

$$d\tau^2 = \left(1 - \frac{2M}{r}\right)dt_{rain}^2 - 2\left(\frac{2M}{r}\right)^{1/2} dt_{rain}dr - dr^2 - r^2 d\phi^2 \qquad \text{[15]}$$

This metric can be used anywhere around a nonrotating black hole, not just inside the horizon. Our ability to write the metric in a form without infinities at $r = 2M$ is an indication that no jerk or jolt is felt as the plunger passes through the horizon.

5 One-Way Motion Inside the Horizon

Why can't we escape from a black hole? Why is travel inside the horizon inevitably a one-way trip to the center? This section treats these questions from the point of view of one special diver: a "raindrop," namely, a plunger that falls freely from rest at a great distance. The basic conclusion, however, is true for *all* travelers inside the horizon: *Inside the horizon you can move only toward the center!*

The raindrop is a local inertial free-float frame. As in any inertial frame, no object or message can move forward or backward faster than light. The worldline of a passing particle must lie within the so-called **forward light cone,** moving no faster than light in either the inward or the outward radial direction. In the following analysis we trace out the local worldlines of light moving in the raindrop frame.

Preview: The rain observer launches light pulses toward the center and away from the center. We shall find that inside the horizon, light shot out the front and light shot out the back both move inward. Hence *all* particles, shot out the front or out the back (necessarily with speeds less than that of light), must also move inward. These facts are important because the only true proof that a horizon exists is the demonstration that worldlines can run through it only in the inward direction, not outward.

QUERY 9 **Light cones in rain coordinates**

A. Multiply out the following expression to show that it is equivalent to the rain-frame metric [15]:

$$d\tau^2 = -\left[dr + \left(1 + \sqrt{\frac{2M}{r}}\right)dt_{\text{rain}}\right]\left[dr - \left(1 - \sqrt{\frac{2M}{r}}\right)dt_{\text{rain}}\right] - r^2 d\phi^2 \qquad [16]$$

B. For light ($d\tau = 0$) moving radially ($d\phi = 0$), show that equation [16] has two solutions, one for the "headlight flash" shot inward and one for the "taillight flash" shot out the back of the raindrop.

C. Show that velocities for both the headlight and taillight flashes can be summarized in the equation

$$\frac{dr}{dt_{\text{rain}}} = -\left(\frac{2M}{r}\right)^{1/2} \pm 1 \qquad [17.\ \text{light}]$$

D. Show that inside the horizon (for $r < 2M$), equation [17] predicts that even the taillight flash, shot out the back of the raindrop, moves *inward*.

E. Fill out the argument that, once the raindrop has crossed the horizon, anything launched from the raindrop in either radial direction, no matter at what speed relative to the raindrop, inevitably moves inward.

Figure 5 shows these initial light cones outside and inside the horizon. Note that these light cones display only the *initial* motion of inward and "outward" light flashes. In the following section we will trace these flashes for some time after their emissions.

QUERY 10 **Initial motion of taillight light flash outside and at the horizon.** As the raindrop falls inward, find the radial location r/M at which the taillight flash emitted from the back of the descending the raindrop moves outward with the following proper velocities:

A. $dr/dt_{\text{rain}} = 0.99$

B. $dr/dt_{\text{rain}} = 0.9$

C. $dr/dt_{\text{rain}} = 0.5$

D. $dr/dt_{\text{rain}} = 0$

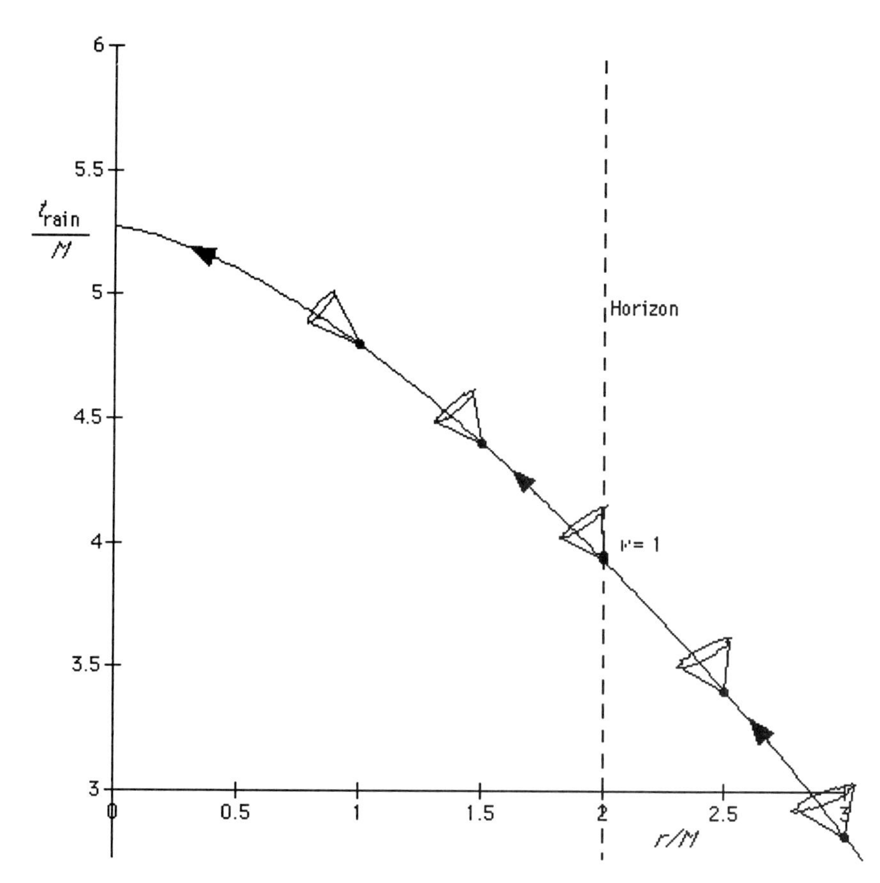

Figure 5 *Computer plot: Worldline of a raindrop emitting flashes as it passes inward through the horizon of a black hole. Arrowheads show the direction of motion of the raindrop along its worldline. Little cones represent light spreading out in all directions from flash emissions along the worldline. The lower line segment leaving each dot represents the initial motion of the portion of the flash sent inward (minus sign chosen in equation [17]). The upper line segment represents the initial motion of the portion of the flash aimed radially outward (plus sign chosen in equation [17]). Inside the horizon even the portion of the flash aimed radially outward moves inward, toward the center. Note that the figure shows only the* ***initial*** *motion of these light flashes.*

QUERY 11 **Initial motion of taillight light flash inside the horizon.** As the raindrop falls inward, find the value of the quantity r/M at which the taillight flash actually moves inward with the following "proper velocities":

E. $dr/dt_{rain} = -0.1$

F. $dr/dt_{rain} = -0.5$

G. $dr/dt_{rain} = -1$

H. $dr/dt_{rain} = -9$

QUERY 12 **Initial motion of headlight flash.** Given each of the eight radial values *r/M* found in Queries 10 and 11, find the proper velocity dr/dt_{rain} of the headlight flash aimed inward at each of these radial values.

6 Radial Trajectories of Light

You are given the task of providing occasional radio news bulletins for the divers. Each of these bulletins, covering the latest news and scientific reports from outside the horizon, will be broadcast radially inward from a fixed station on a shell external to the horizon.

To prepare for this task, you want to know how long it takes light to move from radius r_1 outside the horizon to radius r_2 inside the horizon. And since no clock can be stationary inside the horizon, you choose to measure "how long" for this trip in rain time t_{rain}. The following query is rather technical. You may choose (or may be instructed) to carry it out, or you may simply use the result. In this query and in much of the remainder of this project, we use dimensionless variables that not only simplify the analysis but also make the results independent of the mass M of the black hole being considered. For every domain of radius r and time t we define the dimensionless variables

$$r^* \equiv \frac{r}{M} \qquad \text{and} \qquad t^* \equiv \frac{t}{M} \qquad [18]$$

QUERY 13 **Rain time for light to move from one radius, r_1, to another, r_2.** Rewrite equation [17] of Query 9 to read, for inward-moving light or radio waves,

$$dt^*_{rain} = \frac{-dr^*}{\left(\frac{2}{r^*}\right)^{1/2} + 1} = \frac{-r^{*1/2} dr^*}{\sqrt{2} + r^{*1/2}} \qquad [19.\ light]$$

(For a flash sent "outward," the plus sign becomes a minus sign in both denominators in equations [19].) Make the substitution

$$u = \sqrt{2} + r^{*1/2} \qquad [20]$$

Integrate the resulting equation from u_1 to u_2, recall that $\ln A - \ln B = \ln(A/B)$, then resubstitute equation [20] for u_1 and u_2 to show that for the signal moving inward from r^*_1 to r^*_2, the integrated rain time is

$$t^*_{r2\ rain} - t^*_{r1\ rain} = (r^*_1 - r^*_2) - 2\sqrt{2}(r^{*1/2}_1 - r^{*1/2}_2) + 4\ln\left(\frac{\sqrt{2} + r^{*1/2}_1}{\sqrt{2} + r^{*1/2}_2}\right) \qquad [21\ light]$$

Messy? Sure, but the computer doesn't care and easily plots the results.

QUERY 14 **Horizon-to-crunch rain time for light.** Verify that when $r^*_2 = r^*_1$, the elapsed rain time is zero. Why is it zero? Show that when r^*_1 is at the horizon and $r^*_2 = 0$ (at the crunch point), the elapsed rain time $t^*_{rain} = 0.773$ or $t_{rain} = 0.773\,M$. Compare this result with the horizon-to-crunch wristwatch time $\tau = (4/3)M$ for the rain diver herself, derived in the box on page 3-22. Why is the result using equation [21] less than the result for the rain diver herself?

Figure 6 is a computer plot of some curves derived from equation [21]. For each curve, select r^*_1 and set $t^*_{r1\ rain} = 0$. Then choose a sequence of smaller values of r^*_2 and let the computer calculate $t^*_{r2\ rain}$ and plot the result. For comparison, we include curves showing the worldline of several raindrop divers. It is evident from Figure 6 that our news bulletins can be timed to catch up with the descending diver community.

What about the light flash that the in-falling diver launches radially *outward*? If this launch occurs before the diver falls through the horizon, then the outward flash indeed moves outward. Once the diver has passed inward across the horizon, however, even the "outward" flash moves inward. The little cones sprouting from the worldline in the earlier Figure 5 show the initial motions of inward and "outward" flashes emitted inside the horizon. More complete worldlines for emitted light are shown in Figure 7. You can trace the worldline of the "outward" flash by placing a minus sign in front of the second term in the denominator on the right side of equation [19] and carrying out the integration. Inside the horizon (where r^*_1 is always greater than r^*_2), a convenient form of the solution is

$$t^*_{r2\ rain} - t^*_{r1\ rain} = -(r^*_1 - r^*_2) - 2\sqrt{2}\left(r^{*1/2}_1 - r^{*1/2}_2\right) + 4\ln\left(\frac{\sqrt{2} - r^{*1/2}_2}{\sqrt{2} - r^{*1/2}_1}\right)$$

[22. light]

A diver who has arrived at emission point A shown in Figure 7 can influence only those future events that lie within the shaded region in the diagram. The more time passes—the smaller the radius at which the diver arrives—the fewer events the diver can affect. With unlimited rocket power, the diver could herself be present at any of the events in this shaded region. Indeed, the "outward" flash worldline might be called the *worldline of inevitability* for the rain traveler who has arrived at A; starting from A, no known power in the universe can defer extinction longer than the upper limit traced by this curve.

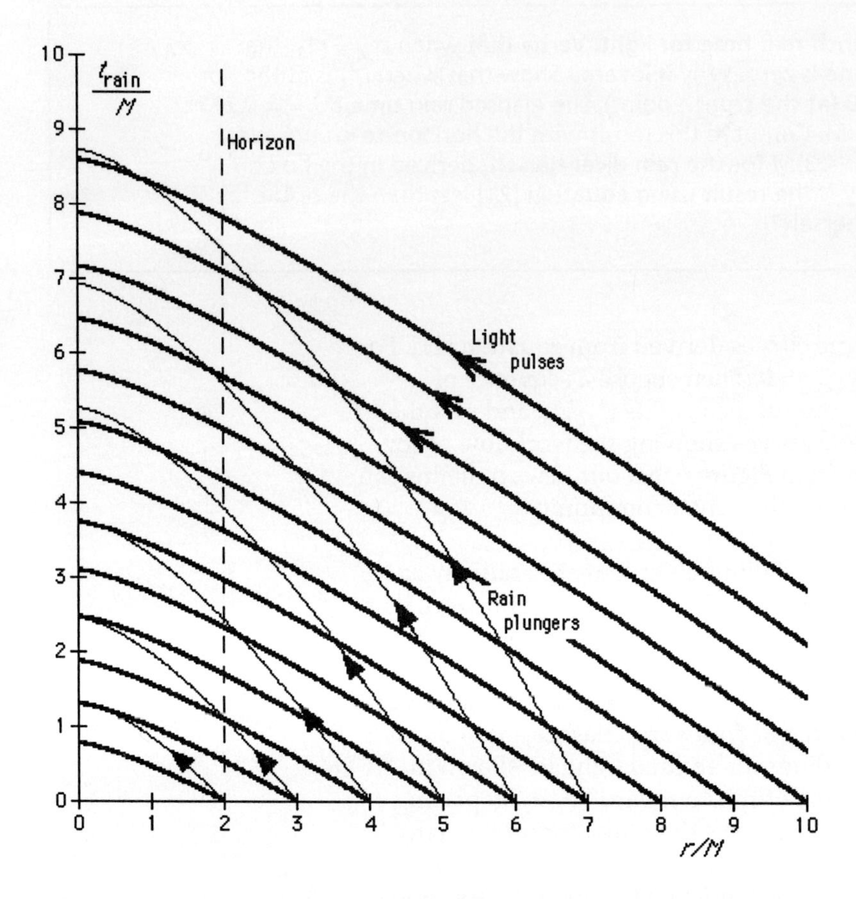

Figure 6 *Computer plot: Worldlines of "rain divers" (thin lines) and inward light flashes (thick lines). The rain diver worldlines all have the same form and can be moved up and down without change. The same is true of the light-flash worldlines. For a given value of r/M, the light pulses are covering more distance per unit time than the rain plunger. From this result it is evident that light or radio waves can be used to communicate from outside the horizon to divers inside the horizon.*

7 A Merciful Ending?

Diving into a black hole is a form of suicide, which may go against religious, moral, or ethical principles. Aside from such considerations, no one will volunteer for your black-hole diver research team if she predicts that her death at the crunch point will be painful. Your task is to estimate the **ouch time** t_{ouch}, defined as the lapse of time on the wristwatch of the diver between the first discomfort and the arrival of the diver at the central crunch point, $r = 0$.

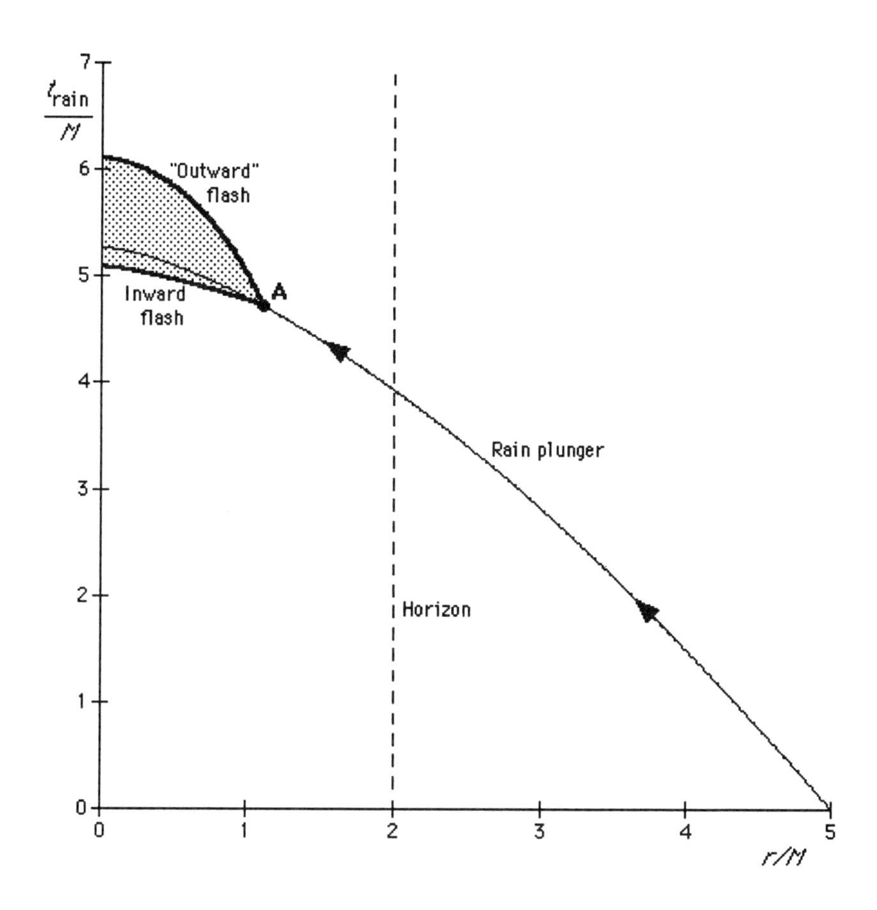

Figure 7 *Computer plot. Thin line: Worldline for rain diver that passes r = 5M at t_{rain} = 0 and emits headlight and taillight flashes at event A inside the horizon. Thick lines: Worldlines for these inward and "outward" flashes. Both flashes move inward. The shaded region contains events the diver can still influence in the future as she passes event A.*

QUERY 15 **Preliminary: Acceleration *g* in geometric units.** Carry out part A of Exercise 9 in Chapter 3 to show, first, that in Newtonian mechanics the acceleration at the surface of Earth is given by the following expression in geometric units,

$$g_{Earth} = g_E = \frac{M}{r_E^2} \qquad \text{[23. Newton]}$$

Show that the acceleration of gravity g_E at Earth's surface has the following approximate value in geometric units:

$$g_{Earth} = g_E \approx 10^{-16} \text{ meter}^{-1} \qquad \text{[24. Newton]}$$

QUERY 16 **Acceleration of the rain frame.** Start with equation [2] with $dt_{rain} = d\tau$:

$$\frac{dr}{d\tau} = -\left(\frac{2M}{r}\right)^{1/2} \qquad [25]$$

Take the derivative of this expression with respect to wristwatch time τ to find the acceleration of the rain observer in the mixed coordinates r and τ. Show that the result has the same form as the Newtonian case (with proper time replacing Newtonian universal time):

$$g_{rain} \equiv \frac{d^2r}{d\tau^2} = \frac{1}{2}\frac{(2M)^{1/2}}{r^{3/2}}\frac{dr}{d\tau} = -\frac{M}{r^2} \qquad [26]$$

where we substituted equation [25] in the next-to-last step.

QUERY 17 **Relative tidal acceleration in the rain frame.** We want to know how much this initial acceleration *differs* between the head and the feet of an in-falling raindrop observer. Start by taking the differential of g_{rain} in equation [26] with respect to r. Show that

$$dg_{rain} = \frac{2M}{r^3}dr \qquad [27]$$

In Query 3 you showed that (for the rain frame only) the rain observer measures the separation between shells to be dr, the same as their separation in the Schwarzschild r-coordinate. The consequence is that the separation between head and feet of the in-falling observer has the same value dr_{rain} in the rain frame as their separation dr in r. Hence equation [27] can be used to find the difference in acceleration between the head and feet of the rain frame diver. Take this separation to be a generous 2 meters. However, we will wait until the last minute to substitute numerical values, hoping that some quantities will cancel.

What are the conditions for discomfort? Let us assume that the diver first becomes uncomfortable when the *difference* in local acceleration between her head and feet, tending to stretch her, is equal to the acceleration at Earth's surface. In other words, if her stomach is in free float, she will become uncomfortable when her feet are pulled downward with half their weight on Earth and her head is pulled upward with half its weight on Earth. The *difference* in acceleration between head and feet is thus equal to the acceleration of gravity on Earth.

QUERY 18 **The beginning of "ouch."** Show that, in equation form, *ouch* first occurs when

$$dg_{\text{rain ouch}} \equiv g_{\text{E}} \qquad [28]$$

so

$$r_{\text{ouch}} = \left(\frac{2Mdr}{g_{\text{E}}}\right)^{1/3} \qquad [29]$$

QUERY 19 **The r-value for "ouch."** Compare the ouch radius [29] with the r-value of the horizon, $r = 2M$. Find the value of the ratiò $r_{\text{ouch}}/2M$ for the following cases:

A. A black hole with ten times the mass of Sun.

B. The "20-year black hole" under study in this project.

C. Suppose the ouch radius is at the horizon. What is the mass of the black hole, in multiples of the Sun's mass?

QUERY 20 **The ouch time t_{ouch}**

A. Show that the ouch time (the total length of time the diver will be uncomfortable) is *independent of the mass of the black hole* and is given by the expression

$$\tau_{\text{ouch}} = \frac{2}{3}\left(\frac{dr}{g_{\text{E}}}\right)^{1/2} \qquad [30]$$

Here, recall, dr is the height of the astronaut, about 2 meters, and g_{E} = Earth-surface gravitational acceleration = 10^{-16} meter^{-1}.

B. Calculate the ouch time τ_{ouch} in seconds—the same value for *all* Schwarzschild black holes.

QUERY 21 **A merciful death?** Doctors tell us that pain signals travel through the nerves at the speed of approximately 1 meter/second. Draw your conclusion about the merciful nature of death at the center of a black hole.

8 Trajectories of Particles Inside the Horizon

Reminder: Sections 8 –10 depend on the contents of Chapters 4 and 5. You may want to return to these sections after completing those chapters.

Starting with the rain metric, equation [15], you can develop the equations for trajectories of stones and light flashes using the procedures of Chapters 3 through 5.

QUERY 22 **Energy.** Follow the steps in Section 3 of Chapter 3, but use the rain metric, to derive the expression for the energy per unit mass of a stone in rain coordinates. Show that the resulting constant of the motion is

$$\frac{E_{rain}}{m} = \left(1 - \frac{2M}{r}\right)\frac{dt_{rain}}{d\tau} - \left(\frac{2M}{r}\right)^{1/2}\frac{dr}{d\tau} \qquad [31.\ m > 0]$$

Here $d\tau$ is the wristwatch time of the moving stone.

QUERY 23 **Angular momentum.** Follow the steps in Section 2 of Chapter 4, but again use the rain metric, to derive the expression for angular momentum per unit mass, the second constant of the motion of the stone. Show that the result is exactly the same as for the Schwarzschild metric, namely,

$$\frac{L}{m} = r^2\frac{d\phi}{d\tau} \qquad [32.\ m > 0]$$

One can go on to describe orbits of particles and light both outside and inside the horizon employing coordinates r, t_{rain}, ϕ. Derivations are similar to those in Chapters 4 and 5, but the new derivations are complicated by the second term in the equation for energy, above. Rather than pursuing these algebraically complicated derivations here, we look instead at what the plunging rain observer sees as she approaches the crunch point at $r = 0$.

9 The Final View

What is the last view of the stars our rain diver sees as she approaches the crunch point? We now have the tools to answer this question.

Outside the horizon, the shell speed of the rain diver is given by equation [1]:

$$\frac{dr_{shell}}{dt_{shell}} = v_{shell} = -\left(\frac{2M}{r}\right)^{1/2} \qquad [33]$$

For a light flash, the relation between energy and the angles ψ at which a shell or rain viewer looks to see the flash is given by equations [40] and [41] on page 5-26. Adapted to the present notation they read:

$$\cos\psi_{rain} = \frac{\cos\psi_{shell} + v_{rel}}{1 + v_{rel}\cos\psi_{shell}} \qquad [34.\ \text{light}]$$

$$E_{\text{rain}} = \frac{E_{\text{shell}}(1 + v_{\text{rel}}\cos\psi_{\text{shell}})}{(1 - v_{\text{rel}}^2)^{1/2}} \qquad \text{[35. light]}$$

Here the rain diver takes the place of the rocket frame while the shell serves as laboratory, and v_{rel} is the velocity of the rocket with respect to the laboratory (shell). The angle ψ is measured with respect to the direction of rocket motion in the laboratory frame. Then, according to these conventions of special relativity, for the rain diver the angle ψ is zero along a direction radially *inward* and v_{rel} is also positive for motion radially *inward*. In other words,

$$v_{\text{rel}} = -v_{\text{shell}} = \left(\frac{2M}{r}\right)^{1/2} \qquad [36]$$

Here is the relation between the energy E_{far} of a light flash at a location far from the black hole and the energy E_{shell} measured by the shell observer, from equation [27] on page 3-17 or Exercise 6 of Chapter 5:

$$E_{\text{far}} = \left(1 - \frac{2M}{r}\right)^{1/2} E_{\text{shell}} \qquad \text{[37. anything]}$$

Equation [37] is true for *anything* moving toward or (when possible) away from a black hole. Moreover, it is true of light flashes (or anything) moving in *any* direction, not just radially. (Proof of these two statements is an *optional Query*: Show that if they were not so, you could build a perpetual motion machine, at least outside the horizon.)

Finally, there is a relation between the angle ψ_{shell} at which the shell observer looks to see a light flash and the impact parameter b of that flash. (The impact parameter b is defined in Figure 2, page 5-6.) From equation [17] and Figure 3 on page 5-9, we obtain an expression for cos ψ_{shell} for use in equations [34] and [35]. The cosine below is written with a negative sign because there is a 180-degree difference between the direction in which the light *moves* and the direction in which one *looks* to *see* the light. For the direction in which one *looks* (with zero angle ψ radially inward), we have

$$\cos\psi_{\text{shell}} = -\left[1 - \left(1 - \frac{2M}{r}\right)\frac{b^2}{r^2}\right]^{1/2} \qquad [38]$$

Now we want to apply these equations to determine the view seen by the rain diver just before she reaches the crunch point at $r = 0$. Combine the above equations and *assume* that the results can be applied to motion inside the horizon. *Note:* For impact parameter with the value $b = 0$, that is, for a star *exactly* radially outward, equation [38] says that shell observers at all radii will see this star in the radially outward direction. This result propagates forward in what follows, so the star exactly radially outward from the plunger will be observed radially outward right down to the crunch point. The analysis below holds for stars in all other directions.

QUERY 24 **Ring around the sky.** Substitute expressions [38] and [36] into [34] and begin to take the limit as $r \longrightarrow 0$. First, neglect all terms except those with some power of r in the denominator (because terms with r in the denominator will be much larger than all others as $r \longrightarrow 0$). Show that the result of this first step is

$$\cos\Psi_{\text{rain}} \to \frac{-\left(\frac{2M}{r}\right)^{1/2}\frac{b}{r} + \left(\frac{2M}{r}\right)^{1/2}}{-\frac{2Mb}{r^2}} \qquad \text{[39. as } r \longrightarrow 0]$$

Next, in the numerator of the fraction on the right side of [39], retain only the first term, with the highest power of r in the denominator (because the first term grows to the largest value as the limit is approached). Show that the result is

$$\cos\Psi_{\text{rain}} \to \frac{\left(\frac{2M}{r}\right)^{1/2}\frac{b}{r}}{\frac{2Mb}{r^2}} \to \left(\frac{r}{2M}\right)^{1/2} \qquad \text{[40. as } r \longrightarrow 0]$$

Equation [40] tells us that as the rain diver approaches the crunch point (as $r \longrightarrow 0$), the cosine of the direction in which she looks to see a star goes to zero, meaning that her viewing angle goes to 90 degrees. *And the 90-degree viewing angle does not depend on the value of the impact parameter b. Hence she sees at 90 degrees EVERY star above her* (except for a star directly outward radially). Hence she sees almost all the stars in a "ring around the sky" transverse to her direction of motion.

What is the energy of light the plunger sees in her last instants?

QUERY 25 **Energy of light in the final view.** Substitute expressions [36], [37], and [38] into equation [35] and take the limit of small r in the same way as in Query 24. Show that the result is

$$E_{\text{rain}} \to \frac{b}{r}E_{\text{far}} \qquad \text{[41. as } r \longrightarrow 0]$$

From Figure 8, show that the energy of light E_{rain} has the value

$$E_{\text{rain}} \to E_{\text{far}}\frac{r_{\text{far}}}{r}\sin\theta_{\text{far}} \qquad \text{[42. as r } \longrightarrow 0]$$

where the subscript *far* refers to values far from the black hole.

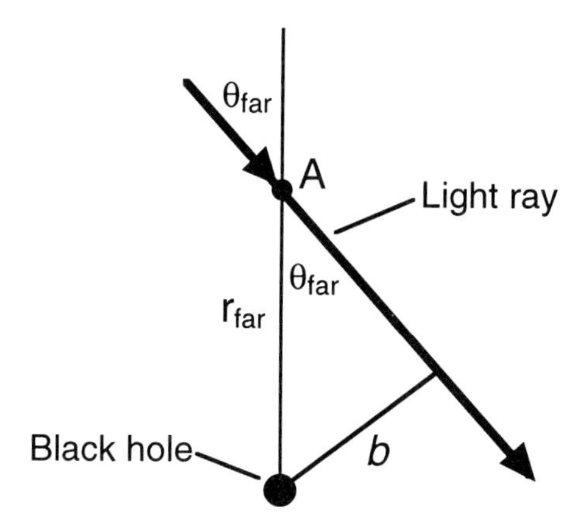

Figure 8 *An observer at A, a great distance r_{far} from a black hole (a distance so great that light is not deflected by the hole), sees a star at an angle θ_{far} from the radially outward direction. This figure shows that the impact parameter b then has the value $b = r_{far} \sin \theta_{far}$, where r and θ are measured at the same point A.*

In summary, just before the rain diver reaches the center of the black hole she views a brilliant ring of high-energy radiation at 90 degrees from her direction of motion, this radiation coming from (almost) all the stars in the sky.

Our description of what the plunger sees does not include images of stars from light paths that wrap once or many times around the black hole. These multiple light paths illuminate the description of the final scene recounted in Section 11 of Chapter 5, starting on page 5-27.

10 Additional Projects

Every possible question about (nonquantum) features of spacetime and orbits of particles and light in the equatorial plane around a spherically symmetric, nonspinning black hole can be answered using the metric and the Principle of Extremal Aging. We have barely begun the description of life inside the horizon of the black hole. Some of the following questions can be answered only after mastering the material in Chapters 4 and 5.

1. Frequency shift. News bulletins for the plunging astronauts are broadcast inward at frequency *f* from a stationary shell outside the horizon. What will be the frequency of arriving signals as received by the diver? Does the broadcast frequency *f* need to be changed as time goes by so that the astronauts can use a receiver tuned to a single frequency?

2. Starlight. What does the canopy of stars look like when viewed from inside the horizon (and before reaching the crunch point)? By what amount does light from the stars change frequency as observed by the diver? How do these effects vary with time as the diver moves inward between horizon and crunch point?

3. Orbits in rain coordinates. In Query 22 you derived the expression for the energy per unit mass of a plunger. After completing Chapter 4, you will be able to verify (Query 23) that the expression for angular momentum in rain coordinates has the same form as the one derived in Section 2 of Chapter 4 from the regular form of the Schwarzschild metric. After completing Chapter 5, you will be able to verify that the equations of motion for a light flash have the same form in rain coordinates as they do in Schwarzschild coordinates.

4. The insertion problem. Astronauts jump off a shell at what radius to start their trip to the center? From a shell on which the local acceleration of gravity does not crush them? Putting in the numbers, you discover that the "insertion" free drop from such a shell to the horizon of a "20-year black hole" will last for hundreds of years. So what is a *practical* plan for beginning a trip to the center of such a black hole? Perhaps start the plunge from a station in the minimum-radius stable circular orbit at $r = 6M$, as described in Figure 13 and exercise 2 of Chapter 4? But how long is the trip into that orbit? And what are the tidal forces in that free-float orbit?

5. Alternative in-falling frames. We have described observations from only one set of reference frames for the diver, the set we call *rain frames*, which fall from rest at a great distance. There are alternative frames, such as the *drip frame* and the *hail frame* described in Section 2. What will be the experience of an observer in one of these alternative frames?

The Galactic Technical Center eagerly awaits your analysis of these problems!

11 References and Acknowledgments

The idea for the rain frame comes from the book *Black Holes: The Membrane Paradigm*, edited by Kip S. Thorne, Richard H. Price, and Douglas A. Macdonald, Yale University Press, New Haven, 1986, page 22 and elsewhere. Their name for the rain observer is "freely falling observer" or FFO.

The box titled "Eggbeater Spacetime " on page B-5 comes, in part, from Chapter 13 of *Black Holes and Time Warps* by Kip Thorne. A description of mass-inflation singularities can be found in E. Poisson and W. Israel, "Inner-Horizon Instability and Mass Inflation in Black Holes," *Physical Review Letters*, Volume 63, Number 16, pages 1663–1666 (16 October 1989). Lior M. Burko and William A. Hiscock made detailed suggestions about the contents of the box. The following references detail current research on the subject: Rhett Herman and William A. Hiscock, "Strength of the Mass Inflation Singularity," *Physical Review D*, Volume 46, Number 4, pages 1863–1865 (1992); Lior M. Burko and Amos Ori, "Introduction to the Internal Structure of Black Holes," and Amos Ori, "On the Traversability of the Cauchy Horizon: Herman and Hiscock's Argument Revisited," both in *Internal Structure of Black Holes and Spacetime Singularities*, Volume XIII of the *Annals of the Israel Physical Society,* coedited by L. M. Burko and A. Ori, Institute of Physics, Bristol, 1997.

Charles W. Misner derived the form of the metric for the global rain frame, equation [15] (private communication). Apparently it was originally derived by Paul Painleve and Allvar Gullstrand. See Werner Israel, "Dark Stars: The Evolution of an Idea" in *Three Hundred Years of Gravitation,* edited by S. W. Hawking and W. Israel, Cambridge University Press, Cambridge, 1987, page 234.

Chapter 4 Orbiting

- *How are orbits around a black hole different from planetary orbits around Sun?*
- *Does a stone orbiting a black hole reach the speed of light?*
- *How close to a black hole can I move in a circular orbit?*
- *Can I use a black-hole orbit to time travel?*

CHAPTER 4

Orbiting

How happy is the little Stone
That orbits a Black Hole alone,*
And doesn't care about Careers
And Exigencies never fears —
Whose Coat of elemental Brown
A passing Universe put on,
And independent as the Sun
Associates or glows alone,
Fulfilling absolute Decree
In casual simplicity —

—Emily Dickinson, about 1881 (poem 1510)
*Line two in the original reads:
That rambles in the Road alone,

1 Step or Orbit?

"Go straight!" implies extremal aging. Extremal aging implies that energy and angular momentum are constants of the motion.

Nature shouts at the stone "Go straight!"

A stone in orbit streaks around a black hole—or around Earth. What tells the stone how to move? Spacetime grips the stone, giving it the simplest possible command: "Follow a straight worldline in the local inertial frame." From instant to instant this directive is enough to tell the stone what to do next, what next step to take in its motion.

Knowledge of its next step is enough for the stone, but it is not enough for us. We want more. We seek a global description of the trajectory of the stone through spacetime, a trajectory that carries it far beyond the confines of a single local inertial frame. The Principle of Extremal Aging answers our need. It converts the local ordinance "Go straight!" into a general executive order spanning spacetime from fixed initial event to fixed final event. "Between initial and final events," the Principle decrees, "follow the worldline of extremal aging." This worldline—the straightest possible worldline through flat or curved spacetime—is called the **geodesic**.

Constants of the motion: energy and angular momentum

In this chapter we win from the Principle of Extremal Aging a simplified account of the orbit of a stone around a spherically symmetric center of attraction. This simplification uses quantities that do not change as the motion progresses. These unchanging quantities—constants of the motion—are energy and angular momentum. The expression for energy for a general trajectory is the same as that derived in Chapter 3 for the

What Then Is Time?

What then is time? If no one asks me, I know what it is.
If I wish to explain it to him who asks me, I do not know.

The world was made, not in time, but simultaneously with time. There was no time before the world.

—St. Augustine (354–430 A.D.)

Time takes all and gives all.

—Giordano Bruno (1548–1600 A.D.)

Everything fears Time, but Time fears the Pyramids.

—Anonymous

Philosophy is perfectly right in saying that life must be understood backward. But then one forgets the other clause—that it must be lived forward.

—Søren Kierkegaard

As if you could kill time without injuring eternity.

Time is but the stream I go a-fishing in.

—Henry David Thoreau

I do not define time, space, place and motion, [because they are] well known to all.

—Isaac Newton

Time is defined so that motion looks simple.

—Misner, Thorne, and Wheeler

Nothing puzzles me more than time and space; and yet nothing troubles me less, as I never think about them.

—Charles Lamb

Either this man is dead or my watch has stopped.

—Groucho Marx

"What time is it, Casey?"
"You mean right now?"

—Casey Stengel

It's good to reach 100, because very few people die after 100.

—George Burns

Time is Nature's way to keep everything from happening all at once.

—Graffito, men's room, Pecan St. Cafe, Austin, Texas

What time does this place get to New York?

—Barbara Stanwyck, during trans-Atlantic crossing on the steamship *Queen Mary*

special case of radial plunge. The expression for angular momentum is new. As the stone orbits, its energy does not change, nor does its angular momentum. How do we know? The Principle of Extremal Aging tells us so!

2 Energy and Angular Momentum from Extremal Aging

Fix the end events of a short path; vary the middle event to find constants of motion.

Special-relativistic expressions become Newtonian at low velocity.

When studying special relativity (Chapter 1), we were forced to use relativistic expressions for energy and linear momentum, expressions different from those of Newton. Why did we accept the unfamiliar formulas? Because only relativistic expressions satisfy laws of conservation of both total energy and total linear momentum in high-speed collisions (and other interactions) in an isolated system. What consolation did we have for leaving the old familiar territory? The consolation that in the limit of low velocity the relativistic expressions give the same values as the Newtonian expressions (provided we include the rest energy—the mass m—in the total energy of each isolated particle).

Next we will derive general-relativistic expressions for the energy and angular momentum of a stone moving near a black hole. Why accept these new expressions? Because as the stone dips and swoops around the uncharged, nonspinning center of attraction, only these formulas reflect conservation of energy and angular momentum. What consolation do we have for leaving the familiar territory of special relativity? The consolation that in the limit of large radius—in the flat spacetime a great distance from the black hole—the new expressions become equal to those of special relativity.

General-relativistic expressions become special-relativistic far from black hole.

The box on page 4-4 uses the now-familiar argument to derive the expression for angular momentum from the Schwarzschild metric and the Principle of Extremal Aging. The resulting constant of the motion is

$$r^2 \frac{d\phi}{d\tau} = \text{constant} \qquad [1]$$

The constant in equation [1] is the **angular momentum *L*** of the particle divided by its mass:

Expression for angular momentum

$$\frac{L}{m} = r^2 \frac{d\phi}{d\tau} \qquad [2]$$

Recognition that equation [2] expresses the angular momentum follows from noticing that this equation has the same form as in Newtonian mechanics (Figure 1) except for the wristwatch time $d\tau$ in the denominator. The presence of the wristwatch time is not surprising, since the relativistic expression for linear momentum (of which angular momentum is composed) also has the wristwatch time in the corresponding position.

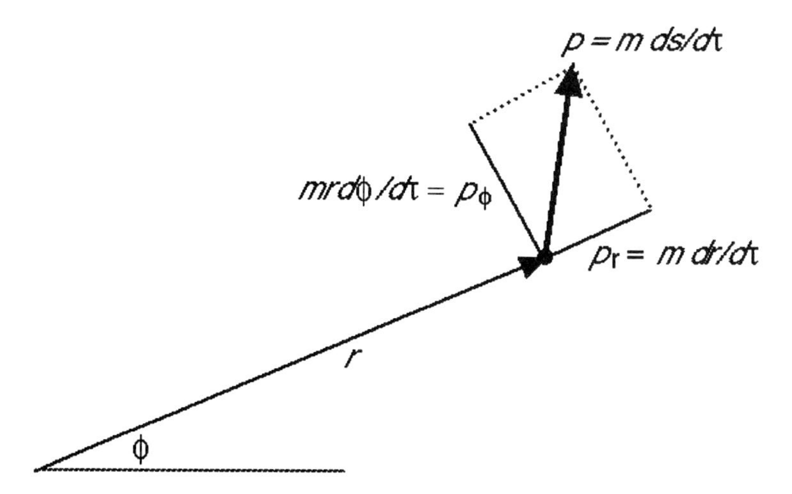

Figure 1 *Angular momentum L is the product of r and the component of linear momentum* p_ϕ *in the tangential or* ϕ *direction, yielding* $L = mr^2 d\phi/d\tau$ *(equation [2]). The radial component of linear momentum is* p_r*. The linear momentum vector* **p** *has its special-relativity magnitude* $mds/d\tau$*, where* $d\tau$ *is the advance of wristwatch time of the particle whose momentum is being determined.*

Expression for energy

In Chapter 3 we used a similar derivation to find an expression for total energy per unit mass, E/m. That derivation of energy has exactly the same form, whether or not the angle changes as the stone moves. Therefore that earlier expression is correct for an orbiting stone as well as for a plunging one:

$$\frac{E}{m} = \left(1 - \frac{2M}{r}\right)\frac{dt}{d\tau} \qquad [3]$$

Derivation of Expression for Angular Momentum

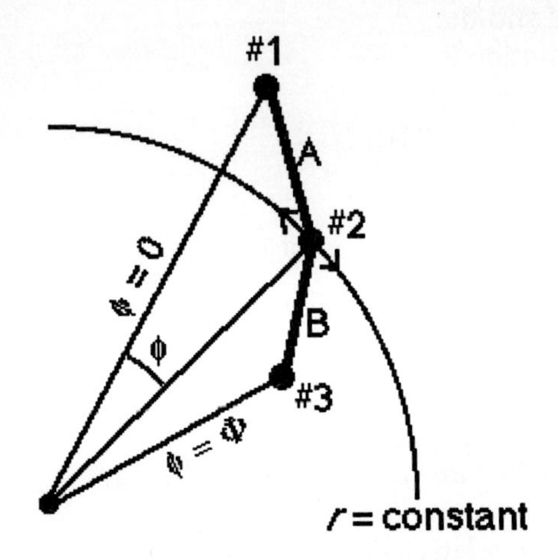

Figure 2 *Choose the intermediate angle φ so that the wristwatch time from event #1 to event #3 is an extremum.*

Strategy: Use the Principle of Extremal Aging and the Schwarzschild metric. A moving particle emits flashes #1, #2, and #3, marking off segments A and *B* along the trajectory. Fix the times of all three flashes and the positions of flashes #1 and #3. Flash #2 is emitted somewhere along a circle (Figure 2), because we vary only ϕ in this derivation to maximize the total wristwatch time racked up between flashes #1 and #3. The result is an expression that is a constant of the motion. Now the details.

Set the fixed angle of event #1 equal to zero and call the fixed angle Φ for event #3. Change the angle ϕ of event #2 by moving it either way along the circle lying between fixed events #1 and #3. (See the arrows in the figure.) Let r_A and r_B be appropriate average values of the radii for segments A and B, respectively, and τ_A and τ_B be the corresponding wristwatch times of the stone moving along these segments. With these substitutions, the timelike version of the Schwarzschild metric for segment A becomes (see equation [12]):

$$\tau_A = [-r_A^2\phi^2 + (\text{terms without } \phi)]^{1/2} \qquad [4]$$

To prepare for the derivative that leads to extremal aging, take the derivative of this expression with respect to ϕ:

$$\frac{d\tau_A}{d\phi} = -\frac{r_A^2\phi}{\tau_A} \qquad [5]$$

Similarly for segment B,

$$\tau_B = [-r_B^2(\Phi - \phi)^2 + (\text{terms without } \phi)]^{1/2} \qquad [6]$$

$$\frac{d\tau_B}{d\phi} = \frac{r_B^2(\Phi - \phi)}{\tau_B} \qquad [7]$$

The total wristwatch time for both segments is

$$\tau = \tau_A + \tau_B \qquad [8]$$

Take the derivative of this expression with respect to ϕ, substitute from equations [5] and [7], and set the resulting derivative equal to zero, thus applying the Principle of Extremal Aging:

$$\frac{d\tau}{d\phi} = \frac{d\tau_A}{d\phi} + \frac{d\tau_B}{d\phi} = -\frac{r_A^2\phi}{\tau_A} + \frac{r_B^2(\Phi - \phi)}{\tau_B} = 0 \qquad [9]$$

The condition for extremal lapse of proper time becomes

$$\frac{r_A^2\phi}{\tau_A} = \frac{r_B^2(\Phi - \phi)}{\tau_B} \qquad [10]$$

The left side contains quantities for segment **A** only; the right side contains quantities for segment **B** only. We have discovered a quantity that is the same for both segments, a constant of the motion for the free particle. In deriving this quantity we assume that each segment of the worldline is small in space and time. To witness this microscopy, restore the incrementals $d\phi$ for ϕ (or for $\Phi - \phi$) and $d\tau$ for τ. Then the constant of motion is written

$$r^2\frac{d\phi}{d\tau} = \text{a constant of the motion} \qquad [11]$$

In the text this constant of the motion is identified as L/m, the angular momentum of the particle per unit mass.

Note: These particular expressions for energy and angular momentum as constants of the motion depend not only on the Principle of Extremal Aging but also on the unique curvature of spacetime around a nonspinning, uncharged center of attraction described by the Schwarzschild metric. These expressions do *not* describe conserved quantities around a *spinning* black hole, for example. (See Project F, The Spinning Black Hole.)

Energy and angular momentum expressions are different for different metrics.

3 Properties of Angular Momentum

The new meaning of angular momentum

The spacetime view of the orbit of a satellite around a spherically symmetric center of gravitational attraction shares an insight with Newton: There are two "constants of the motion" that remain unchanged as the satellite zooms along, namely, angular momentum and total energy.

Two "constants of the motion"

From Newton's viewpoint, angular momentum of the satellite remains constant if and only if zero force acts on the satellite in the direction of increasing or decreasing azimuthal angle, ϕ.

From the spacetime-physics point of view, in contrast, there is no force on the satellite at all, not even a force toward the center of attraction. The satellite moves through spacetime on the absolutely straightest worldline it knows—along a *geodesic* in the local free-float frame. Instead of looking for the absence of a force along the direction of increasing ϕ, we look for symmetry of spacetime geometry with respect to the azimuthal angle ϕ. The Schwarzschild metric (equation [A] in Selected Formulas at the end of the book) depends on the radius r but not on the angle ϕ:

$$d\tau^2 = \left(1 - \frac{2M}{r}\right)dt^2 - \frac{dr^2}{\left(1 - \frac{2M}{r}\right)} - r^2 d\phi^2 \qquad [12]$$

The metric is not different for different values of the angle ϕ. (Only the *change* $d\phi$ appears in the metric.) A constant of the motion, namely angular momentum, corresponds to this symmetry. Note that the same can be said for the time coordinate t: only the *change* dt appears in the metric. We already know that there is another constant of the motion that corresponds to this symmetry in t, namely the energy E. And note, finally, that the same *cannot* be said for the coordinate r. The coordinate r itself appears in the metric, along with dr. The lack of symmetry in r is evidence that there is no r-related third constant of the motion.

Symmetries in t and ϕ predict constants of the motion.

Our way of describing angular motion—that is, zero change in spacetime geometry with ϕ—by no means implies a constant rate of change of azimuth for a moving particle, a constant value for $d\phi/dt$. Constant angular momentum, yes; constant angular velocity, $d\phi/dt$, no. A striking example of this truth lies close at hand (Figure 3). Let the center of attraction lose, in everything but name, the character of a center of attraction by being deprived of all mass. Then the satellite zooms past it—at whatever distance of closest approach we may have selected—in a straight line with a uniform velocity.

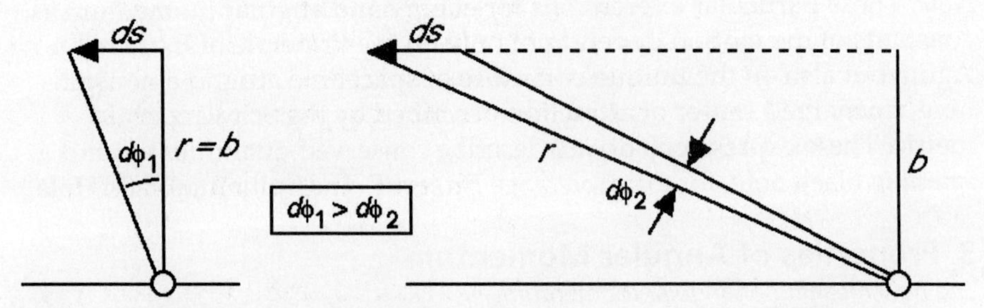

Figure 3 *Changing r means changing $d\phi/d\tau$ when angular momentum is a constant of the motion. An object moving along a straight line in flat spacetime covers equal lengths ds in equal proper times. However, the change in angle $d\phi_1$ through which we turn our telescope to watch the object as it covers this length is great if that object is near the minimum radius (left figure). In contrast, as the object disappears in the distance (large r—right figure), the rotation of our telescope $d\phi_2$ is much slower, even though the length ds covered by the particle is the same. If the center of attraction has no effect on the path of the moving object, then the distance of closest approach is equal to the value of the so-called* ***impact parameter b****.*

Angular momentum being constant, it is impossible for angular velocity also to be constant when r is changing. We see the change in angular velocity most directly by rewriting equation [2] in the form

$$\begin{pmatrix}\text{angular}\\ \text{velocity}\\ \text{sensed by}\\ \text{the satellite}\end{pmatrix} = \frac{d\phi}{d\tau} = \frac{\begin{pmatrix}\text{angular}\\ \text{momentum}\\ = \text{a constant}\end{pmatrix}}{(\text{mass})\ r^2} \qquad [13]$$

We, standing at a center of attraction deprived of all attractive mass, and watching the receding satellite, turn our telescope at an ever-slower rate as the remoteness increases (Figure 3). This finding, so natural and so familiar to any bystander watching a car speeding down the highway, is perhaps the best known of all everyday consequences of the law of conservation of angular momentum.

Smaller rate of change of ϕ for greater *r*.

Now install a nonspinning spherically symmetric uncharged mass M at the center of attraction. No longer is it necessarily true that the satellite will fly off to infinite distance. Indeed, we normally do not even give the name "satellite" to an object unless it goes round and round the center of attraction. In that motion its distance from the center of attraction typically oscillates from a smaller to a larger value. Smaller distance from the center? Then a higher angular velocity. Greater distance from the center? Then a lower angular velocity.

But beauty of beauties, this correlation between angular velocity and radius r still obeys the relations [2] and [13] connecting distance from the center of attraction with angular velocity. Angular velocity remains inversely proportional to r^2.

Rather than using L alone, we use the ratio L/m in the relativistic law of conservation of angular momentum. Why? Because we recognize that par-

SAMPLE PROBLEM 1 Escape from the Vicinity of the Black Hole?

The governor of the spherical shell at radius $r_0 = 10M$ fires a satellite horizontally (that is, at 90° from the radially outward direction) at a speed $v_{o\ shell} = 0.500$ (half the speed of light), both quantities measured with his local shell clocks and rods. What is the angular momentum per unit mass, L/m, of the satellite? What is its energy per unit mass, E/m? Will this satellite escape to great distance? (*Careful:* Here the word "shell" refers to the spherical shell and its observers, not to the launched satellite.)

SOLUTION

The satellite is fired horizontally with initial shell speed $v_{o\ shell} = dx_{shell}/dt_{shell} = 0.500$. In order to compute the angular momentum L (equation [2]), we need the proper time $d\tau$, rather than shell time dt_{shell}, in the denominator. To obtain $d\tau$, think of two flashes emitted by the satellite. The proper time $d\tau$ between these two emissions is measured directly on the clock carried by the satellite. The relation between dt_{shell} and $d\tau$ is just the special-relativity expression:

$$\frac{dt_{shell}}{d\tau} = \gamma_{o\ shell} \equiv (1 - v_{o\ shell}^2)^{-1/2} = (1 - 0.5^2)^{-1/2} = 1.155 \qquad [14]$$

Substitute this result into the definition of angular momentum, recognizing that $dx_{shell} = r_0\, d\phi$:

$$\frac{L}{m} = r_o^2 \frac{d\phi}{d\tau} = r_o \frac{dt_{shell}}{d\tau}\left(\frac{r_o d\phi}{dt_{shell}}\right) = r_o \gamma_{o\ shell} \frac{dx_{shell}}{dt_{shell}} = r_o \gamma_{o\ shell} v_{o\ shell} \qquad [15]$$

from which we find the numerical value of the angular momentum per unit mass:

$$\frac{L}{m} = r_o \gamma_{o\ shell} v_{o\ shell} = 10M \times 1.155 \times 0.500 = 5.775M \qquad [16]$$

To calculate energy (equation [3]), we need the value of $dt/d\tau$, the ratio of far-away time to wristwatch time of the satellite. The relation between shell time and far-away time t (equation [C] in the Selected Formulas at the end of the book) comes ultimately from the Schwarzschild metric:

$$\frac{dt}{dt_{shell}} = \left(1 - \frac{2M}{r}\right)^{-1/2} \qquad [17]$$

From these equations and the expression [3] for energy per unit mass, we find

$$\frac{E}{m} = \left(1 - \frac{2M}{r_o}\right)\frac{dt}{d\tau} = \left(1 - \frac{2M}{r_o}\right)\frac{dt}{dt_{shell}}\frac{dt_{shell}}{d\tau} = \left(1 - \frac{2M}{r_o}\right)\left(1 - \frac{2M}{r_o}\right)^{-1/2}\gamma_{o\ shell} = \left(1 - \frac{2M}{10M}\right)^{1/2}\gamma_{o\ shell} = (0.8)^{1/2} \times 1.155 = 1.033 \qquad [18]$$

The answer to the question "Will this satellite escape to great distance?" is "Maybe." The total energy E is slightly greater than the rest energy m. If the particle were launched radially outward, it would make it to great distances and still be moving. In contrast, if the projectile were launched radially inward, it surely would plunge into the black hole. The problem states that the projectile is launched perpendicular to the radius as measured in the local shell frame. Tracing out the orbit with a computer will allow us to answer the question. Using the effective potential (Section 7) provides an instant answer (Sample Problem 2, page 4-21).

ticles of different mass m follow the same worldline through spacetime. What describes the motion is neither the mass of the soaring particle by itself nor its angular momentum by itself but only the ratio of the two, the angular momentum per unit mass, L/m. Equation [2], with proper time τ expressed in meters and angle ϕ having no units, shows us that the units of L/m are meters.

How does angle ϕ change with satellite time τ?

For now we bypass the question "What is the angular velocity at a point on the orbit where r has a given value?" by asking instead "How great is the advance, $d\phi$, of the azimuth of the satellite in a given microscopic increment, $d\tau$, in its wristwatch time?" The answer follows from the law of conservation of angular momentum when we multiply both sides of equation [2] by $d\tau$ and divide both sides by r^2; thus,

$$d\phi = \frac{(L/m)}{r^2} d\tau \qquad [19]$$

We have in equation [19] half the story of the orbit, the part focused on angular momentum and advance in azimuth ϕ. Now for the other half, the connection between energy and advance in r.

4 Forecasting the Orbit

Satellite wristwatch ticks off $d\tau$. From $d\tau$ reckon the resulting changes dr, $d\phi$, and dt.

How do ϕ and r change with change in satellite time τ?

We now have in hand the tools we need to figure the step-by-step advance of the satellite through the world of space and time. Advance? Yes, (1) advance dt of far-away time, (2) advance $d\phi$ of azimuth, and (3) advance dr of r-coordinate, all orchestrated—at our choice and for our convenience—to the tick $d\tau$ of every wristwatch aboard the satellite. This analysis is carried out in the box on page 4-9.

5 "Knife-Edge" Orbit

Teetering on the brink!

Compute orbit around black hole at center of galaxy.

We want to know whether that satellite whirring around the 2.6-million-solar-mass black hole at the center of the Milky Way is going to fall in. Originally this satellite was orbiting at very high speed very close to the black hole. Suddenly a neutron star zoomed through the system and then disappeared, leaving our satellite in a perturbed state.

For the perturbed satellite orbiting the black hole at the center of our galaxy, we feed the numbers into the computer. In a matter of seconds the computer has drawn the **Schwarzschild map** of Figure 4. Amazing motion! It differs fantastically from the circles and ellipses that we know from the orbits of the planets about our Sun. Yes, both our satellite and any planet coast out to a maximum distance from the center of attraction. But there the similarity ends. After moving outward, the planet turns around and comes in to a closest distance of approach and then turns back. The satellite, too, reduces its inward speed as it approaches the attracting mass, reduces it greatly, reduces it almost to zero. But now the difference: The inward speed of the satellite stays low for a long time. The inward velocity of the planet, in contrast, soon turns around to become outward. In that short time of planetary turnaround the planet advances not many degrees, tracing out as it goes the inner curve of the familiar ellipse. The satellite, however, dwells so long in a stage of low radial velocity that its azimuth advances not a few degrees but—in our example—four and one-half revolutions! Only then does its future course clearly declare itself: plunge. "Why, oh why," its captain cries, "didn't I carry along a booster rocket! As we teetered between infall and escape, it could have made all the difference in our fate. But now it's too late."

This knife-edge orbit will be explained in Section 7 and especially in Figure 9.

Computing the Orbit

Strategy

From the initial position and directed velocity, find the value of angular momentum per unit mass L/m and energy per unit mass E/m.

Step 1: Express dt and $d\phi$ in terms of satellite wristwatch time $d\tau$.

Step 2: Substitute these results into the Schwarzschild metric to find dr as a function of $d\tau$.

Result: All bookkeeper increments dt, $d\phi$, and dr are now locked to satellite time increment $d\tau$.

Computer: Starting with the initial position, let the computer advance satellite time τ by increments as it updates values of t, ϕ, and r. Now for the details.

Step 1: The first of these movements in spacetime we read off from the law of conservation of energy as seen in equation [3]:

$$dt = \frac{E/m}{1 - \frac{2M}{r}} d\tau \qquad [20]$$

Similarly find the advance of azimuth from the law of conservation of angular momentum, equation [19]:

$$d\phi = \frac{L/m}{r^2} d\tau \qquad [21]$$

Step 2: With dt and $d\phi$ now known in terms of $d\tau$, we lack only dr to specify completely the displacement of the satellite in space and time in one tick, $d\tau$, of satellite time. This one unknown, dr, appears along with the three knowns, dt, $d\phi$, and $d\tau$, in the standard Schwarzschild expression for the metric, equation [12]:

$$d\tau^2 = \left(1 - \frac{2M}{r}\right)dt^2 - \frac{dr^2}{1 - \frac{2M}{r}} - r^2 d\phi^2 \qquad [12]$$

Into this equation substitute dt from [20] and $d\phi$ from [21] and solve for dr. The result is a single equation that relates dr to $d\tau$:

$$dr = \pm\left[\left(\frac{E}{m}\right)^2 - \left(1 - \frac{2M}{r}\right)\left\{1 + \left(\frac{L}{mr}\right)^2\right\}\right]^{1/2} d\tau \qquad [22]$$

Starting with initial values of r and ϕ, equations [21] and [22] tell how each bookkeeper polar coordinate changes as the satellite clock ticks. If the Schwarzschild bookkeeper demands that the increments be expressed also in terms of far-away time t, equation [20] provides the corresponding change, dt.

Two technical details for the programmer: First, the square root in equation [22] gives us the magnitude of dr, not its sign. Therefore whether r initially increases or decreases must be derived from the initial conditions. Second, many orbits include a turnaround, where the radius reaches a maximum or minimum. Exactly at turnaround, radius r does not change with time; dr is zero, which means that the expression in the square bracket of equation [22] (a function of r) goes to zero. Once it reaches zero, the iterative process keeps the square-bracket expression at zero, because $dr = 0$ as long as the bracket stays zero—which it will until r changes. The basic problem here is that we have simplified everything to first differences (first derivatives) and at a maximum or minimum of r we need a second derivative. These technical difficulties in computation we do not consider further here.

Einstein "invented" curved spacetime

Ever since Francis Bacon, it had been believed that the laws of Nature were there to be "discovered," if only one made the right experiments. Einstein taught us differently. He stressed the vital role of human inventiveness in the process. Newton "invented" the force of gravity to explain the motion of the planets. Einstein "invented" curved spacetime and the geodesic law; in his theory there is no force of gravity. If two such utterly different mathematical models can (almost) both describe the same observations, surely it must be admitted that physical theories do not tell us what nature is, *only what it is* like. *The marvel is that nature seems to go along with some of the "simplest" models that can be constructed . . .*

—Wolfgang Rindler

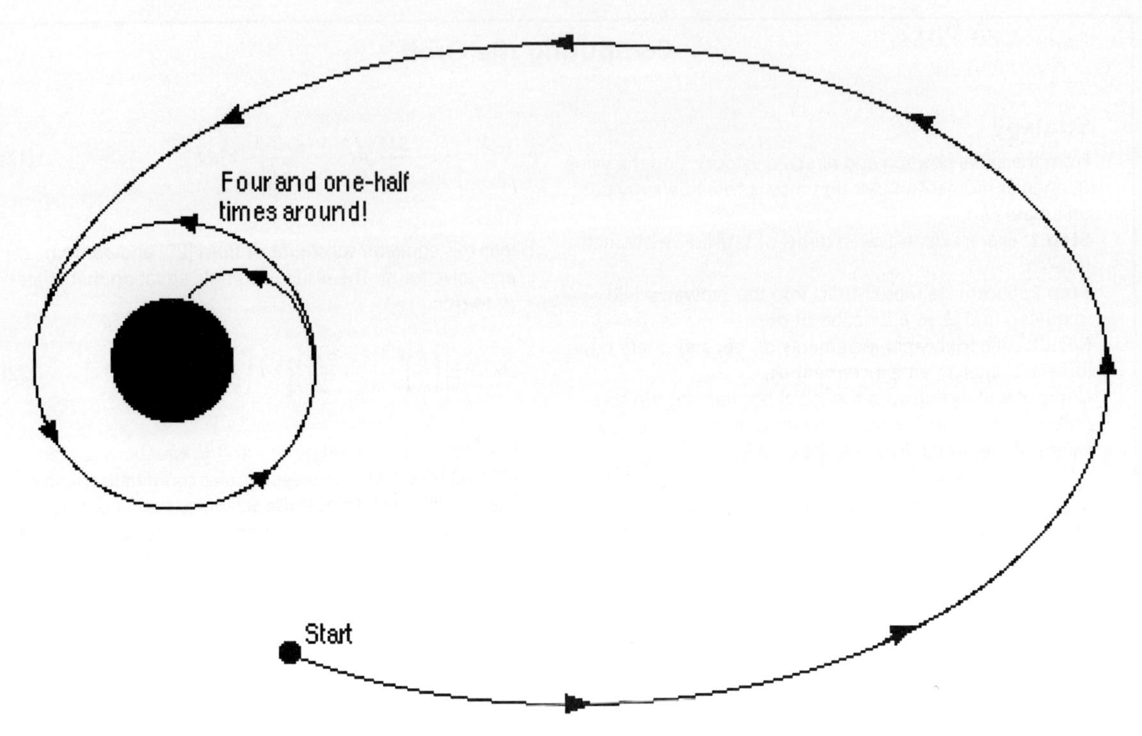

Figure 4 Schematic *Schwarzschild map of the trajectory of a satellite near a black hole after it is perturbed from closed orbit by a passing neutron star. The path in space is strikingly different from the elliptical orbit predicted by Newtonian mechanics, particularly its nonclosed profile and the temporary "knife-edge" unstable circular orbit followed by a plunge through the horizon of the black hole clocked on the wristwatch of the orbiter. This figure is not a computer plot but a qualitative sketch of the major features of the orbit. For an explanation, see Section 7 and Figure 9.*

Where did you get the value of four and one-half revolutions? Why not three revolutions? or thirty? What determines the exact duration of the "teetering on the edge"?

A spaceship teetering on the edge of the abyss near a black hole is in no way different in principle from a pencil balancing on its point on your desktop. Try it! How long does the pencil stay balanced? One second? Ten seconds? That depends. Depends on what? On how nearly vertical the pencil is when released. On whether it is released at rest or with a little motion. On air currents in the room. On vibration of your desktop. On your criterion for deciding when the pencil is no longer balanced. Similar uncertainties describe the spaceship teetering on the edge of a black hole. In the example of Figure 4, we arbitrarily chose four and one-half revolutions as illustrative only.

We now have powerful tools to describe the wealth of orbits around any spherically symmetric center of attraction. Indeed, the possibilities are so great that we need a simplifying scheme that allows us to understand qualitatively many different orbits at once, including the one just described. Such a scheme makes use of the so-called **effective potential**, presented in the Section 7. But first take a look at effective potential in Newtonian mechanics, as outlined in Section 6.

6 Effective Potential in Newtonian Mechanics

Predicting the Newtonian orbit of a particle with a glance at the effective potential

The angular momentum has the same form in Newtonian mechanics as in relativity, equation [2], with "Newtonian universal time" t replacing proper time τ:

$$\frac{L_{\text{conv}}}{m_{\text{conv}}} = r^2 \frac{d\phi}{dt_{\text{conv}}} \qquad \text{[23. Newton]}$$

Here the subscript "conv" reminds us that we are using conventional units, such as seconds and kilograms, so L_{conv} has the units kilogram-meter^2 per second.

The Newtonian energy is a constant of the motion, and for an inverse-square gravitational field the energy is the sum of kinetic energy and potential energy in joules, here divided by m in conventional units:

$$\frac{E_{\text{conv}}}{m_{\text{conv}}} = \frac{1}{2}v_{\text{conv}}^2 - \frac{GM_{\text{kg}}}{r} \qquad \text{[24. Newton]}$$

Simplify energy using conservation of angular momentum.

But the velocity squared is the sum of squares of the radial component and the component in the direction of increasing ϕ. Use equation [23] for angular momentum to simplify the resulting equation:

$$v_{\text{conv}}^2 = \left(\frac{dr}{dt_{\text{conv}}}\right)^2 + \left(r\frac{d\phi}{dt_{\text{conv}}}\right)^2 = \left(\frac{dr}{dt_{\text{conv}}}\right)^2 + \frac{L_{\text{conv}}^2}{m_{\text{conv}}^2 r^2} \qquad \text{[25. Newton]}$$

Make this substitution for v^2 in the energy equation [24]:

$$\frac{E_{\text{conv}}}{m_{\text{conv}}} = \frac{1}{2}\left(\frac{dr}{dt_{\text{conv}}}\right)^2 + \left[-\frac{GM_{\text{conv}}}{r} + \frac{(L_{\text{conv}}/m_{\text{conv}})^2}{2r^2}\right] \qquad \text{[26. Newton]}$$

Even though equation [26] is a Newtonian expression, we prefer the simplification of working in our accustomed geometric units. We want to measure E and m in the same units and the mass M of the attracting center in units of length. Divide both sides of equation [26] by the conversion factor c^2. Recognize that all units cancel in the resulting expression $E_{\text{conv}}/m_{\text{conv}}c^2 = E/m$, while the expression $GM_{\text{conv}}/c^2 = M$ has units of length. Equation [23] shows the conventional units of L_{conv}/m to be meters^2/second. To express this in our simplified units, where time is measured in meters, divide by c to give $L = L_{\text{conv}}/c$. Solve the resulting equation for the square of the radial velocity:

$$\frac{1}{2}\left(\frac{dr}{dt}\right)^2 = \frac{E}{m} - \left[-\frac{M}{r} + \frac{(L/m)^2}{2r^2}\right] \qquad \text{[27. Newton]}$$

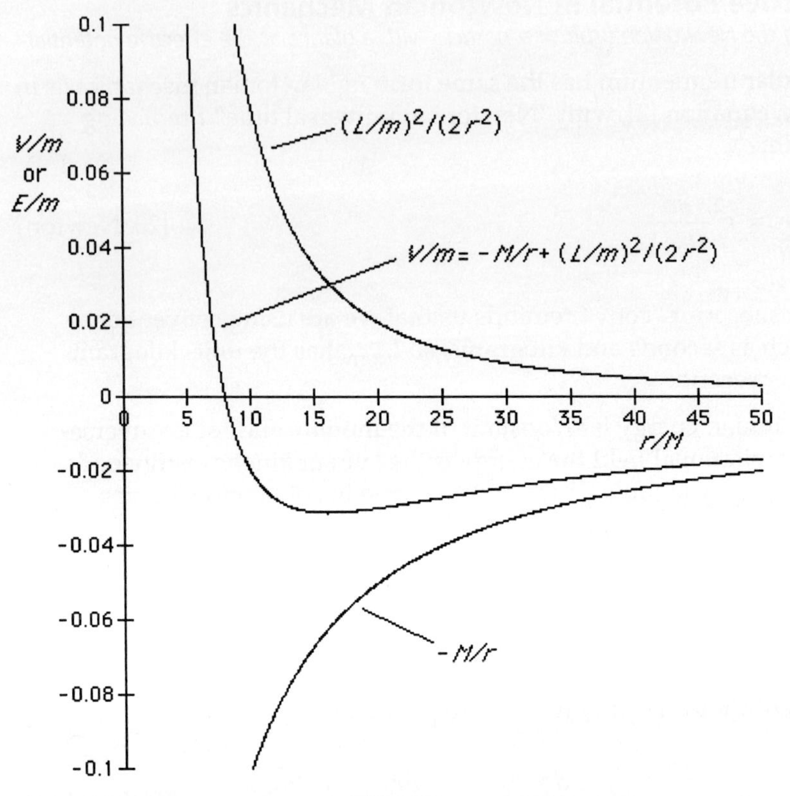

Figure 5 *Computer plot: Newtonian effective potential (middle curve) as a sum of two terms: gravitational potential (lower curve) and a "repulsive" term due to angular momentum, equation [28] for the specific case L/m = 4M (upper curve).*

Newtonian effective potential

Now we *define* the **effective potential** $V(r)$ by what we have to take away from the total energy term to get a measure of the radial velocity,

$$\begin{pmatrix}\text{effective potential}\\ \text{per unit of}\\ \text{satellite mass}\end{pmatrix} \equiv \frac{V(r)}{m} = -\frac{M}{r} + \frac{(L/m)^2}{2r^2} \qquad \text{[28. Newton]}$$

Then the equation for radial velocity squared becomes

$$\frac{1}{2}\left(\frac{dr}{dt}\right)^2 = \frac{E}{m} - \frac{V(r)}{m} \qquad \text{[29. Newton]}$$

Figure 5 illustrates the effective potential of equation [28], showing the separate terms of which it is made up. The first term, decreasing toward the center, corresponds to the attractive "gravitational force." At large distances this term predominates and an orbiting particle is attracted inward. In contrast, the second term, increasing toward the center, corresponds to a repulsive force (repulsive effective potential yielding "centrifugal force"). At close distances this term predominates and an orbiting particle is repelled outward. At intermediate distances the inward and outward slopes of the two dueling terms cancel each other to make a minimum in the effective potential.

What is the source of the repulsive term in the effective potential?

An effective potential—Newtonian equation [28] on page 4-12 and relativistic equation [32] on page 4-18—looks beyond azimuthal motion and focuses on progress along one dimension: the radial direction. But the single *r*-dimension fails to represent reality. As the pole rider pictured in Figure 6 climbs inward along the pole, he is also sweeping around in a circle. The pole-rider experiences an outward force, sometimes called a *centrifugal force* or a *centrifugal pseudo force,* resulting from the circular orbit of his noninertial rotating reference frame. This outward force is the source of the repulsive term in the effective potential.

Can you give me a more concrete feeling for the repulsive term in the effective potential?

Yes. Think of yourself clinging to a lightweight pole pivoted at one end and rotating freely in a horizontal plane (Figure 6). The only force on you toward the center is the force of your hands grasping the pole. It is hard enough to hang on when you revolve far from the pivot; you feel pulled outward by an inertial "centrifugal force" because you are in a rotating reference frame. Now you grasp the pole tenaciously and pull yourself inward, hand over hand. You find that the rate of rotation increases—conservation of angular momentum! The effective outward force on you increases sharply, for two reasons. First, you are closer to the pivot and therefore travel in a tighter circle. Second, you move at greater speed around the smaller circle. The closer to the center you manage to drag yourself along the pole, the more savagely the repulsive centrifugal potential tears at your grip. Try as you will, there is a minimum radius beyond which you are not strong enough to haul yourself inward. Desperately you hang on at that innermost point. Soon your head swims, your hands tire, and you slip outward along the pole, leave the pole, and plunge into the net held by your laughing companions. You have experienced personally the ferocious repulsive term in the effective potential.

That's pretty graphic. It makes my arms ache! But what about me in orbit around Earth? There I feel nothing. I am weightless! I feel no inward pull of gravity. I feel no ferocious repulsive term in the effective potential. Where has your effective potential gone in this case? And what's the good of an effective potential in a situation where I feel no force?

Newton has two complicated explanations of weightlessness in free orbit; Einstein has one simple explanation.

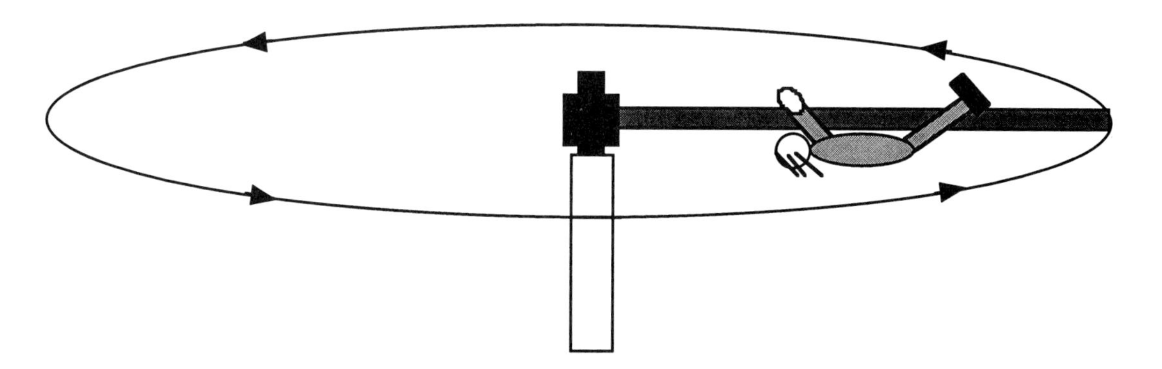

Figure 6 *Crawling inward along a freely rotating pole in order to experience centrifugal force.*

Newton's first complicated explanation of weightlessness: As viewed from the center of attraction, every particle in your body experiences an equal "gravitational acceleration." Different parts experience the same inward acceleration. Therefore there is no *relative* acceleration between adjacent parts; the bonds holding your body together feel no tension or compression. They feel nothing. You feel nothing.

Newton's second complicated explanation of weightlessness: As viewed from your accelerating reference frame, every particle in your body experiences two equal and opposite forces. First force: gravity, directed toward the center of attraction. Second force: centrifugal force, directed away from the center of attraction. In your accelerating frame these two forces balance, so each particle in your body feels zero net force. So does each nearby particle. There is no stress on the bonds holding parts of your body together. You feel nothing.

Einstein's simple explanation: In orbit you are in a free-float (inertial) frame. A frame can be in free float even near a center of attraction. In a free-float frame you float free, feeling nothing. Period.

How is the experience in orbit different from my experience of hauling myself inward along the pole by hand?

Very different! In Einstein's view, the force exerted on you by the pole drags you away from free-float motion. In Newton's view applied to your rotating frame, the outward centrifugal force acts on every particle in your body, while the inward force is applied through your hands. The inward force must be transmitted to every other particle in your body by your bones and muscles. These experience tension; you definitely do not feel weightless!

And when I stand on Earth's surface?

Very similar analysis. Einstein says the force on your feet pushes you away from a free-float frame. Newton says that every particle in your body experiences a downward "gravitational force," while the upward force is applied through your feet. The upward force must be transmitted to every particle in your body by your bones and muscles, which experience compression (legs and torso) or tension (arms hanging by your side). Again, you feel weight.

7 Effective Potential in Schwarzschild Spacetime

The orbit at a single glance!

The box on page 4-9 describes a method of using a computer to reckon the orbit of a stone. In this section we outline a much simpler method. By looking at the so-called **effective potential**, we can describe at a glance the major features of the trajectory of any particle. Clearer than our computer printouts, clearer even than our orbits, the effective potential lets us see at a glance the central features of the motion of a satellite. The basic idea is similar to the Newtonian analysis (Section 6).

Preview: a pit in the potential

The principal new results can be stated simply in terms of an effective potential: In addition to the attractive potential of gravity at great distances and the repulsive effects of angular momentum at intermediate distances (Section 6), Einstein's theory adds at still shorter distances a pit in the potential. This pit (1) captures a particle that comes too close, (2) establishes a critical distance of closest approach for this black-hole capture process, (3) for a particle that approaches this critical point without crossing it, lengthens the turnaround time as compared with

Newtonian expectations. This lengthening of the turnaround time (a) makes the time for in-and-out radial motion longer than the period for one revolution, (b) thus causes the major axis of an otherwise elliptical orbit to rotate, and (c) deflects a fast particle through larger angles than Newtonian theory would predict. Results 3a, b, and c are verified directly by motions observed in our solar system.

The pit comes from constants of the motion.

The potential? A pit in this potential? A potential attractive at large distances, repulsive at intermediate distances, and attractive again at yet smaller distances? Can we get this potential from principles that are simple, clear, and solid? Yes, from two principles: Energy as a constant of the motion! Angular momentum as a constant of the motion!

Anyone rolling a marble down the hard plastic surface illustrated in Figure 7 has before his very eyes such an effective potential. The marble, left to itself, sits at rest at the bottom of the "cup" (open circle in upper diagram). Sitting at a fixed r-value is the decisive feature of a circular orbit. Next, disturb the marble a little. It rolls back and forth in a vibration of small amplitude (solid circles, upper diagram). Changing r-value with time is exactly what marks an elliptic orbit in the Newtonian description of planetary orbits.

New feature: Capture at small radius

But now comes the new feature (Figure 7, lower diagram). Release the marble from a point high enough up on the right-hand slope so that it just makes it over the barrier summit at the left—very very slowly, yes, but definitely—irretrievably captured into territory new to it. The behavior of the marble's motion in the x-direction is exactly analogous to the behavior of the r-value of the satellite orbit in Figure 4: an orbit in which a particle moves on an inward spiral at nearly constant radius, approximating a circle for several orbits, then plunges into the black hole (Figures 8 and 9).

Satellite energy does not vary with radius.

Two features dominate radial motion as described in the language of the effective potential. One is the height of the curve as it depends on the single coordinate to which we give our attention, an x-coordinate for the marble, an r-coordinate for the satellite. The other feature is not a curve described by a function but a constant described by a number: in Figure 7, the height of release of the marble; in the motion of the satellite, the energy. Particle energy does not vary with radius—it is a constant of the motion. As a result the energy-versus-position plot is a horizontal line. Such horizontal lines for different energies are drawn in Figure 8.

Both the constant energy and the effective potential function become apparent when we derive from equation [22] the square of the radial velocity as registered in satellite wristwatch time:

$$\left(\frac{dr}{d\tau}\right)^2 = \left(\frac{E}{m}\right)^2 - \left(1 - \frac{2M}{r}\right)\left[1 + \frac{(L/m)^2}{r^2}\right] \qquad [30]$$

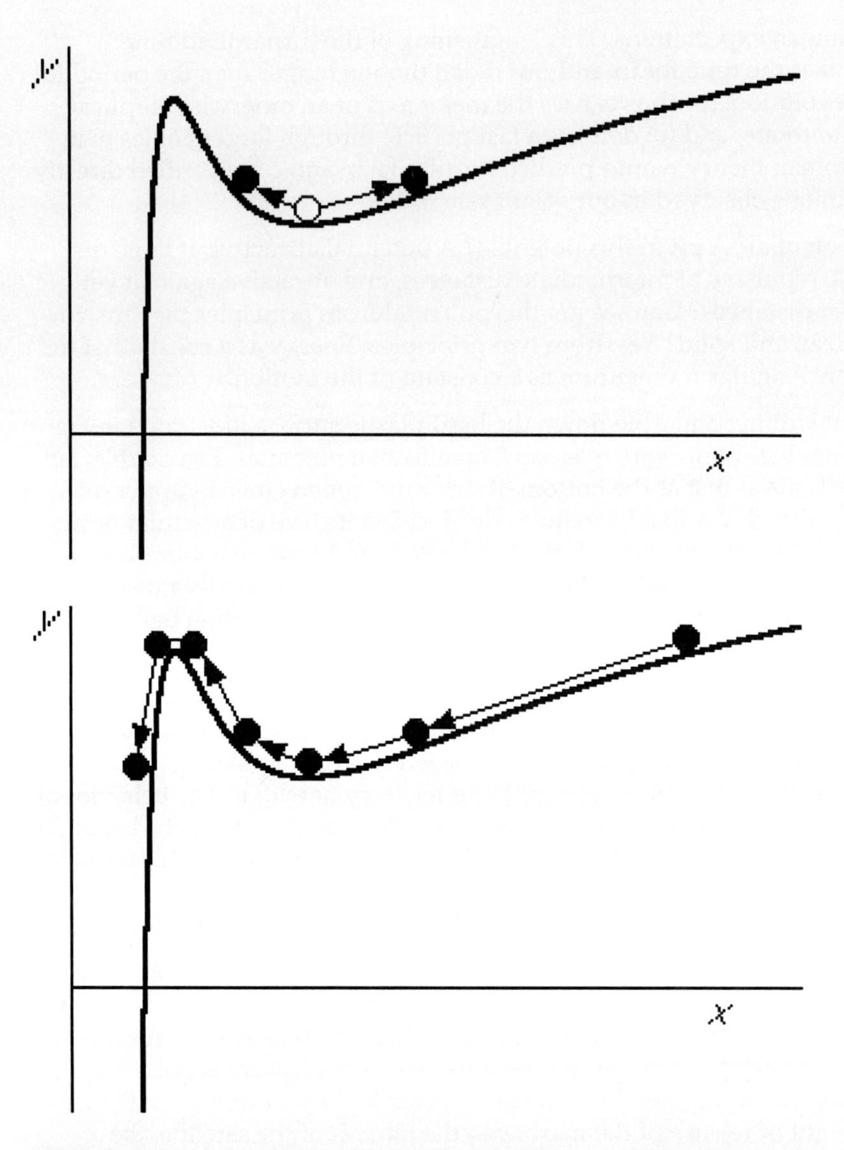

Figure 7 *Computer plot: Marble rolling on a plastic surface, y vs. x.* ***Upper diagram:*** *Placed at rest at the minimum, the marble remains at rest (open circle). Displaced from this minimum and released, the marble oscillates back and forth between inner and outer limits (solid circles).* ***Lower diagram:*** *Released from rest far enough up the right-hand slope, the marble barely makes it over the hill at the left, then plunges into new territory.*

Use equation [30] to introduce a new quantity V/m.

$$\left(\frac{dr}{d\tau}\right)^2 = \left(\frac{E}{m}\right)^2 - \left(\frac{V}{m}\right)^2 \qquad [31]$$

Equation [31] *defines* the effective potential, $V(r)$, in its dependence on r, by what we have to take away from the squared energy term to get the square of the radial velocity.

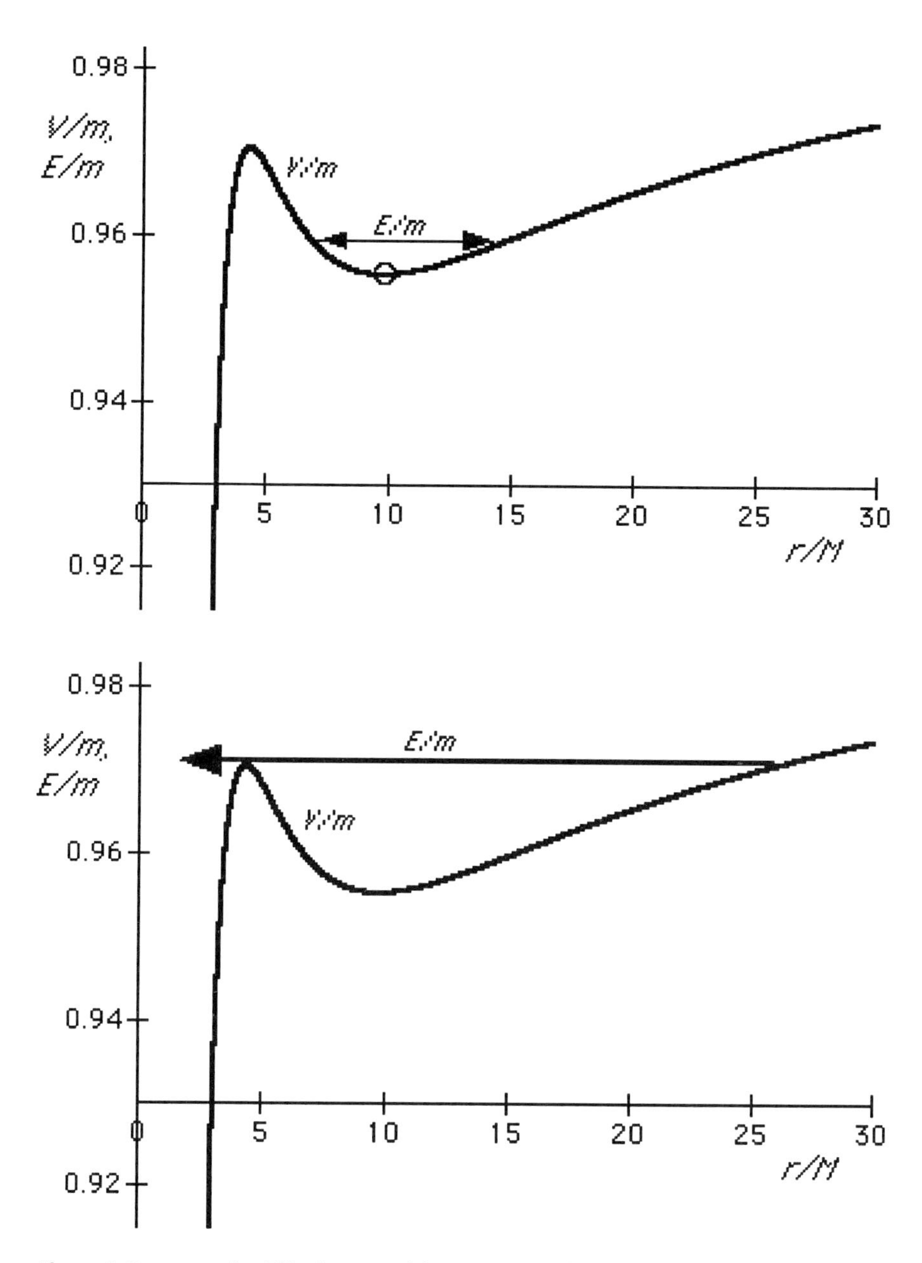

Figure 8 *Computer plot: Effective potential curve versus radius for an object orbiting a black hole for angular momentum L/m = 3.75M. Both effective potential and particle energy are measured along the vertical axis. Particle energy does not vary with radius—it is a constant of the motion. As a result the energy-versus-position plot is a horizontal line.* **Upper diagram:** *When the particle has energy corresponding to the minimum of the potential (point indicated by open circle), the particle remains at a constant radius and orbits the black hole in a circle. When it has a somewhat greater energy (line with double arrow), the particle oscillates back and forth in radius while orbiting around the center of attraction.* **Lower diagram:** *When the particle comes in with energy only slightly greater than the maximum of the potential at the left of the lower diagram, it slows its radial motion to nearly zero at the peak and orbits the black hole on a nearly circular path before plunging, as in the terminal portion of the orbit drawn in Figure 4. For further analysis of Figure 4, see Figure 9.*

Define effective potential.

$$\left(\frac{V(r)}{m}\right)^2 \equiv \left(1 - \frac{2M}{r}\right)\left[1 + \frac{(L/m)^2}{r^2}\right] \qquad [32]$$

The squared effective potential is the product of two simple factors. The first depends entirely on the attracting mass M and the r-coordinate, not at all on the angular momentum. The second factor depends not at all on the attracting mass, but only on the angular momentum per unit mass L/m of the satellite and on its r-coordinate.

There are two obvious difficulties with equation [31]:

First difficulty: *Equation [31] is the difference of SQUARES of the total energy per unit mass and the effective potential per unit mass. Why squares? Why isn't it the simple difference between total energy per unit mass and the effective potential per unit mass, as in the case of the corresponding Newtonian equation [29]?*

Second difficulty: *For very large values of the radius r, the expression [32] for V(r)/m takes on the value unity, In contrast, the Newtonian effective potential V/m in equation [28] approaches the value zero at large r, as it must by the definitions used in Newtonian mechanics.*

Both these features of the effective potential are characteristic of relativity. To take the second objection first, the total energy of the particle at rest at infinity is $E = m$, so $E/m = 1$. Therefore the effective potential per unit mass must also have the value unity ($V/m \longrightarrow 1$ as $r \longrightarrow \infty$) so that the difference of their squares—equal to the square of dr/dt—remains zero in equation [31] for a particle at rest at infinity. In contrast, the Newtonian effective potential is arbitrarily given the value zero at large separation. In essence the relativistic formula provides the new value of unity for this previously arbitrary constant.

Concerning the difference of squares in equation [31], we mentioned earlier that in general relativity one cannot separate different forms of energy. Rest energy, potential energy, and kinetic energy all combine into a larger unity in the term E. Therefore we cannot claim that $V(r)$ defined in equation [32] is the actual potential energy. Instead it is a quantity that helps us to visualize the radial component of the trajectories of particles—it is an "effective potential." This phrase does not explain the difference of squares in equation [31], but it shows that the interpretation of the separate terms is different from the Newtonian case.

Visual measure of radial speed

Plotting both V/m and E/m on the same graph, as in Figure 8, provides a measure of the radial speed, as predicted by equation [31]. The greater the vertical separation in this graph between the E/m line and the V/m curve for a given r, the greater is the radial component of the particle velocity at that r. (The vertical separation does not tell us whether the particle is moving inward or outward.) In particular, the radial motion is zero when the two graphs touch.

"Knife-edge" orbit predicted

The decisive new feature of orbits around the black hole is shown in the lower diagram of Figure 8. Let the particle have energy slightly greater than the peak of the effective potential. Then its radial motion will slow to nearly zero as it creeps down in radius, spiraling in to the radius corresponding to this peak. Very slow radial motion corresponds to nearly

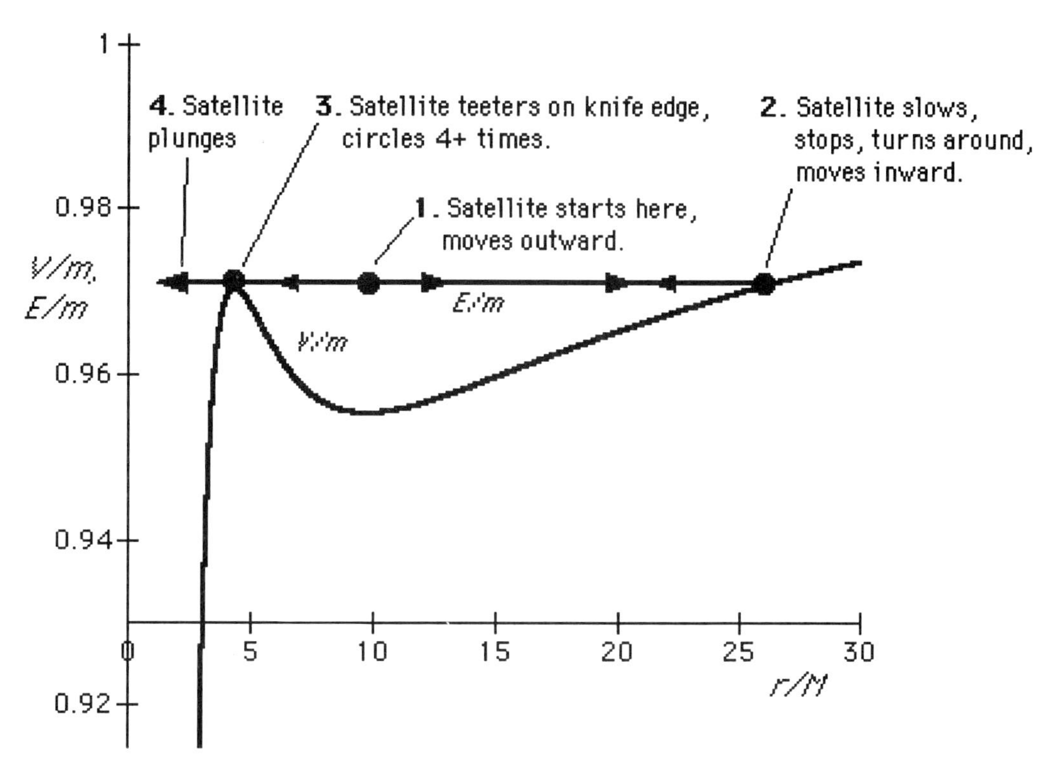

Figure 9 *Computer plot: Effective potential for particle in the orbit exhibited in Figure 4 for L/m = 3.75M. To trace the resulting radial component of motion in Figure 4, follow steps 1 through 4 in the figure.*

circular motion in the plane of the orbit. Exactly such a motion marks the r-value of the satellite orbit we see in Figure 4: an orbit in which a particle moves inward, pauses to circulate in a near-circular "knife-edge" orbit, then plunges into the black hole. But now we can predict this result with a glance at the diagram of effective potential in Figure 9. The summit of the barrier in the effective potential lies at a height slightly less than the energy available for the motion. No wonder that the inward motion slowed down when the satellite came to this point in its motion! But no wonder, either, that it kept on going inward to its dark fate. Evidently all the interesting qualitative features of the orbit let themselves be seen at a glance from the effective potential. No computing of orbits!

Radically different orbits with different energies for a single effective potential

The effective potential has this wonderful feature that for a given m and M it depends only on the angular momentum of the satellite, not at all on its energy, E. Therefore it is enough to look at a reasonable sampling of angular momentum values and the effective potentials that go with them to be able to foresee every kind of motion a satellite may undergo under the influence of a spherically symmetric center of attraction. The trajectories illustrated in Figure 11 (page 4-23) display shapes dramatically different as energy E is assigned this or that value for a given angular momentum. All these details, however, we capture from the central box by a single glance at the relation between these energies and the curve for effective potential.

Figure 12 (page 4-24) compares the Newtonian effective potential with that of a black hole for the same angular momentum. For a bound satellite, the black hole effective potential leads to deeper penetration of the satellite to smaller radii. The additional "dwell time" at the smaller radius—not obvious from the figure—results in a rotation of the major axis of the orbit.

Circular orbit from minimum of effective potential

Circular orbits are those for which the energy of the satellite nestles at the minimum of the effective potential. Different circular orbits lie at the minima of effective potentials with different angular momenta, as shown in Figure 13, page 4-25. That figure shows a stable circular orbit of minimum radius at $r = 6M$. For details, see the exercises at the end of this chapter.

These results spring from Einstein's standard geometric theory of gravity. However, some features were foreshadowed in the Newtonian use of effective potential, as we saw in Section 6.

8 Summary

How can one predict the motion of a free material particle near a spherically symmetric center of gravitational attraction? The Principle of Extremal Aging and the Schwarzschild metric combine to give us two constants of the motion—angular momentum L and energy E:

$$\frac{L}{m} = r^2 \frac{d\phi}{d\tau} \qquad [2]$$

$$\frac{E}{m} = \left(1 - \frac{2M}{r}\right)\frac{dt}{d\tau} \qquad [3]$$

Note the wristwatch time increment $d\tau$ in both of these equations.

From these equations plus the Schwarzschild metric we can find expressions for dr, $d\phi$, and dt as functions of the advance in proper time $d\tau$ (equations [20], [21], and [22], page 4-9). These equations permit a stepwise computer solution for the orbits of a material particle once we know its energy E and angular momentum L. Values for E and L can be computed, for example, from the initial speed and direction of the object launched from a spherical shell (Sample Problem 3).

The resulting orbits around a black hole show some features utterly different from those predicted by Newtonian mechanics. For example, a satellite may sweep in from a distance, then move in a circle for one or many orbits, then either move outward again or plunge into the black hole (Figure 4, page 4-10). As another example, a more remote satellite may move in a nearly elliptic orbit whose major axis changes direction with time (Figure 11, case 2, page 4-23). This is what happens to the orbit of Mercury (Project C, Advance of the Perihelion of Mercury).

SAMPLE PROBLEM 2 Escape from a Black Hole? (Revisited)

Sample Problem 1 on page 4-7 reads: "The governor of the spherical shell at reduced circumference $r_0 = 10M$ fires a satellite horizontally (that is, at 90° from the radially outward direction) at a speed $v_{0\ \mathrm{shell}} = 0.500$ as measured with his local shell clocks and rods. What is the angular momentum (per unit mass), L/m, of the satellite? What is its energy per unit mass, E/m?" These questions were both answered there, so we turn now to the third query: "Will this satellite escape to great distance?"

SOLUTION

In Sample Problem 1 we found for this case that $L/m = 5.775\ M$ and $E/m = 1.033$. Now, a particle launched radially outward (zero angular momentum!) with the lesser energy $E/m = 1.000$ will coast to rest at a very great distance from the black hole. If the satellite of this problem were launched radially outward with the slightly greater energy $E/m = 1.033$, it would arrive at a great distance with some kinetic energy. In contrast, launching the particle radially inward will surely lead to capture by the black hole, no matter what the energy. The prescribed launch at 90° from the outward direction as measured by the shell observer leaves us uncertain: Does the satellite escape from the black hole?

Figure 13 permits us to answer this question. See the dashed horizontal line at $E/m = 1.033$. Notice that a particle traveling inward along this line with the smaller value of angular momentum $L/m = 4.33M$ stops moving inward and returns again outward. From the progression of curves for different angular momenta in the diagram, one sees that the larger value of angular momentum $L/m = 5.775\ M$ will lead to an even higher potential barrier. So a satellite with energy $E/m = 1.033$ and $L/m = 5.775\ M$ will surely not penetrate the effective barrier to the black hole. Instead, it will escape outward to infinity. Further question: Will it start inward first and then turn back? Or start outward immediately?

These effects—and many more—can be predicted quickly and easily using an equation that relates radial motion, energy E, and the so-called effective potential V,

$$\left(\frac{dr}{d\tau}\right)^2 = \left(\frac{E}{m}\right)^2 - \left(\frac{V}{m}\right)^2 \qquad [31]$$

where the squared effective potential is defined as

$$\left(\frac{V(r)}{m}\right)^2 \equiv \left(1 - \frac{2M}{r}\right)\left[1 + \frac{(L/m)^2}{r^2}\right] \qquad [32]$$

For large enough values of the angular momentum L, a plot of this effective potential shows a minimum at an intermediate radius, a maximum for a smaller radius, and a plunge to zero for even smaller radii as r approaches the horizon at $r = 2M$. The rate of change of r with satellite time then follows from equation [31], allowing a quick visual prediction of the essential features of the orbit (Figure 11, page 4-23). In particular, a satellite with energy at the minimum of the effective potential does not change radius and moves in a circular orbit. (More on circular orbits in the exercises of this chapter.)

SAMPLE PROBLEM 3 Angular Momentum and Energy from Shell Velocity

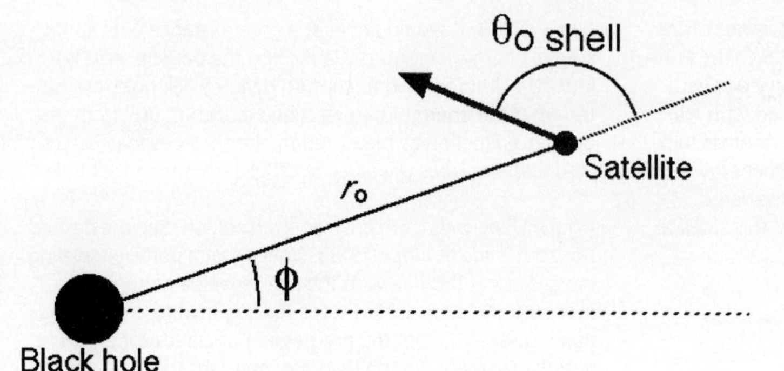

Figure 10 *Schematic diagram of the initial conditions, showing the difference between azimuth ϕ and angle of launch $\theta_{o\ shell}$ measured in shell coordinates.*

The governor of the spherical shell at reduced circumference r_o launches a satellite of mass m with speed $v_{o\ shell}$ at an initial angle $\theta_{o\ shell}$ with respect to the radially outward direction. Both speed and angle are measured with local shell rods and clocks. What is the angular momentum of the satellite? What is its energy?

SOLUTION

The solution to this exercise allows us to calculate the angular momentum *L/m* and energy measured at infinity *E/m* from initial satellite velocity measured in shell coordinates.

Figure 10 distinguishes between the angle of launch $\theta_{o\ shell}$ as measured in shell coordinates and the azimuthal angle ϕ that tracks the satellite on its orbit.

The satellite is launched with initial speed $v_{o\ shell}$ and at an angle $\theta_{o\ shell}$ with respect to the outward direction, both as measured by the shell observer. The satellite's tangential component of velocity, v_ϕ is

$$v_\phi = r\frac{d\phi}{dt_{shell}} = v_{o\ shell}\sin\theta_{o\ shell} \qquad [33]$$

In order to compute the angular momentum L (equation [2]), we want to put the proper time $d\tau$ in the denominator of the derivative in equation [33], replacing dt_{shell}. To find $d\tau$, think of the separation between two flashes emitted by the satellite. The proper time $d\tau$ between these two emissions is measured on the clock carried by the satellite. The relation between dt_{shell} and $d\tau$ is just the special relativity expression for time stretching:

$$\frac{dt_{shell}}{d\tau} = \gamma_{o\ shell} = (1 - v_{o\ shell}^2)^{-1/2} \qquad [34]$$

The value of the angular momentum of the satellite comes from the preceding two equations:

$$\frac{L}{m} = r_o^2\frac{d\phi}{d\tau} = r_o\frac{dt_{shell}}{d\tau}\left(r_o\frac{d\phi}{dt_{shell}}\right) = r_o\gamma_{o\ shell}v_{o\ shell}\sin\theta_{o\ shell} \qquad [35]$$

This last expression for angular momentum has a simple interpretation. Angular momentum has its usual vector cross-product form $\boldsymbol{L} = \boldsymbol{r} \times \boldsymbol{p}$, whose magnitude is $L = r\,p\,\sin\theta_{o\ shell}$. Here p is the expression from special relativity, $p = \gamma mv$. In the present case, simply use on-shell values of shell quantities θ, γ, and v.

To calculate energy we need the value of $dt/d\tau$, the ratio of far-away time to wristwatch time of the satellite. The relation between shell time t_{shell} and far-away time t comes from equation [C] in Selected Formulas at the end of the book:

$$\frac{dt}{dt_{shell}} = \left(1 - \frac{2M}{r}\right)^{-1/2} \qquad [36]$$

Find the expression for energy from these equations and the definition of energy [3]:

$$\frac{E}{m} = \left(1 - \frac{2M}{r_o}\right)\frac{dt}{d\tau} = \left(1 - \frac{2M}{r_o}\right)\frac{dt}{dt_{shell}}\frac{dt_{shell}}{d\tau} = \left(1 - \frac{2M}{r_o}\right)\left(1 - \frac{2M}{r_o}\right)^{-1/2}\gamma_{o\ shell} = \left(1 - \frac{2M}{r_o}\right)^{1/2}\gamma_{o\ shell} \qquad [37]$$

Equations [35] and [37] allow us to determine the satellite's angular momentum L and energy E from initial speed and angle of motion measured by a shell observer at radius r_o. Then energy and angular momentum specify the shape of the entire orbit from its start onward, thanks to equations [21] and [22] that tell us how angle ϕ and radial coordinate r tick ahead.

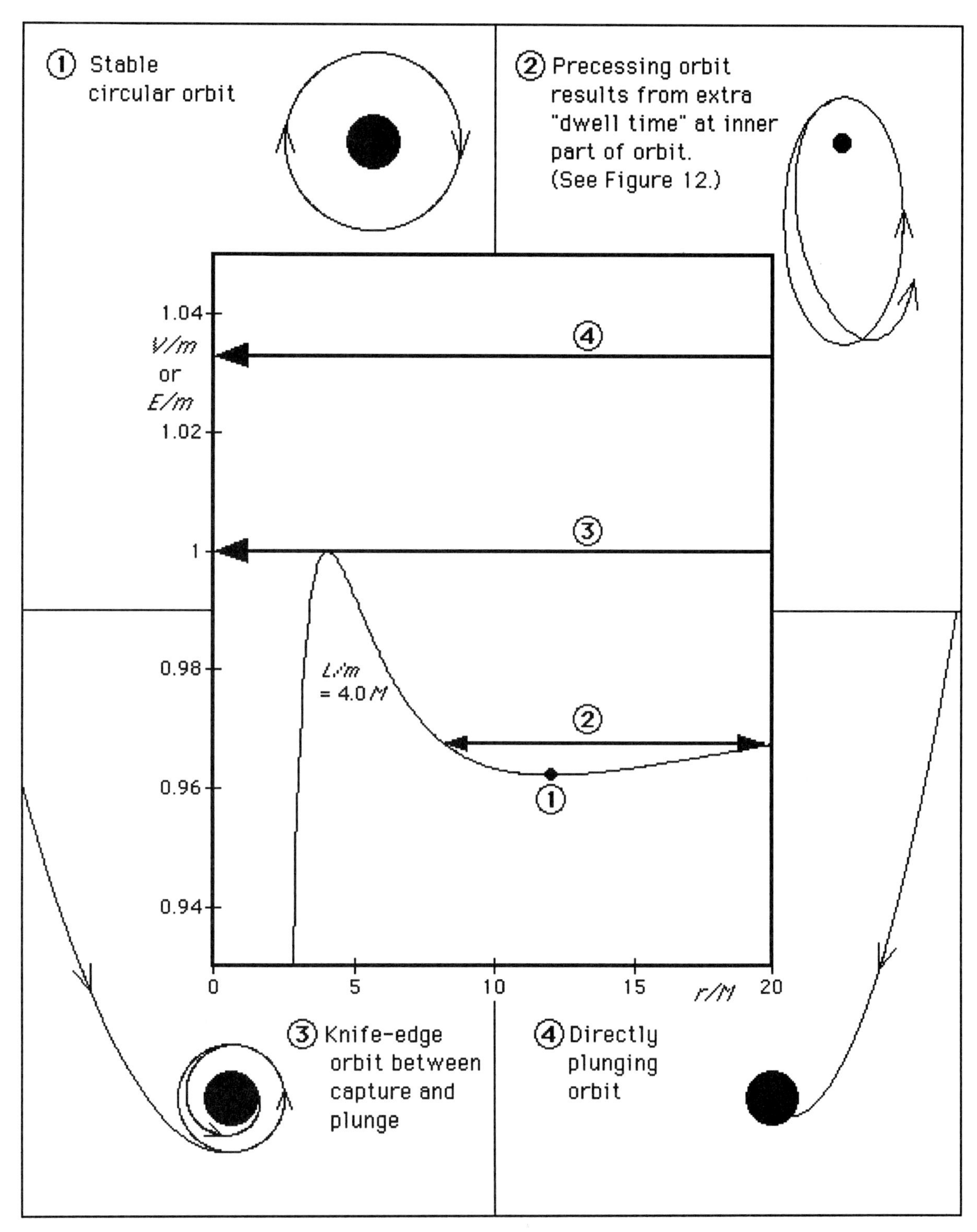

Figure 11 *Computer plots: Predicting trajectories at a glance. The effective potential—shown here for a particular value of angular momentum L/m—plus the value of total energy E/m allow us to make quick predictions about the trajectories of particles orbiting a black hole. Four different cases are numbered in the central plot of a single effective potential; the corresponding trajectories for these cases appear clockwise in the four outer corners of the figure.*

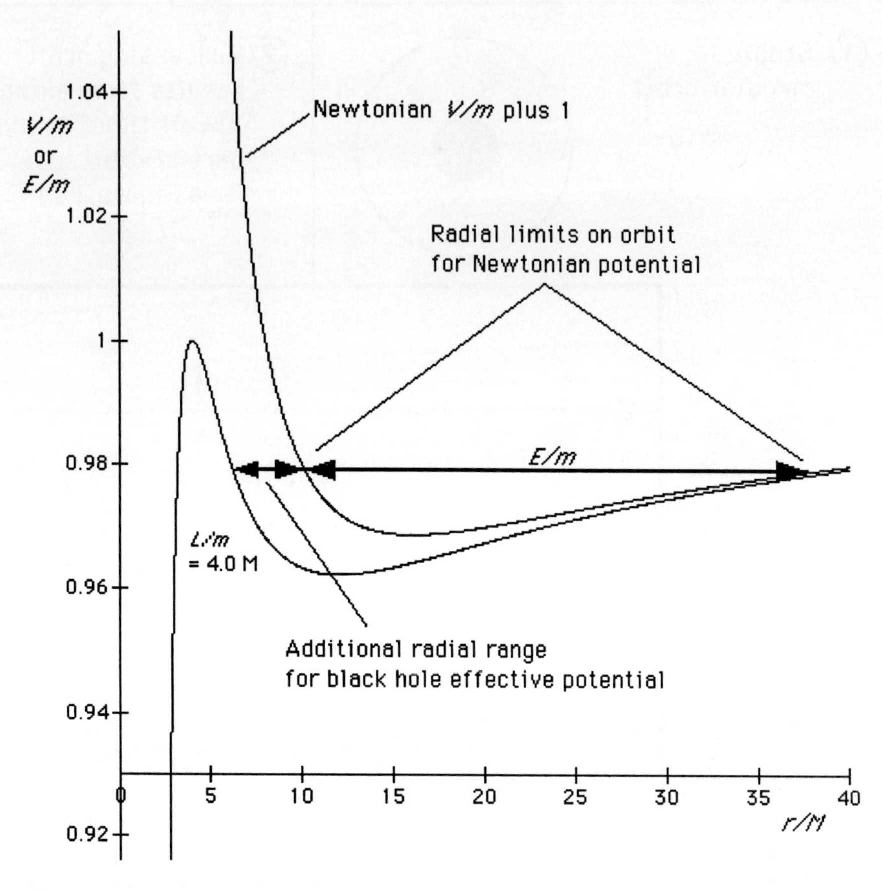

Figure 12 *Computer plot: Case 2 of Figure 11 in more detail. Effective potential for black hole and corresponding effective potential for the Newtonian case (with unity added to the Newtonian figure to include the rest energy E/m =* **1** *for a particle at rest at large radius). Both curves are for angular momentum L/m =* **4.0** *M. The radial excursion in the Newtonian case leads to an elliptic orbit. In contrast, general relativity predicts that r-values extends to smaller radii. The orbiting satellite spends more time at smaller radii than the Newtonian model would predict—a fact not directly obvious from this effective potential diagram. During the additional "dwell time" near the inner edge of the orbit—and elsewhere on the orbit—the angle ϕ keeps changing according to the conservation of angular momentum. In consequence, the orbit swings out to maximum radius at a changed angle instead of at the same angle predicted by Newtonian mechanics. The result is described as an elliptic orbit whose major axis rotates, or "advances" (case 2 in Figure 11). For the planet Mercury, this effect—though very much smaller in magnitude than for the case shown here—results in the advance of the major axis of the orbit by nearly 43 seconds of arc (0.0119 degrees) per century (Project C, following this chapter).*

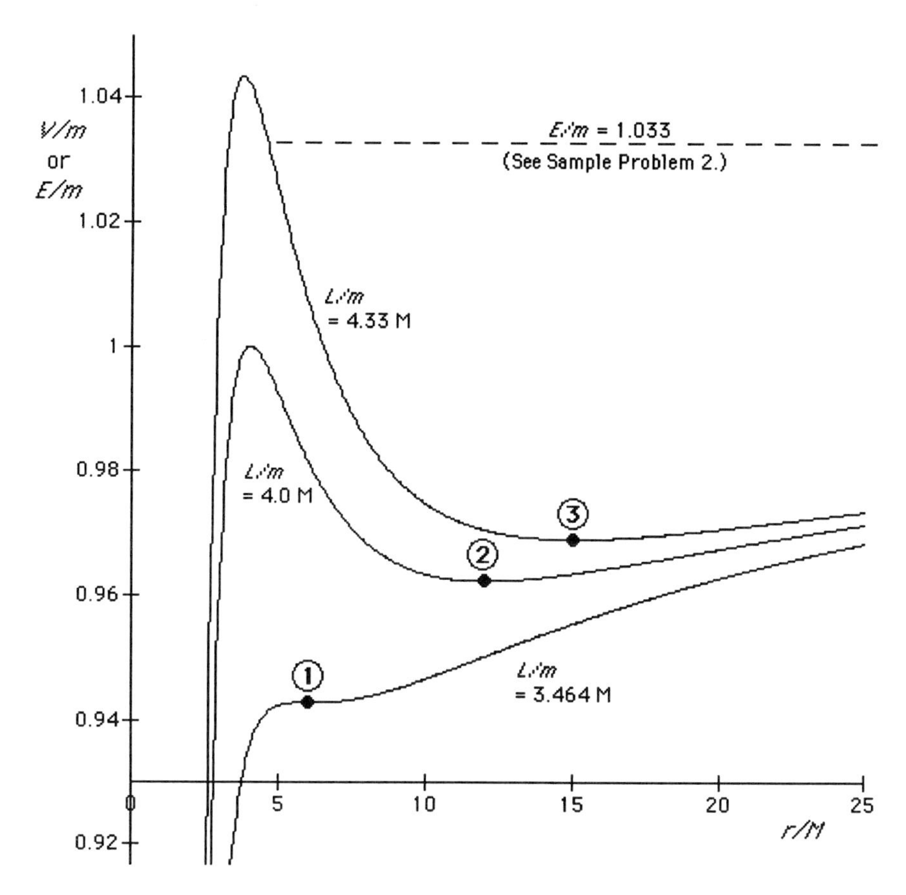

Figure 13 *Computer plot: Radii (circled numbers 1, 2, and 3) of different* **circular orbits**, *each of which lies at the radius of the effective potential minimum. The stable circular orbit of smallest radius lies at r = 6M. For details, see the exercises of this chapter. With rockets blasting, you can still explore from r = 6M down to r = 2M, but you cannot be in a stable circular orbit in this region.*

I frame no hypothesis

Hitherto we have explained the phenomena of the heavens and of our sea by the power of gravity, but have not yet assigned the cause of this power. This is certain, that it must proceed from a cause that penetrates to the very centres of the sun and planets, without suffering the least diminution of its force; that operates not according to the quantity of the surfaces of the particles upon which it acts (as mechanical causes used to do), but according to the quantity of the solid matter which they contain, and propagates its virtue on all sides to immense distances, decreasing always as the inverse square of the distances. Gravitation towards the sun is made up out of the gravitations towards the several particles of which the body of the sun is composed; and in receding from the sun decreases accurately as the inverse square of the distances as far as the orbit of Saturn, as evidently appears from the quiescence of the aphelion of the planets; nay, and even to the remotest aphelion of the comets, if those aphelions are also quiescent. But hitherto I have not been able to discover the cause of those properties of gravity from phenomena, and I frame no hypothesis; for whatever is not deduced from the phenomena is to be called an hypothesis; and hypotheses, whether metaphysical or physical, whether of occult qualities have no place in experimental philosophy. In this philosophy particular propositions are inferred from the phenomena, and afterwards rendered general by induction. Thus it was that the impenetrability, the mobility, and the impulsive force of bodies, and the laws of motion and of gravitation, were discovered. And to us it is enough that gravity does really exist, and acts according to the laws which we have explained, and abundantly serves to account for all the motions of the celestial bodies, and of our sea.

—Isaac Newton
about 1686

9 References and Acknowledgments

The Emily Dickinson poem (1510) at the beginning of the chapter is from *The Complete Poems of Emily Dickinson*, edited by Thomas H. Johnson, Little Brown and Company, Boston, 1960. Reprinted and modified with permission of Harvard University.

Quotes about time on page 4-2: St. Augustine, source of first quote unknown. Second quote from *Confessions*, quoted in *A Dictionary of Scientific Quotations*, edited by Alan L. MacKay, Institute of Physics Publishing, Bristol and Philadelphia, 1991, page 17. Giordano Bruno, *The Candle Bearer*, 1582, translated by J. B. Halle. Charles Lamb, Letter to Thomas Manning, 2 January 1806, quoted in *A Dictionary of Scientific Quotations*, page 145. Soren Kierkegaard, *Journals and Papers*, 1843, translated by Howard V. Hong and Edna H. Hong. H. D. Thoreau, *Walden* (1854). Isaac Newton, Scholium to Definition 8 at the beginning of *Principia Mathematica*. Charles W. Misner, Kip S. Thorne, and John Archibald Wheeler, *Gravitation*, W. H. Freeman and Company, San Francisco (now New York), 1973. George Burns, quoted in *New York Times*, 23 August 1993, page A14. John Archibald Wheeler, *American Journal of Physics*, Volume 46, page 323 (1978). Barbara Stanwyck, from *A Dictionary of Scientific Quotations*, page 229.

Quote, page 4-9 by Wolfgang Rindler, *American Journal of Physics*, Volume 60, Number 10, page 893 (October 1994).

Final quote from *Sir Isaac Newton's Mathematical Principles of Natural Philosophy and His System of the World*, translated by Andrew Motte, Edited by Florian Cajori, University of California Press, 1946, pages 546-7.

Lester Clendenning helped to simplify the analysis of circular orbits in the exercises of this chapter.

Chapter 4 Exercises

1. Three Views of a Circular Orbit

A shell observer on the shell of radius r compares measurements with a free-float observer moving past with speed v_{shell}. *Assume* that these comparisons can be made using the laws of special relativity. That is, assume that locally we can use the usual Lorentz transformations, calling the shell frame "laboratory," the free-float frame "rocket," and the relative speed v_{shell}. *Be careful* to have the x-axis of relative motion for the Lorentz transformation lie along the direction of motion of the "rocket" frame with respect to the "laboratory" (shell) frame.

Now consider a free-float, unpowered spaceship of mass m in circular orbit of radius r around a black hole. This rocket skims around the shell of radius r with speed v_{shell}.

A. The orbiter does one circuit and returns to the same shell clock. What time lapse T_{shell} does this shell clock record for one circuit, in terms of v_{shell}?

B. Assume that during a complete circuit of the shell, the orbiting clock "runs slow" by the usual stretch factor γ of special relativity when compared with the shell clock to which it returns. What time lapse $T_{orbiter}$ does the circular orbiter's clock record between sequential passes over the recording shell clock, in terms of v_{shell}?

C. What is the angular momentum per unit mass, L/m, of the spaceship, in terms of v_{shell}? (The answer is *not* rv_{shell})

D. What time lapse T_{bkkpr} does the bookkeeper record for one circuit of the orbiter, in terms of v_{shell}?

E. A later exercise gives the opportunity to show that the smallest radius for a stable circular orbit is $r = 6M$ and that in this orbit the satellite moves with speed $v_{shell} = 0.5$, that is at half the speed of light. Assume the central attractor to be Black Hole Alpha, with $M = 5000$ meters. Following, to one significant figure, are L/m and the times of one orbit for the shell observer, orbiter, and bookkeeper. Find the value of L/m and these times to three significant digits. (Notice that this orbiter completes one circuit in approximately 1 millisecond!)

$$T_{shell} \approx 4 \times 10^5 \text{ meters}$$

$$T_{orbiter} \approx 3 \times 10^5 \text{ meters}$$

$$L/m \approx 2 \times 10^4 \text{ meters}$$

$$T_{bkkpr} \approx 5 \times 10^5 \text{ meters}$$

2. Radii of Circular Orbits Around a Black Hole

Note: Exercises 2 through 8 build on one another. Later ones in the sequence require answers from earlier ones.

Find the radii of circular orbits around a black hole as a function of the angular momentum of the satellite, using the following outline or some other method.

An object has known angular momentum per unit mass, L/m. This value of L/m fixes the effective potential (equation [32]) for all r-coordinate values. The particle moves in a stable circular orbit if its energy is equal to the minimum in the effective potential (examples: dots with circled numbers in Figure 13, page 4-25). The circular orbit is unstable if the particle energy is equal to the maximum of the effective potential (the peaks of the barriers in the effective potentials of Figures 8, 9, 11, and 12). A satellite in an unstable orbit is like a pencil balanced on its point and will leave this orbit if perturbed in the slightest.

A. Start with the equation for the square of the effective potential from equation [32]:

$$\left[\frac{V(r)}{m}\right]^2 = \left(1 - \frac{2M}{r}\right)\left[1 + \frac{(L/m)^2}{r^2}\right] \qquad [38]$$

Make computation simpler by converting to dimensionless units, shown with an asterisk.

$$V^* \equiv \frac{V(r)}{m}$$

$$r^* \equiv \frac{r}{M} \qquad [39]$$

$$L^* \equiv \frac{L}{mM}$$

(Recall that m is the mass of the satellite in whatever units you choose for energy and M is the mass of the black hole in meters.) Show that in these dimensionless units, the square of the effective potential is

$$V^{*2} = \left(1 - \frac{2}{r^*}\right)\left(1 + \frac{L^{*2}}{r^{*2}}\right) \qquad [40]$$

B. Find the maximum and minimum of this (squared) effective potential by taking its derivative with respect to r^* and setting this derivative equal to zero. (The maximum or minimum of a positive function occurs at the same coordinate as the maximum or minimum of its square.) Show that the result is

$$r^{*2} - L^{*2} r^* + 3L^{*2} = 0 \qquad [41]$$

C. Solve this equation for the radius r^* of circular orbits. Show that the result can be written

$$r^* = \frac{L^{*2}}{2}\left[1 \pm \left(1 - \frac{12}{L^{*2}}\right)^{1/2}\right] \qquad [42]$$

or, in our regular units,

$$r = \frac{(L/m)^2}{2M}\left[1 \pm \left(1 - \frac{12M^2}{(L/m)^2}\right)^{1/2}\right] \qquad [43]$$

The plus sign yields the radius at the minimum of effective potential and therefore the radii of stable circular orbits. See equation [32]. The minus sign yields the radius at the maximum of effective potential and thus the radii of unstable circular orbits of smaller radius. Refer to Figures 8, 9, 11, and 12. (*Optional:* Verify the statements about minima and maxima by taking the second derivative of V^{*2} with respect to r^* to determine whether this second derivative is positive or negative at the given values of r^*.)

D. Show that there are no circular orbits of any kind for angular momentum given by $L/m < (12)^{1/2}M$. Show that for the minimum angular momentum $L/m = (12)^{1/2}M$, the radius of the circular orbit is $r = 6M$. This is the stable circular orbit of smallest radius. See the case with the circled number 1 in Figure 13.

3. Satellite Speed in Circular Orbit

Compute the speed of the satellite in a circular orbit. Consider the satellite speed (1) as measured by the shell observer at the reduced radius of the orbit, (2) as measured by an observer on the orbiting satellite, and (3) as computed by the remote Schwarzschild bookkeeper from her records.

A. Find the satellite speed measured by the shell observer. Consider two ticks of the satellite clock, separated by proper time $d\tau$ and by zero distance in the satellite frame, separated by time dt_{shell} and by distance $rd\phi$ in the shell frame. The relation between dt_{shell} and $d\tau$ is just the special-relativity expression

$$dt_{shell} = \gamma_{shell} d\tau = (1 - v_{shell}^2)^{-1/2} d\tau \qquad [44]$$

Knowing angular momentum, we can now reckon shell speed:

$$v_{shell} = \frac{rd\phi}{dt_{shell}} = (1 - v_{shell}^2)^{1/2} \frac{r^2 d\phi}{r d\tau}$$
$$= (1 - v_{shell}^2)^{1/2} \frac{L/m}{r} = (1 - v_{shell}^2)^{1/2} \frac{L^*}{r^*} \qquad [45]$$

From this equation, show that

$$v_{shell}^2 = \frac{1}{\frac{r^{*2}}{L^{*2}} + 1} \qquad [46]$$

From equation [41] show that

$$\frac{r^{*2}}{L^{*2}} = r^* - 3 \qquad [47]$$

Substitute this into [46] to find

$$v_{shell}^2 = \frac{1}{r^* - 2} = \frac{M}{r - 2M}$$

$$= \frac{M}{r\left(1 - \frac{2M}{r}\right)} \qquad [48]$$

B. Remember that the stable circular orbit of smallest radius is at $r = 6M$ (part D of Exercise 2). What is the value of the speed v_{shell} in this smallest stable orbit? (You will show in Exercise 6 that the *unstable* circular orbit of smallest radius is $r = 3M$.)

C. Present an argument that the speed of the passing shell measured by the satellite observer has the same value as the speed of the passing satellite measured by the shell observer.

D. Find the speed of the satellite as recorded by the remote Schwarzschild bookkeeper as follows: Use the expression for the angular momentum to replace $d\tau$ in the Schwarzschild metric by an expression in $d\phi$. Find an expression for the tangential speed as recorded by the Schwarzschild bookkeeper $v_{bkkpr} = r\,d\phi/dt$

$$v_{bkkpr}^2 = \frac{1 - \frac{2M}{r}}{\left(\frac{mr}{L}\right)^2 + 1} = \frac{1 - \frac{2}{r^*}}{\frac{r^{*2}}{L^{*2}} + 1} \qquad [49]$$

Use equation [47] to simplify [49]:

$$v_{bkkpr}^2 = \frac{1 - \frac{2}{r^*}}{r^* - 2} = \frac{1 - \frac{2}{r^*}}{r^*\left(1 - \frac{2}{r^*}\right)} \qquad [50]$$

$$= \frac{1}{r^*} = \frac{M}{r}$$

(In Exercise 6 we show that $3M \le r < 6M$ for unstable circular orbits.)

An aside: Equation [50] can be derived directly from equation [48] using equation [C] in Selected Formulas at the end of this book that relates shell time and bookkeeper time for a clock at rest on the shell:

$$dt_{shell}^2 = \left(1 - \frac{2M}{r}\right)dt^2 \qquad [51]$$

Here is how. Look at the two expressions for speed in circular orbit:

$$v_{shell} = \frac{r\,d\phi}{dt_{shell}} \qquad [52]$$

$$v_{bkkpr} = \frac{r\,d\phi}{dt} \qquad [53]$$

Both have the same numerator, $r\,d\phi$. Now look at the denominators. The shell time lapse in the denominator of v_{shell} is recorded by two clocks, both of which are on the same shell at the same radius r. Therefore these two clocks can be synchronized and will give the same result (equation [51]) as if the timepieces were one and the same shell clock.

E. What is the value of the bookkeeper speed of a satellite in the stable circular orbit of smallest radius?

4. Newton's Circular Orbits

Under what circumstances are circular orbits predicted by Newton indistinguishable from circular orbits predicted by Einstein? Answer this question using the following outline or some other method.

A. Find the Newtonian expression similar to equation [43] for the radius of a stable circular orbit, starting with equation [28], page 4-12.

B. Recast equation [43] for the general-relativistic prediction of r for stable orbits in the form

$$r = r_{Newt}(1 + q)$$

where r_{Newt} is the radius of the orbit predicted by Newton and q is a small quantity. This expression neglects differences between the Newtonian and relativistic values of L/m when expressed in the same units. Use the approximation [E] in Selected Formulas at the end the book to derive a simple algebraic expression for q in terms of M and r_{Newt}.

C. Set your expression for q equal to –0.01 as a criterion for good-enough equality of the radius according to both Newton and Einstein. Find an expression for r_{min}, the smallest value of the radius for which this approximation is valid. The result is what multiple of M?

D. Find a numerical value for r_{min} in meters for our Sun. Compare the value of r_{min} with the radius of Sun.

E. What is the value of q for the radius of the orbit of the planet Mercury, whose orbit has an average radius 0.387 times that of Earth? (See Selected Physical Constants inside the back cover.)

5. Kepler's Laws of Planetary Motion

Johannes Kepler (1571-1630) provided a milestone in the history of mechanics with his **Three Laws of Planetary Motion**, deduced from a huge stack of planetary observations made by Tycho Brahe.

1. The planets circulate around Sun in elliptical orbits with Sun at one focus.

2. The radius vector from Sun to any planet sweeps out equal areas in equal times.

3. The square of the period of any planet is proportional to the cube of the planet's mean distance from Sun.

A. Show by example that Kepler's first law is *not* true for orbits near a black hole. (Hint: Look at the figures in this chapter.) The planet Mercury departs slightly from this law (Project C, Advance of the Perihelion of Mercury).

B. Show by a simple symmetry argument that Kepler's Second Law is true for circular orbits around a black hole.

C. From equation [50] show that for *circular* orbits the period squared is given by the expression

$$T_{bkkpr}^2 = \frac{(2\pi)^2}{M} r^3 \qquad [54]$$

so that Kepler's Third Law is also valid for circular orbits around a black hole.

D. Kepler's Third Law is sometimes called **The 1-2-3 Law** from the exponents in the following equation. Show that for circular orbits

$$M \equiv M^1 = \omega^2 r^3 \qquad [55]$$

where ω is the bookkeeper angular velocity, given by the expression

$$\omega \equiv \frac{2\pi}{T_{bkkpr}} \qquad [56]$$

6. Satellite Speed in *Unstable* Circular Orbit

Examine more closely the unstable circular orbits, those with energy E at the top of the peak of effective potential. In order to stay dependably in such an orbit, the satellite needs an automatic control device that fires small rockets to keep the satellite on the knife edge at this peak.

A. Begin with equation [42] for the radius of circular orbits, replacing the plus/minus symbol in that expression by minus for the radius of the unstable orbits:

$$r^* = \frac{L^{*2}}{2}\left[1 - \left(1 - \frac{12}{L^{*2}}\right)^{1/2}\right] \qquad [57. \text{ unstable}]$$

(Recall that $r^* = r/M$ and $L^* = L/(mM)$.) Show that expressions [48] and [50] for satellite speed are valid regardless of the choice of sign (plus or minus) in equation [42].

B. What is the unstable circular orbit of smallest possible radius? Look at equation [48], which gives the speed of an object in a circular orbit as measured by the observer on the shell of radius r. In principle this measurement can actually be carried out. From our eternal requirement that no object can have a locally-measured speed greater than the speed of light, find the smallest possible radius for this unstable circular orbit.

C. From earlier expressions, find the value of angular momentum for a particle in the unstable orbit of smallest possible radius. Show that this angular momentum is infinite.

Discussion: How can the angular momentum go to infinity? Recall that the angular momentum is equal to $L = m\, r^2\, d\phi/d\tau$. The relation between proper time $d\tau$ and shell time dt_{shell} is given by the usual time-stretch formula of special relativity. As the satellite speed approaches the speed of light, the wristwatch time, the proper time $d\tau$ goes to zero. Then it takes zero proper time to circulate once around the black hole. Because $d\tau$ is in the denominator of the expression for angular momentum, the angular momentum L grows large without limit.

D. For the foregoing unstable circular orbit of minimum radius, what is the speed of the satellite reckoned by the Schwarzschild bookkeeper? Believe it or not, this maximum satellite speed is the *speed of light* in this orbit as reckoned by the Schwarzschild bookkeeper. For more on bookkeeper values of the speed of light, see Chapter 5.

7. Time Travel Using the Black Hole: *Stable* Circular Orbits

You are on a panel of experts called together to evaluate a proposal to travel forward in time using the difference in rates between a clock in a stable circular orbit around a black hole and our far-away clocks remote from the black hole. Give your advice about the feasibility of the scheme, based on the following analysis or some other that you devise.

A. Consider two sequential ticks of the clock of a satellite in a stable circular orbit around a black hole. We want to find the ratio $d\tau/dt$. The numerator in this fraction is equal to the wristwatch time $d\tau$ between the ticks in the frame of the satellite; the denominator is the far-away time lapse dt recorded by the Schwarzschild bookkeeper. Use the expression for angular momentum to eliminate $d\phi$ from the Schwarzschild metric in this case to obtain

$$\left(\frac{d\tau}{dt}\right)^2 = \frac{1-\dfrac{2M}{r}}{1+\left(\dfrac{L}{mr}\right)^2} = \frac{1-\dfrac{2}{r^*}}{1+\left(\dfrac{L^*}{r^*}\right)^2} \qquad [58]$$

Substitute from equation [47] and simplify to show that

$$\frac{d\tau}{dt} = \left(1-\frac{3}{r^*}\right)^{1/2} = \left(1-\frac{3M}{r}\right)^{1/2} \qquad [59]$$

B. What is the value of the ratio $d\tau/dt$ for the stable circular orbit of smallest possible radius, $r = 6M$?

C. What rocket speed in flat spacetime gives the same ratio of rocket clock time to "laboratory" time"?

D. Based on this analysis, do you recommend in favor of—or against—the proposal for time travel using stable circular orbits around a black hole?

8. Time Travel Using the Black Hole: *Unstable* Circular Orbits

The proposal for travel forward in time (Exercise 7) is modified to use an unstable circular orbit, with the assumption that an automatic device controls correcting rockets to keep the satellite safely on its knife-edge orbit.

A. What is the value of the ratio $d\tau/dt$ for the unstable circular orbit of smallest possible radius? Based on this result, would you recommend in favor of using an unstable orbit around a black hole for time travel?

B. With a glance at effective potential curves for different values of angular momentum in Figure 13, draw a conclusion about the value of the total energy per unit mass, E/m, for a particle that is to be put into the unstable orbit of minimum radius $r = 3M$. While it is still far from the black hole, what is the time-stretch factor γ_{far} for a spaceship with this energy? Does this result alter your recommendation about using unstable circular orbits around a black hole for time travel?

9. Turning Around Using the Black Hole

The starship *Enterprise* is headed toward a black hole of known mass M. As captain, you want to use this black hole to reverse your direction of motion with a minimum expenditure of rocket fuel. You look at Figure 11, page 4-23 and decide to use Case 3, but with energy E/m smaller than the peak of effective potential so that you will return outward immediately rather than enter the risky knife-edge orbit that has claimed the lives of so many graduates of Starfleet Academy.

A. Will the value of E/m change as the *Enterprise* descends toward the black hole?

B. What is the *linear* momentum per unit mass p_{far}/m of the spaceship while far from the black hole, expressed as a function of E/m of the spaceship? Will the value of the linear momentum of the spaceship change as it descends toward the black hole?

C. What is the magnitude of the *angular* momentum L/m per unit mass of the spaceship about the black hole, expressed in terms of E/m of the spaceship? See Figure 3, page 4-6, and take the impact parameter b to be the would-have-been distance of closest approach if the black hole had zero mass. (For a more careful definition of b, see Figure 2, page 5-6.) Will the value of the angular momentum change as the spaceship descends toward the black hole of mass M?

D. What happens to you and your crew if E/m for the *Enterprise* is greater than the peak of the effective potential?

E. You do not want to change E/m for the *Enterprise*, because that requires expenditure of considerable rocket fuel. Instead, you change the value of your angular momentum L/m by slightly altering your direction of motion as you approach the black hole from a distance. What is the minimum value of the impact parameter b that brings the peak of the effective potential above the value of E/m?

F. Explain the advantages of the strategy outlined in this exercise compared with rocket-propelled velocity changes for interstellar round trips or for time travel into the future. What complications does the black hole introduce compared to the use of rockets in flat spacetime?

Project C Advance of the Perihelion of Mercury

- *How much do Newton and Einstein disagree about the orbit of Mercury?*
- *How do theory and experiment decide between them?*
- *Why Mercury?*

Project C

Advance of the Perihelion of Mercury

This discovery was, I believe, by far the strongest emotional experience in Einstein's scientific life, perhaps in all his life. Nature had spoken to him. He had to be right. "For a few days, I was beside myself with joyous excitement." Later, he told Fokker that his discovery had given him palpitations of the heart. What he told de Haas is even more profoundly significant: when he saw that his calculations agreed with the unexplained astronomical observations, he had the feeling that something actually snapped in him.

—Abraham Pais

1 Joyous Excitement

What discovery sent Einstein into "joyous excitement" in November of 1914? It was the calculation showing that his brand new (actually not quite completed) theory of general relativity gave the correct value for one detail of the orbit of the planet Mercury that had previously been unexplained.

Mercury circulates around Sun in a not quite circular orbit: The planet oscillates in and out radially while it circles tangentially. The result is an elliptic orbit. Newton tells us that if we consider only the interaction between planet and Sun, then the time for one circular orbit is *exactly* the same as one in-and-out radial oscillation. Therefore the orbital point closest to Sun, the so-called **perihelion**, stays in the same place; the elliptical orbit does not shift around with each revolution—according to Newton. In this project you will begin by verifying this nonrelativistic result. Why bother calculating something that does not change? Because observation shows that Mercury's orbit does, in fact, change. The innermost point, the perihelion, moves around the Sun a *little*; it *advances* with each orbit (Figure 1). The long (major) axis of the ellipse rotates at the tiny rate of 531 seconds of arc (0.147 degrees) *per century*. (One degree equals 3600 seconds of arc.) Newtonian mechanics accounts for most of this advance by computing the perturbing influence of the other planets. But a stubborn 43 seconds of arc (0.0119 degree) per century (called a **residual**) remains after all these effects are accounted for. This discrepancy (though not its modern value) was computed from observations by LeVerrier as early as 1859.

Simon Newcomb

Simon Newcomb
Born March 12, 1835, Wallace, Nova Scotia
Died July 11, 1909, Washington, D.C.
(Photo courtesy of Yerkes Observatory)

From 1901 until 1959 and even later, the tables of locations of the planets (so-called **ephmerides**) used by most astronomers were those compiled by Simon Newcomb and his collaborator, George W. Hill. By the age of five Newcomb was spending several hours a day making calculations and before the age of seven was extracting cube roots by hand. He had little formal education but avidly explored many technical fields in the libraries of Washington, D. C. He discovered the *American Ephemeris and Nautical Almanac*, of which he said, "Its preparation seemed to me to embody the highest intellectual power to which man had ever attained."

Newcomb became a "computer" (someone who computes) in the American Nautical Almanac Office and, by stages, rose to become its head. The greater part of the rest of his life was spent calculating the motions of bodies in the solar system from the best existing data. Newcomb collaborated with Q. M. W. Downing to inaugurate a worldwide system of astronomical constants, which was adopted by many countries in 1896 and officially by all countries in 1950.

The advance of the perihelion of Mercury computed by Einstein in 1914 would have been compared to entries in the tables of Simon Newcomb.

The advance of the perihelion of Mercury is sometimes called the **precession of the perihelion**.

Newtonian mechanics says that there should be *no residual* advance of the perihelion of Mercury's orbit and so cannot account for the 43 seconds of arc per century which, though tiny, is nevertheless too large to be ignored or blamed on observational error. But Einstein's general relativity hit it on the button. Result: joyous excitement!

In this project we review Newton's incorrect prediction and then carry out a general-relativistic approximate calculation of the advance of the perihelion of Mercury adapted from that of Robert M. Wald (*General Relativity,* University of Chicago Press, 1984, pages 142–143). This approximation describes the angular motion of the planet as if it were in a nearly circular orbit. From this assumption we calculate the time for one orbit. The approximation also describes the small inward and outward radial motion of the planet as if it were a harmonic oscillator moving back and forth radially about the minimum in a potential well (Figure 2). We calculate the time for one round-trip radial oscillation. These two times are equal, according to Newton, if one considers only the planet-Sun interaction. In that case the planet goes around once in the same time that it oscillates radially inward and back out again. The result is an elliptical orbit that closes on itself, so the planet repeats its elliptic path forever. In contrast, these two times—the angular and the radial—are *not quite* equal according to the Einstein approximation. The radial oscillation takes place more slowly. From the difference we reckon the approximate rate of advance of Mercury's perihelion around Sun.

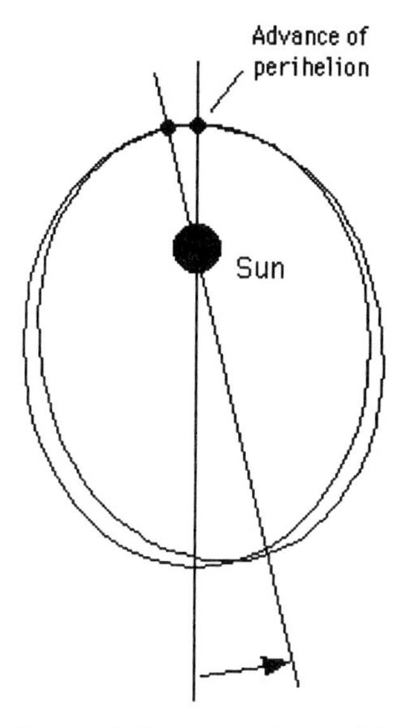

Figure 1 *Exaggerated view of the change in orientation of Mercury's orbit during one century.*

2 Linear Harmonic Oscillator

Why should the satellite oscillate in and out radially? Look at the effective potential for Newtonian motion, the heavy line in Figure 2. This heavy line has a minimum, the location at which a particle can rest and ride around at constant r, executing a circular orbit. But it can also oscillate radially in and out, as shown by the two-headed arrow.

How long will it take for one in-and-out oscillation? That depends on the shape of the effective potential curve near the minimum shown in Figure 2. If the amplitude of the oscillation is small, then the important part of the curve is very close to this minimum, and we can use a well-known mathematical theorem: If a continuous, smooth curve has a minimum, then near that minimum the curve can be approximated by a parabola with its vertex at the minimum point. Such a parabola is shown superimposed on the effective potential curve of Figure 2. From the diagram it is apparent that the parabola is a good approximation of the potential near that minimum. In fact Mercury's orbit swings from a minimum radius (the perihelion) of 46.04 million kilometers to a maximum radius (the so-called **aphelion**) of 69.86 million kilometers.

From introductory physics we know how a particle moves in a parabolic potential. The motion is called **harmonic oscillation** and follows a formula of the kind

$$x = A \sin \omega t \qquad [1]$$

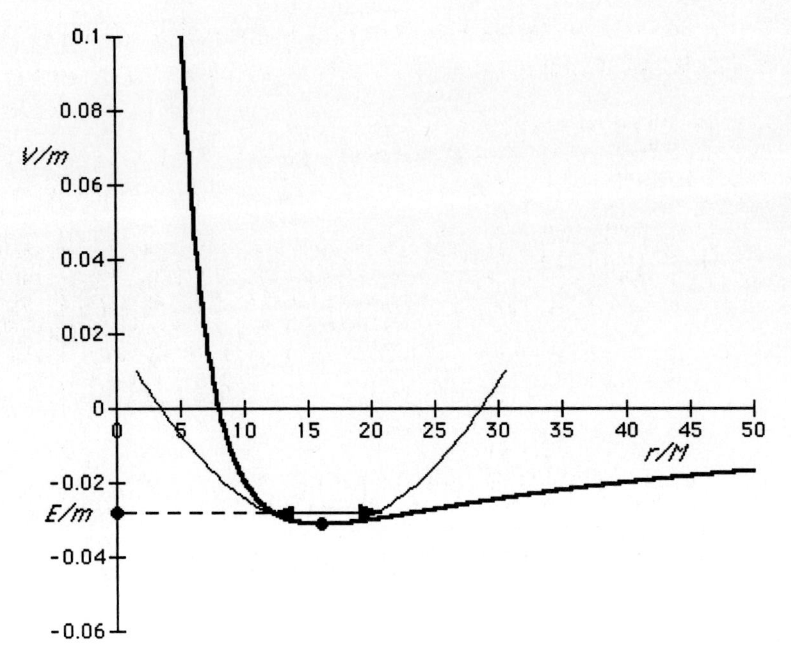

Figure 2 Computer plot: The Newtonian effective potential (thick curve), copied from Figure 5, page 4-12, on which is superimposed the parabolic potential of the simple harmonic oscillator (thin curve). The two curves conform to one another only near the minimum of the effective potential. We use a similar set of curves to approximate the radial oscillation of Mercury in its orbit as an harmonic oscillation of small amplitude.

Here A is the amplitude of the oscillation and ω (Greek lower-case omega) tells us how rapidly the oscillation occurs. The potential energy per unit mass V/m of a particle oscillating in a parabolic potential is given by the formula

$$V/m = \frac{1}{2}\omega^2 x^2 \qquad [2]$$

From equation [2] we can find an expression for ω by taking the second derivative of both sides with respect to the displacement x:

$$\frac{d^2(V/m)}{dx^2} = \omega^2 \qquad [3]$$

In general, if we have the expression for the potential, we can find the rate ω of harmonic oscillation around a minimum by taking the second derivative of the curve and evaluating it at that minimum where $d(V/m)/dx = 0$.

3 Radial Harmonic Oscillation of Mercury: Newton

The trouble with the in-and-out radial oscillation of Mercury is that it does not take place around $x = 0$ but around the average radius r_0 of its orbit.

What is the value of r_0? It is the radius for which the effective potential has a minimum. For Newtonian orbits the radial motion is given by equations [27], page 4-11, and [29], page 4-12:

$$\frac{1}{2}\left(\frac{dr}{dt}\right)^2 = \frac{E}{m} - \left[-\frac{M}{r} + \frac{(L/m)^2}{2r^2}\right] = \frac{E}{m} - \frac{V(r)}{m} \qquad \text{[4. Newton]}$$

From this equation we define the effective potential (equation [28] on page 4-12):

$$\frac{V(r)}{m} = -\frac{M}{r} + \frac{(L/m)^2}{2r^2} \qquad \text{[5. Newton]}$$

QUERY 1	**Finding the potential minimum.** Take the derivative with respect to r of the potential per unit mass, V/m given in equation [5]. Set this first derivative aside for use in Query 2. As a separate calculation, equate this derivative to zero in order to determine the radius r_o at the effective potential minimum. Use the result to write down an expression for the unknown quantity $(L/m)^2$ in terms of the known quantities M and r_o.
QUERY 2	**Oscillation rate ω_r for radial motion.** We want to use equation [3] to find the rate of radial oscillation. Accordingly, continue by taking a second derivative of V/m in equation [5] with respect to r. Set $r = r_o$ in the resulting expression and substitute your value for $(L/m)^2$ from Query 1. Use equation [3] to find an expression for the rate at which Mercury oscillates in and out radially—according to Newton!

4 Angular Velocity of Mercury in Its Orbit

We want to compare the rate ω_r of in-and-out radial motion of Mercury with its rate ω_ϕ of round-and-round tangential motion. Use the Newtonian definition of angular momentum, with increment dt of Newtonian universal time, similar to equation [2], page 4-3.

$$L/m = r^2\frac{d\phi}{dt} = r^2\omega_\phi \qquad \text{[6. Newton]}$$

We want to find the value for the angular velocity $\omega_\phi = d\phi/dt$ of Mercury along its almost circular orbit.

QUERY 3	**Angular velocity of Mercury in orbit.** Into equation [6] substitute your value for L/m from Query 1 and set $r = r_o$. Find an expression for $d\phi/dt$ in terms of M and r_o.

QUERY 4 **Comparing radial oscillation rate with orbital angular velocity.** Compare your value of angular velocity ω_ϕ from Query 3 with your value for radial oscillation rate ω_r from Query 2. State your conclusion about the advance of the perihelion of Mercury's orbit around Sun (when only the Sun-Mercury interaction is considered), according to Newton.

5 Effective Potential: Einstein

Now we repeat the analysis for the general relativistic case, using the Newtonian analysis as our model. Equation [30], page 4-15 gives a measure of the radial motion of the orbiting planet. Multiply through by 1/2 to obtain an equation similar to equation [4] above for the Newtonian case:

$$\frac{1}{2}\left(\frac{dr}{d\tau}\right)^2 = \frac{1}{2}\left(\frac{E}{m}\right)^2 - \frac{1}{2}\left(1-\frac{2M}{r}\right)\left[1+\frac{(L/m)^2}{r^2}\right] = \frac{1}{2}\left(\frac{E}{m}\right)^2 - \frac{U}{m} \qquad [7]$$

Equations [4] and [7] are of similar form, and we use this similarity to make a harmonic analysis of the radial motion of Mercury in orbit in general relativity similar to the Newtonian analysis of Sections 3 and 4. Begin by assigning the name *effective potential* and the symbol U/m to the term subtracted from the squared energy in [7], as indicated on the right end of the equation.

Before proceeding further, note first that the time in equation [7] is the proper time τ, the wristwatch time of the satellite, not Newton's universal time t. This different time standard is not necessarily fatal, since in Newtonian mechanics there is only one universal time, and we have not yet had to decide which relativistic time should replace it. You will show that for Mercury the choice of which time to use (wristwatch time, bookkeeper far-away time, or even shell time at the radius of the orbit) makes a negligible difference in our predictions about the rate of advance of the perihelion.

Second, note that the relativistic expression $(1/2)(E/m)^2$ in equation [7] stands in the place of the Newtonian expression (E/m) in equation [4]. Do we dare replace an energy with a squared energy? Both represent a constant of the motion and, strange as it may seem, the difference does not affect our analysis. Evidence that we are on the right track follows from multiplying out the second term of the middle equality in equation [7]. We have assigned the symbol U/m to this second term.

$$\begin{aligned}\frac{U}{m} &= \frac{1}{2}\left(1-\frac{2M}{r}\right)\left[1+\frac{(L/m)^2}{r^2}\right] \\ &= \frac{1}{2} - \frac{M}{r} + \frac{(L/m)^2}{2r^2} - \frac{M(L/m)^2}{r^3}\end{aligned} \qquad [8]$$

On the right side of the second line are the two effective potential terms that made up the Newtonian expression [5]. In addition, the first term (1/2) assures that far from the center of attraction the radial speed in [7] will have the correct value. For example, let the total energy equal the rest energy ($E/m = 1$). Then for large r, the radial speed $dr/d\tau$ (equation [7]) goes to zero, as it must in this case. The potential U/m is plotted in Figure 3.

The final term on the right of the second line of [8] describes an attractive potential arising from general relativity. This causes the slight deviation of the orbit of Mercury from that predicted by Newton. Because of the r^3 in the denominator, near a black hole this negative term overwhelms all others at small radii, leading to the downward plunge in the effective potential at the left side of Figure 3.

In summary, the forms of equations [7] and [8] allow us to use the tools of Newtonian mechanics to analyze the radial component of the satellite's motion predicted by general relativity, provided that we are satisfied with the wristwatch time of the satellite and with an "energy term" equal to $(1/2)(E/m)^2$. Of course, we are trying to solve a relativistic problem. Nevertheless, because of its form we can use the Newtonian manipulation to carry out a general relativistic calculation.

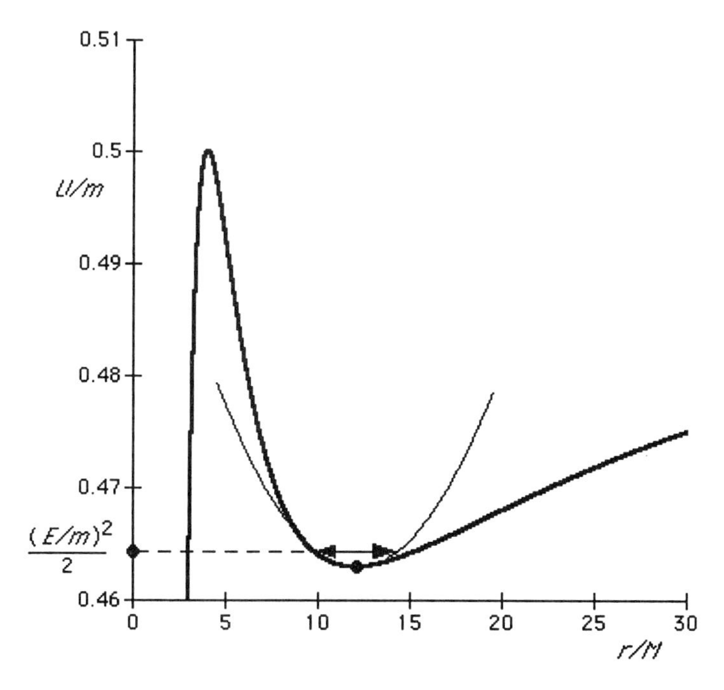

Figure 3 *Computer plot: Approximation of the general-relativistic effective potential U/m (heavy curve) at the minimum with a parabola (light curve) in order to analyze the radial excursion (double-headed arrow) as simple harmonic motion. The heavy effective potential curve is for a black hole, not for Sun, whose effective potential would be indistinguishable from the Newtonian function on the scale of this diagram.*

6 Radial Harmonic Oscillation of Mercury: Einstein

Now analyze the radial oscillation of Mercury according to Einstein.

QUERY 5 **Finding the potential minimum.** Take the derivative of the effective potential [8] with respect to *r*. Set this first derivative aside for use in Query 6. As a separate calculation, equate this derivative to zero, set $r = r_o$, and solve the resulting equation for the unknown quantity $(L/m)^2$ in terms of the known quantities M and r_o.

QUERY 6 **Radial oscillation rate.** We want to use equation [3] to find the rate of oscillation in the radial direction. Accordingly, continue to the second derivative of U/m from equation [8]. Set $r = r_o$ in the result and substitute the expression for $(L/m)^2$ from Query 5 to obtain

$$\left.\frac{d^2(U/m)}{dr^2}\right|_{r=r_0} = \omega_r^2 = \frac{M(r_o - 6M)}{r_o^3(r_o - 3M)} \qquad [9]$$

QUERY 7 **Newtonian limit of radial oscillation.** The radius of Mercury's orbit around Sun has the value $r_o = 5.80 \times 10^{10}$ meters. Compare this radius with the value M for the mass of Sun in geometric units. If one of these can be neglected in equation [9] compared with the other, demonstrate that the resulting value of ω_r is the same as your Newtonian expression derived in Query 2.

7 Angular Velocity in Orbit: Einstein

We want to compare the rate of in-and-out radial oscillation of Mercury with the angular rate at which Mercury moves tangentially in its orbit. The rate of change of azimuth ϕ springs from the definition of angular momentum, equation [2], page 4-3:

$$\frac{L}{m} = r^2 \frac{d\phi}{d\tau} \qquad [10]$$

Note that the time here, too, is the wristwatch (proper) time τ of the satellite.

QUERY 8 **Angular velocity.** Square both sides of equation [10] and use your result from Query 5 to eliminate $(L/m)^2$ from the resulting equation. Show that the result can be written

$$\omega_\phi^2 \equiv \left(\frac{d\phi}{d\tau}\right)^2 = \frac{M}{r_o^2(r_o - 3M)} \qquad [11]$$

According to the relativistic prediction, does the round-and-round tangential motion of Mercury take place in step with the in-and-out radial oscillation, as it does in the Newtonian analysis?

QUERY 9 **Newtonian limit of angular velocity.** Make the same kind of approximation as in Query 7 and demonstrate that the resulting value of ω_ϕ is the same as your Newtonian expression derived in Query 3.

8 Predicting Advance of the Perihelion

The advance of the perihelion of Mercury springs from the difference between the frequency at which the planet sweeps around in its orbit and the frequency at which it oscillates in and out radially. In the Newtonian analysis these two frequencies are equal if one considers only the interaction between planet and Sun. But Einstein's theory shows that these two frequencies are *not quite* equal, so Mercury reaches its maximum (or minimum) radius at a slightly different angular position in each orbit. This results in the advance of the perihelion. The rate of advance is the difference between the orbital angular frequency ω_ϕ and the radial angular frequency ω_r.

QUERY 10 **Difference in squared oscillation rates.** From equations [11] and [9] construct and simplify an expression for the difference of squares $\omega_\phi^2 - \omega_r^2$ in terms of M, r_o, and ω_ϕ plus numerical constants.

QUERY 11 **Difference in oscillation rates.** The two angular rates ω_ϕ and ω_r are *almost* identical in value, even in the Einstein analysis. Therefore write the result of Query 10 in the following form:

$$\omega_\phi^2 - \omega_r^2 = (\omega_\phi + \omega_r)(\omega_\phi - \omega_r) \approx 2\omega_\phi(\omega_\phi - \omega_r) \quad [12]$$

Use outcomes of earlier queries to show that this approximation can be written

$$\omega_\phi - \omega_r \approx \frac{3M}{r_o}\omega_\phi \quad [13]$$

Equation [13] gives us the difference in angular rate between the tangential motion and the radial oscillation. From this rate difference we can calculate the rate of advance of the perihelion of Mercury.

All of the ω-expressions are of the form $d(\text{angle})/d\tau$ or $d(\text{phase angle})/d\tau$. Since $d\tau$ is in the denominator everywhere, it can be canceled out and the angle increments added to give angles. The resulting adaptation of equation [13] has the following form:

$$\begin{pmatrix}\text{predicted}\\ \text{angle of}\\ \text{advance}\end{pmatrix} = \begin{pmatrix}\text{total angle}\\ \text{covered in}\\ \text{orbital motion}\end{pmatrix} - \begin{pmatrix}\text{total phase angle}\\ \text{covered in}\\ \text{radial motion}\end{pmatrix} \qquad [14]$$

$$= \frac{3M}{r_0}\begin{pmatrix}\text{total angle}\\ \text{covered in}\\ \text{orbital motion}\end{pmatrix}$$

Moreover, we can use any measure of angle we wish—degrees or radians or seconds of arc—as long as we are consistent. Numerical prediction based on this equation must be compared with results of observation.

9 Comparison with Observation

QUERY 12 **Mercury's orbital period.** The period of Mercury's orbit is 7.602×10^6 seconds and that of Earth is 3.157×10^7 seconds. What is the value of Mercury's period in Earth-years?

QUERY 13 **Mercury's revolution in one century**. How many revolutions around Sun does Mercury make in one century (100 Earth-years)? How many degrees of angle are traced out by Mercury in one century?

QUERY 14 **Correction factor.** The mass M of Sun is 1.477×10^3 meters and the radius r_0 of Mercury's orbit is 5.80×10^{10} meters. Calculate the value of the correction factor $3M/r_0$ in equation [14].

QUERY 15 **Advance angle per century in degrees.** Using equation [14], multiply your answers from Queries 13 and 14 to obtain a prediction of the advance of the perihelion of Mercury's orbit per century in degrees.

QUERY 16 **Advance angle per century in seconds of arc.** There are 60 minutes of arc per degree and 60 seconds of arc per minute of arc. Multiply your result from Query 15 by $60 \times 60 = 3600$ to obtain your prediction of the advance of the perihelion of Mercury's orbit per century in seconds of arc.

A more careful analysis predicts a value of 42.98 seconds of arc (0.0119 degrees) per century (see Table 1). The observed rate of advance of the perihelion is in perfect agreement with this value: 42.98 ± 0.1 seconds of arc per century. (See references.) How close was your prediction?

10 Advance of the Perihelia of the Inner Planets

Do the *perihelia* (plural of *perihelion*) of other planets in the solar system also advance as described by general relativity? Yes, but these planets are farther from Sun, so the predicted advance is less than that of Mercury. In this section we compare our estimated advance of the parahelia of the inner planets Mercury, Venus, Earth, and Mars with results of an accurate calculation.

The Jet Propulsion Laboratory (JPL) in Pasadena, California, supports an active effort to improve our knowledge of the positions and velocities of the major bodies in the solar system. For the major planets and the moon, JPL maintains a database and set of computer programs known as the Solar System Data Processing System (SSDPS). The input database contains the observational data measurements for current locations of the planets. Working together, more than 100 interrelated computer programs use these data and the relativistic laws of motion to compute locations of planets at times in the past and future. The equations of motion take into account not only the gravitational interaction between each planet and Sun but also interactions among all planets, Earth's moon, and 300 of the most massive asteroids, as well as interactions between Earth and Moon due to nonsphericity and tidal effects.

To help us with our project on perihelion advance, Myles Standish, Principal Member of the Technical Staff at JPL, kindly used the numerical integration program of the SSDPS to calculate orbits of the four inner planets over four centuries, from A.D. 1800 to A.D. 2200. In an overnight run he carried out this calculation twice, once with the full program including relativistic effects and a second time "with relativity turned off." Standish "turned off relativity" by setting the speed of light to 10^{10} times its measured value, effectively making light speed infinite. (By combining equation [5], page 2-14, with equation [10], page 2-19, we can show that the Schwarzschild curvature factor in conventional units is written $(1 - 2GM_{kg}/rc^2)$; the value of this expression approaches unity for a large value of c.) For each of the two runs, the perihelia of the four inner planets were computed for a series of points in time covering the four centuries. The results from the nonrelativistic run were subtracted from those of the relativistic run, revealing advances of the perihelia per century accounted for only by general relativity. The second column of Table 1 shows the results, together with the estimated computational error. Later columns show additional data on these planets.

Table 1 Advance of the perihelia of the inner planets

Planet	Advance of perihelion in seconds of arc per century (JPL calculation)	Radius of orbit in AU*	Period of orbit in years
Mercury	42.980 ± 0.001	0.38710	0.24085
Venus	8.618 ± 0.041	0.72333	0.61521
Earth	3.846 ± 0.012	1.00000	1.00000
Mars	1.351 ± 0.001	1.52368	1.88089

*Astronomical Unit (AU): average radius, Earth's orbit; inside back cover.

QUERY 17 **Perihelia advance of the inner planets.** Compare the JPL-computed advances of the perihelia of Venus, Earth, and Mars with results of the approximate formula developed in this project.

11 Checking the Standard of Time

QUERY 18 **Difference between shell and wristwatch times.** Use special relativity to find the fractional difference between satellite wristwatch time τ and the time t_{shell} read on shell clocks at the same radius r_o at which Mercury moves in its orbit at the average velocity 4.8×10^4 meters/second. By what fraction could a change of time from τ to t_{shell} change the total angle covered in the orbital motion of Mercury in one century (equation [14])? Therefore by what fraction could it change the predicted angle of rotation of the major axis?

QUERY 19 **Difference between shell and far-away times**. Find the fractional difference between shell time t_{shell} at radius r_o and bookkeeper far-away time t for r_o equal to the radius of the orbit of Mercury. By what fraction could a change of time from t_{shell} to t change the total angle covered in the orbital motion of Mercury in one century (equation [14])? Therefore by what fraction will it change the predicted angle of axis rotation?

QUERY 20 **Does time standard matter?** From your results of Queries 18 and 19, say whether or not the choice of a time standard (planet proper time τ, t_{shell}, or far-away time t) would make a significant difference in the numerical prediction of the advancement of the perihelion of Mercury in one century. Would your answer differ if the time were measured with clocks on Earth's surface?

12 References and Acknowledgments

Initial quote: Abraham Pais, *Subtle Is the Lord: The Science and the Life of Albert Einstein*, Oxford University Press, New York, 1982, page 253.

Observed value of the advance of the perihelion of Mercury: Irwin Shapiro in *General Relativity and Gravitation, 1989*, edited by N. Ashby et al., Cambridge University Press, New York, 1990, page 313.

Periods and orbital radii in Table 1 from two sources: Kenneth R. Lang, *Astrophysical Data, Planets and Stars*, Springer-Verlag, New York, 1991, and *Landholt-Bornstein, Numerical Data and Functional Relationships in Science and Technology, Group VI: Astronomy and Astrophysics*, Volume 34, *Astronomy and Astrophysics*, Extension and Supplement to Volume 2, Subvolume a, *Instruments, Methods, Solar System*, edited by H. H. Voigt, Springer-Verlag, New York, 1993.

Myles Standish of the Jet Propulsion Laboratory ran the programs on the inner planets presented in Section 10. He also made useful comments on the project as a whole.

Chapter 5 Seeing

- *Can I see a black hole at all?*
- *Does light have its usual speed near a black hole?*
- *Can light orbit a black hole?*
- *What do the stars above me look like as I fall through the horizon?*

CHAPTER 5

Seeing

Tell all the Truth but tell it slant—
Success in Circuit lies
Too bright for our infirm Delight
The Truth's superb surprise

As Lightning to the Children eased
With explanation kind
The Truth must dazzle gradually
Or every man be blind—

—Emily Dickinson, about 1868 (poem 1129)

1 Motion of Light

What do we see when we (finally!) look around?

What can we say about the motion of light around, past, or into a spherically symmetric nonspinning black hole? We ask here no small question. Almost everything that we learn about happenings out in space comes to us by way of light or signals that travel with the speed of light. When we examine happenings near any compact massive object, however, we cannot assume that the evidence we see comes to us without distortion.

Curved spacetime results in a distorted visual view.

Is light deflected as it passes near a center of attraction? If so, how can we trust what we see? Stand on the spherical shell surrounding a black hole. Or ride an unpowered spaceship that orbits the black hole. What do we see? How big does the black hole *look*? What is the apparent position and color of a particular distant star? The entire canopy of stars: Do we see it spread uniformly overhead? We answer these questions by tracing out the trajectory of a light pulse as it approaches a center of gravitational attraction. The results apply to an electromagnetic wave of any frequency.

But first, in the following section, we describe pieces of light trajectories in Schwarzschild geometry as tracked by shell observers, by free-float observers, and by recorded data of the bookkeeper who uses increments of Schwarzschild coordinates dr, $d\phi$, and dt. We recognize that each element of the path of a light pulse racks up zero proper time ($d\tau = 0$) and use this feature of the trajectory to analyze pieces of it. The resulting analysis stops short of what is needed to plot a full trajectory. To do that, it turns out, we need to find a constant of the motion for light. We find this constant of the motion in Section 3.

Look first at "pieces" of light trajectory.

2 Alternative Speeds of Light I

When does light move slower than speed unity?

Light moves at $v = 1$ for free-float observer.

Free-Float Observer Measures $v = 1$ for Light. Our unpowered spaceship plunges radially toward the center of a black hole. Or comfortably speeds around in a stable circular orbit while on-board astrophysicists make observations. While they do so, we measure the speed of light along round-trip paths entirely within our spaceship: port to starboard, fore to aft, floor to ceiling. What will be our experimental results? What is the speed of light in our free-float frame?

Why do you ask such an elementary question this late in the game? The measured speed of light is unity, as always in a free-float (inertial) frame—of course! From the beginning of our study of relativity, the invariance of the speed of light as clocked in overlapping free-float frames and the constancy of its value over time have been key-stones of our analysis.

Yes, $v = 1$ will be the result of all our light-speed experiments carried out inside the unpowered spaceship, whatever its orbit around the black hole. Of course, tidal accelerations within our spaceship require us to limit the spatial extent and time duration of our light-speed experiments. As we plunge and swoop, these tidal accelerations vary, forcing us repeatedly to redefine the spacetime region that we call *inertial* or *free-float*, according to the sensitivity of our experimental equipment (Section 8 of Chapter 1). Nevertheless, with these limitations the speed of light remains unity for us, whatever our free-float orbit around the black hole.

Why bring up the speed of light now? Because our assumption about light speed is about to suffer a jolt. This jolt to our expectations occurs when we examine the predicted bookkeeper speed of light, the changes in reduced circumference dr and tangential displacement $rd\phi$ as a light pulse moves during an increment dt of far-away time.

Schwarzschild Bookkeeper Reckons $v < 1$ for Light. A pulse of laser light flies along a slanting arc toward and around a black hole. Along its path it brilliantly illuminates two detonators lying close to one another. With the touch of light, each detonator instantly explodes. Examine these explosion events. Since both detonators are ignited by the same pulse of light, the proper time between explosion events is zero. Light follows a **null geodesic**, a locally straight worldline with zero wristwatch time between any two events along it. The Schwarzschild metric (equation [10], page 2-19) lets us apply this condition to our two neighboring detonations:

$$d\tau^2 = 0 = \left(1 - \frac{2M}{r}\right)dt^2 - \frac{dr^2}{1 - \frac{2M}{r}} - r^2 d\phi^2 \qquad \text{[1. light]}$$

(We place the label [light] to the right of equations that describe the motion of light. In contrast, from now on the label [$m > 0$] goes to the right of equations that apply to material particles, that is, particles with nonzero mass.)

First, analyze the laser pulse when it is fired radially outward from a spherical shell centered on the black hole. In that case $d\phi = 0$ and equation [1] can be written

Bookkeeper speed of radial light pulse

$$\frac{dr^2}{1 - \frac{2M}{r}} = \left(1 - \frac{2M}{r}\right)dt^2 \qquad \text{[2. light]}$$

or

$$\frac{dr}{dt} = \pm\left(1 - \frac{2M}{r}\right) \qquad \text{[3. light, radial motion]}$$

Carry out for yourself a similar analysis for tangential motion. Note that $rd\phi$ is the tangential displacement, so $rd\phi/dt$ is the tangential bookkeeper velocity. Show that

Bookkeeper speed of tangential light pulse

$$r\frac{d\phi}{dt} = \pm\left(1 - \frac{2M}{r}\right)^{1/2} \qquad \text{[4. light, tangential motion]}$$

Notice the square root on the right side of equation [4] and *no* square root on the right side of [3]. These equations tell a strange story! They show that the bookkeeper's value for the speed of light is less than unity and even depends on the direction of motion: radial or tangential. At great distance r from the black hole, both radial and tangential computed light speeds approach unity, the directly observed speed of light in flat spacetime. However, closer to the black hole this (r, ϕ, t) value for light speed can be less than unity. For the Schwarzschild bookkeeper, one can say, "The speed of light is less than the (unity) speed of light!" In particular the speeds of light as measured by the bookkeeper in both the radial and the tangential directions go to zero at the horizon, $r = 2M$.

Bookkeeper records $v < 1$ for light near the horizon.

WHAT? Throughout the study of relativity everyone claims that the speed of light is an invariant, with value unity, the same for all observers. This principle has given me a lot of trouble, but I have finally accepted it. Now you assert that the speed of light differs from unity near a black hole and, in particular, goes to zero at the horizon. You have succeeded in confusing me all over again. Please make up your mind!

No nearby observer makes a *direct local* measurement of the slowed light speed. Light speeds figured using equations [3] and [4] are *bookkeeper's* accounting entries, not direct reports from local experiments. Actually, these predictions should not be too surprising in view of earlier results. Recall the particle of mass m streaking radially inward across the horizon after starting its fall from a remote point (Chapter 3, Section 5). That particle moves with the speed of light at the horizon as recorded by nearby shell observers. Yet the Schwarzschild bookkeeper—shell data though she uses, converting to dr and dt—records the stone as coming gently to rest at the horizon (equation [21], page 3-13).

Bookkeeper slow light speeds are not measured locally.

The present case is similar. No nearby local observer measures directly the smaller speed of a passing light flash expressed in equations [3] and [4]. The in-falling free-float observer in her flat-spacetime inertial special-relativity capsule can check as often as she wants—and at every stage of her journey—to find that the speed of light is unity, as always in a free-float frame. The shell observer also measures the speed of light to have the value unity (see the following paragraphs). So is the bookkeeper's smaller speed of light totally unmeasurable? And if so, how can it be part of a physics theory?

Bookkeeper light speeds can be measured from a distance.

True, none of our local observers measures directly the smaller speed of light near a center of gravitational attraction as evaluated by the bookkeeper. However, a remote observer can measure this smaller speed—and has already done so! Irwin Shapiro recognized this prediction of smaller light speed past our Sun and in 1979 reported time delays in the propagation of radio waves that graze the surface of Sun, described in Project E, Light Slowed Near Sun.

Incidentally, the lowered bookkeeper speed of light resolves the paradox how an in-faller can cross the horizon at the speed of light (as she measures her speed past local shells) while a remote observer, looking inward, sees the in-faller slowing over a period of time, never reaching the horizon (Section 5 of Chapter 3). The outward-moving signals continue to come out for a long time because of the "slowed bookkeeper speed of light" near the horizon. Kip Thorne calls these signals "relics of the past."

Light speed is $v = 1$ for shell observer.

Shell Observer Measures $v = 1$ for Light. The observer on the local shell through which the light is currently passing also measures the light to have its usual speed: unity. To verify that our mathematics correctly predicts speed unity, recall that the local metric for shell observers is identical in form to the metric for flat spacetime (equation [33], page 2-33):

$$d\tau^2 = dt_{\text{shell}}^2 - dr_{\text{shell}}^2 - r^2 d\phi^2 = dt_{\text{shell}}^2 - ds_{\text{shell}}^2 \qquad [5]$$

For light, $d\tau = 0$ and we simplify with $ds^2_{\text{shell}} = dr^2_{\text{shell}} + r^2 d\phi^2$ to find

$$\frac{ds_{\text{shell}}}{dt_{\text{shell}}} = \pm 1 \qquad [6.\ \text{light}]$$

Hence the speed of light is unity as measured locally on *all* shells, down to the horizon. Typically a light beam curves near a black hole, as we see later in this chapter. Still, along a small segment of path every shell observer measures every passing light flash to move at speed unity.

Measure shell distances with radar.

Shell light speed equal to unity has an important consequence for measurements by shell observers: They can use radar to measure local shell distances. Send a flash of light (or a radar pulse) in some direction, time its return after reflection from a nearby object, and assume that the distance in meters to the reflecting object equals half the round-trip time in meters as recorded on the local shell clock. No more worry about stretching or flexing meter sticks! Note, however: Radar pulses that do not move radially will also return the correct *magnitude* of distance of nearby objects along the curved radar trajectory in space, but the *direction* to the reflecting object may be distorted by bending of the radar path in curved spacetime geometry, as described in Sections 5 and 6.

Shell light speed equal to unity can also help us to understand better the slowed light speed reckoned using Schwarzschild coordinates. Think of two events along the path of a light flash. The tangential bookkeeper velocity of light between these two events is $(1 - 2M/r)^{1/2}$ because for the bookkeeper the light covers the same tangential distance $rd\phi$ as for the shell observer, but in a longer lapse of time, $dt = dt_{\text{shell}}/(1 - 2M/r)^{1/2}$. For the radial motion of light not only is the bookkeeper lapse of time longer between these two events than for the shell observer, but also the radial distance between them is smaller: $dr = dr_{\text{shell}}(1 - 2M/r)^{1/2}$. The result is a bookkeeper radial speed of $(1 - 2M/r)$.

3 Orbiting Light

Treat a light flash as a very fast particle with vanishingly small mass.

Where do we look to see the familiar stars?

The stars! The stars! What do the stars look like as we stand on a spherical shell near the horizon of a black hole? Where do we look to see Sirius, the Dog Star, as we careen in free float through the spherical shells around a black hole? The Schwarzschild bookkeeper tells us to look in what direction to see remnants of the latest supernova in Andromeda? And if we turn our telescope in the direction she indicates, will we see it?

To answer these questions we trace inward from each star, each quasar, each supernova the long trajectory, the single thin pencil of light that spirals in toward the black hole so precisely as to enter the iris of our eye as we stand on a spherical shell near the horizon. The goal now is to connect the pieces of light paths described in Section 2 into trajectories girdling the black hole and plunging across its horizon. Describing complete trajectories allows us—at last!—to predict what we see when we raise our eyes to behold the heavens from the vicinity of a black hole.

Total aging along light trajectory is zero.

To track the path of light, we return to the Principle of Extremal Aging for material particles, the fundamental rule that each particle with mass moves from an initial fixed event to a final fixed event along a worldline chosen to maximize (or in general to make extreme) the time lapse on the wristwatch carried by the particle. This extremum principle we cannot apply directly to light, and for an elementary reason—the aging of a light pulse is automatically zero! Each element of the worldline has equal space and time parts as observed in a local free-float frame (Figure 1). Therefore the aging along each element of its worldline is zero. Aging is zero also along the entire worldline! The total aging of a light flash from creation to annihilation is zero. How can we possibly apply the Principle of Extremal Aging to a light pulse whose aging is automatically zero?

Treat light pulse as fast particle with vanishingly small mass.

Answer: Sneak up on it! There is nothing about the motion of a light pulse in a vacuum that cannot be discovered by analyzing a sequence of material particles, each one moving faster than the previous one, each with less mass than the previous one. Arrive in this way at a particle of vanishingly small mass but with speed approaching that of light. We shall see that, in

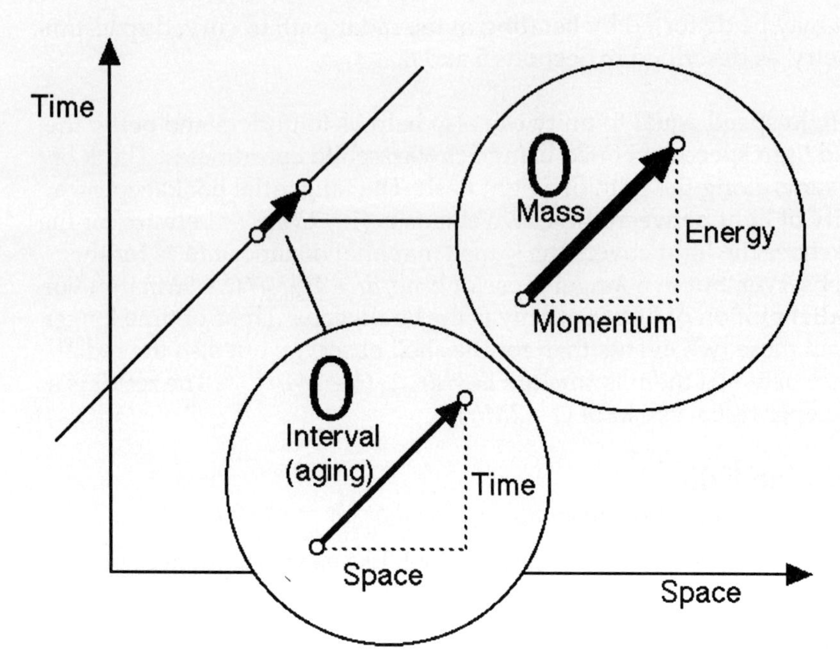

Figure 1 *Worldline of a light flash with respect to a free-float (inertial) frame, showing its unit slope in spacetime. Insets: Unit slope of worldline means equal space and time separations between events along this worldline, hence zero interval between them—and zero aging for the light flash (equation [30], page 1-12, with $d\tau = 0$). Momentum-energy of the same light flash, also with unit slope, symbolizing three properties of light: It has zero mass, it travels with light speed, and it has a momentum identical in magnitude to its energy (equation [32], page 1-13, with $m = 0$).*

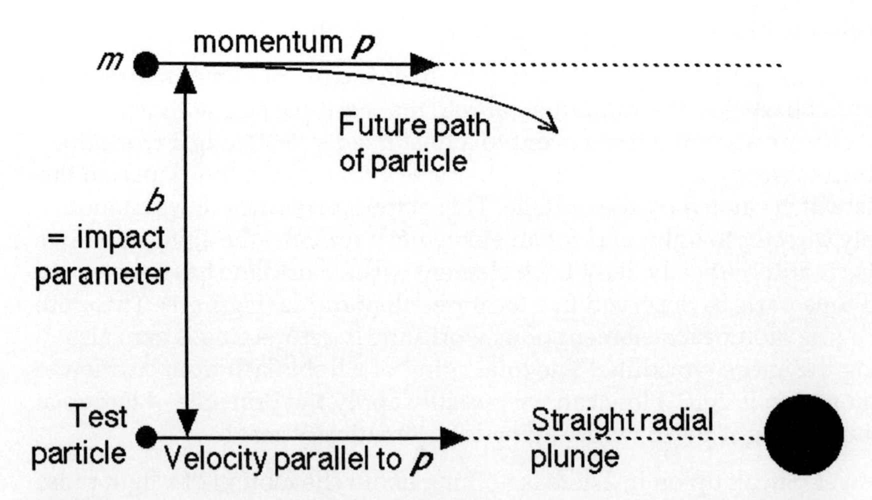

Figure 2 *Impact parameter b defined. A fast particle (mass m) approaches a black hole from a great distance with vector momentum **p**. Find a test particle with a parallel velocity that plunges radially—without deflection—into the black hole. The perpendicular distance b between their initially parallel paths (at this great distance) is called the impact parameter b.*

this limit, motion of such a particle mimics that of the light flash. Now for the details.

A very fast particle—still of mass m—about to pass a black hole is initially at a great distance from it (Figure 2). Spacetime in the local neighborhood of the particle is effectively flat. The particle has a momentum vector $\boldsymbol{p}$ along its line of motion. By trial and error, find a second line of motion, initially parallel to the first, such that a test particle moving along this second line plunges radially into the black hole, deviating neither one way nor the other as it does so. Our original particle and the test particle start along parallel tracks from our remote location in flat spacetime. The perpendicular distance between these tracks at this remote location is called the **impact parameter** and given the symbol b (Figure 2).

Impact parameter b defined

The box on page 5-8 uses the impact parameter and the sneak-up-on-it strategy to derive the equations of motion for light.

4 Alternative Speeds of Light II

Bookkeeper: Different light speeds in different directions.
Shell observer: Light speed always v = 1.

Equations [14] and [15] in the box on page 5-8 give the radial and tangential components of velocity as calculated by the remote bookkeeper, who reckons everything in reduced circumference r, far-away time t, and azimuthal angle ϕ. (For radial motion, $b = 0$, compare these results with equation [3].) Both radial and tangential velocity components get smaller and smaller as the light flash gets closer and closer to the event horizon at $r = 2M$. Now we can find the speed of light for any impact parameter. Square both sides of equations [14] and [15] in that box, add their respective sides, and take the square root of the result, giving equation [16]:

Bookkeeper light speed is a function of impact parameter and r.

$$\begin{pmatrix}\text{light speed}\\ \text{reckoned by}\\ \text{bookkeeper}\end{pmatrix} = \left[\left(\frac{dr}{dt}\right)^2 + \left(r\frac{d\phi}{dt}\right)^2\right]^{1/2}$$

$$= \left(1 - \frac{2M}{r}\right)\left[1 + \frac{2Mb^2}{r^3}\right]^{1/2} \qquad \text{[16. light]}$$

At great distance r from the black hole (in the limit $2M/r$ —> 0), this computed light speed approaches unity, the directly observed speed of light in flat spacetime. However, closer to the black hole this r, ϕ, t coordinate light speed decreases, approaching zero at the horizon.

No local observer measures directly the slower light speed. It is a bookkeeper's accounting entry. Nevertheless the prediction has been verified by remote observers employing radio waves that graze the surface of Sun, as described in Project E, Light Slowed Near Sun.

Two observers who *do* measure the speed of light directly are the shell observer and the plunging observer. We have already seen (equation [6]) that the shell observer measures the speed of light to be unity. We can find

Shell and plunging observers measure v = 1 for light.

Motion of Light in Schwarzschild Geometry

By definition, the magnitude of the angular momentum L for the particle in flat spacetime, shown as the upper particle in Figure 2, is its linear momentum p_{far} far from the black hole multiplied by the impact parameter b, thus: $L = b\, p_{far}$. Hence the ratio of angular momentum to linear momentum L/p_{far} is just the impact parameter b. In this region of flat spacetime we can use the expressions of special relativity to find b for a particle of mass m (equation [32], page 1-13). In flat spacetime, $m^2 = E^2 - p^2$ and thus $p = (E^2 - m^2)^{1/2}$. Energy in this expression is just the total energy, the energy measured at infinity E, because the particle is effectively *at* infinity. Therefore

$$b = \frac{\begin{pmatrix}\text{angular}\\ \text{momentum}\end{pmatrix}}{\begin{pmatrix}\text{linear}\\ \text{momentum}\end{pmatrix}} = \frac{L}{p} = \frac{L}{[E^2 - m^2]^{1/2}} \qquad [7.\ m > 0]$$

Now, both angular momentum L and energy measured at infinity E are constant (Chapter 4) as this ultrafast particle (this almost-but-not-quite light flash) moves from far away and approaches the black hole. Therefore the right side of equation [7] is constant during this motion. Hence the left side, the numerical value of b, the impact parameter measured at infinity, also remains constant as the particle swoops in toward the center of attraction.

Note: Both the impact parameter measured at infinity of a satellite and its energy measured at infinity can be measured directly only at infinity. Closer in, where spacetime is curved, a shell observer must infer their at-infinity values from local measurements (see Sample Problem 1 on page 5-10 and Sample Problem 3 on page 4-22).

For the very fast particle we have chosen to mimic the light flash, hold the energy and the angular momentum constant while letting the mass m approach zero. Equation [7] becomes

$$b_{light} = \lim_{m \to 0} \frac{\begin{pmatrix}\text{angular}\\ \text{momentum}\end{pmatrix}}{\begin{pmatrix}\text{linear}\\ \text{momentum}\end{pmatrix}} = \frac{L}{E} \qquad [8.\ \text{light}]$$

The original equations of motion for a particle of mass m are equations [20], [21], and [22] on page 4-9:

$$\left(\frac{dr}{d\tau}\right)^2 = (E/m)^2 - \left(1 - \frac{2M}{r}\right)\left[1 + \frac{(L/m)^2}{r^2}\right] \qquad [9.\ m > 0]$$

$$\frac{d\phi}{d\tau} = \frac{(L/m)}{r^2} \qquad [10.\ m > 0]$$

$$\frac{d\tau}{dt} = \frac{\left(1 - \frac{2M}{r}\right)}{(E/m)} \qquad [11.\ m > 0]$$

We must rewrite equations [9] and [10] without the proper time $d\tau$ because elapsed proper time is zero for the light flash. Eliminate proper time by multiplying equations [9] and [10] through by the appropriate power of $d\tau/dt$ from equation [11], yielding the two equations

$$\left(\frac{dr}{dt}\right)^2 = \left(\frac{dr}{d\tau}\right)^2\left(\frac{d\tau}{dt}\right)^2$$
$$= \left(1 - \frac{2M}{r}\right)^2 - \left(1 - \frac{2M}{r}\right)^3\left[\frac{m^2}{E^2} + \frac{1}{r^2}\left(\frac{L}{E}\right)^2\right] \qquad [12.\ m>0]$$

$$\frac{d\phi}{dt} = \frac{d\phi}{d\tau}\frac{d\tau}{dt} = \left(\frac{L}{E}\right)\frac{\left(1 - \frac{2M}{r}\right)}{r^2} \qquad [13.\ m > 0]$$

Now let $m \longrightarrow 0$, the limit indicated by equation [8]. In the limit of vanishingly small mass, the resulting equations describe the motion of light:

$$\frac{dr}{dt} = \pm\left(1 - \frac{2M}{r}\right)\left[1 - \left(1 - \frac{2M}{r}\right)\frac{b_{light}^2}{r^2}\right]^{1/2} \qquad [14.\ \text{light}]$$

$$r\frac{d\phi}{dt} = \pm\frac{b_{light}}{r}\left(1 - \frac{2M}{r}\right) \qquad [15.\ \text{light}]$$

We write the last of these equations with the multiplier r on the left because the product $r(d\phi/dt)$ is the component of velocity perpendicular to the radial direction as reckoned by the Schwarzschild bookkeeper. Equations [14] and [15] are the equations of motion of a light flash in the neighborhood of a nonspinning black hole, described using bookkeeper coordinates r, ϕ, and t. Describe the complete orbit of the light flash by starting with its initial radius r and angle of motion ϕ, advancing far-away time t and computing corresponding changes in r and ϕ.

Notation: Hereafter we omit the subscript "light" from b, counting on the equation label **["light"]** to remind us that the impact parameter b is for light.

the shell observer's radial and tangential components of the velocity of light by using equations [C] and [D] in Selected Formulas at the end of the book to convert equations [14] and [15] to shell measures of time and space increments:

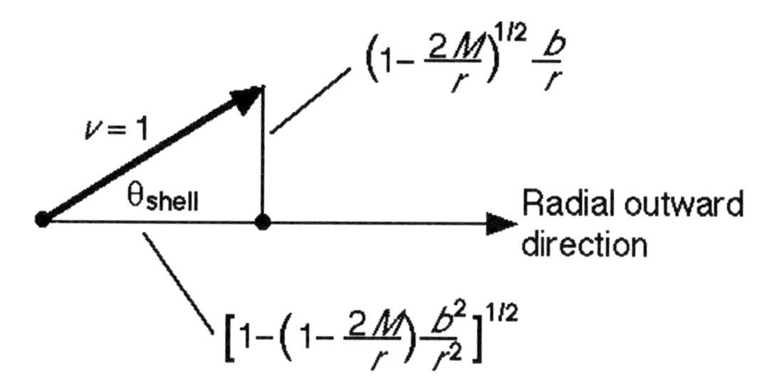

Figure 3 *Light velocity vector and its components as observed by the shell viewer located at radius r. The shell observer measures the speed of light to have the standard value unity.*

$$\frac{dr_{shell}}{dt_{shell}} = \pm\left[1-\left(1-\frac{2M}{r}\right)\frac{b^2}{r^2}\right]^{1/2} \qquad \text{[17. light]}$$

$$r\left(\frac{d\phi}{dt_{shell}}\right) = \pm\left(1-\frac{2M}{r}\right)^{1/2}\frac{b}{r} \qquad \text{[18. light]}$$

Note that the sum of the squares of these components is equal to unity, as shown in Figure 3.

Speed of light equal to unity is also measured by the free-float orbiting or plunging observer, safe inside her capsule of flat spacetime, an unpowered spaceship that orbits the black hole or hurtles radially inward.

5 Forecasting the Trajectory of Light

Give the computer the impact parameter b—and let it crank!

Equations [14] and [15] can be solved for dr and $d\phi$, respectively, and integrated by computer to yield the trajectory of a light pulse described in bookkeeper coordinates r, ϕ, and t. For light coming from a remote source, the impact parameter b is known directly. In contrast, if the light flash is launched from a shell we must derive the value of b indirectly. Sample Problem 1 sets up this procedure.

Plot light trajectory with computer.

Will a laser pulse fired from a shell escape from the black hole? *Any* light flash has enough *energy* to escape if it starts outside the event horizon. However, the pulse may move along a trajectory that crosses the horizon, in which case it will be captured. Escape or capture? So far we have only one way to decide: integrate equations [14] and [15]. Another method, described in Section 6, uses an effective potential to make an instant qualitative judgment about the fate of any flash of light.

Quick analysis of orbits requires effective potential.

SAMPLE PROBLEM 1 The Laser Cannon

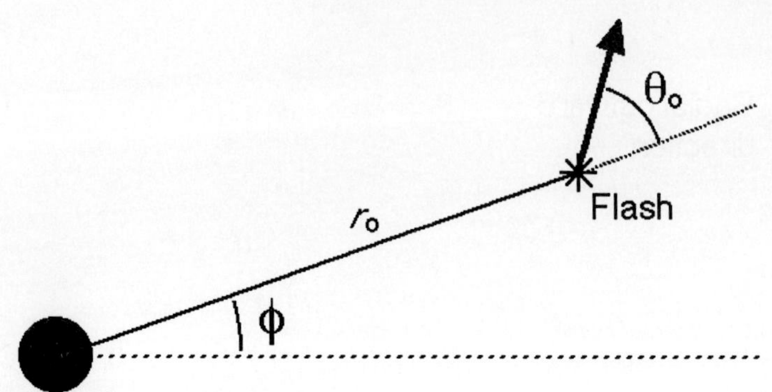

***Figure 4** A flash of light is launched at angle θ_0 = 30 degrees from the radially outward direction as measured in shell coordinates.*

Shell Governor on the spherical shell with reduced circumference $r_0 = 10M$ fires a laser pulse at angle θ_0 = 30 degrees from the radially outward direction as measured in his shell coordinates. Set up the equations for computing the trajectory of the light pulse.

SOLUTION

We need the value of the impact parameter b. The problem statement does not tell us its value measured at infinity (Figure 2). Instead, we obtain the value of b by equating the tangential component of the speed of light (perpendicular to the radius) from equation [18] to that computed from the angle θ_0 in Figure 4. With a shell speed of light equal to unity, we have

$$1 \times \sin\theta_0 = r_0\left(\frac{d\phi}{dt_{\text{shell}}}\right) = \left(1 - \frac{2M}{r_0}\right)^{1/2}\frac{b}{r_0} \qquad \text{[19. light]}$$

from which

$$b = \frac{r_0 \sin\theta_0}{\left(1 - \frac{2M}{r_0}\right)^{1/2}} = \frac{10M \sin 30^\circ}{\left(1 - \frac{2M}{10M}\right)^{1/2}} \qquad \text{[20. light]}$$

$$= \frac{10M \times 0.500}{(0.8)^{1/2}} = 5.59M$$

Adapt equations [14] and [15] to compute the orbit step by step:

$$dr = \pm\left[\left(1 - \frac{2M}{r}\right)^2 - \left(1 - \frac{2M}{r}\right)^3 \frac{b^2}{r^2}\right]^{1/2} dt \qquad \text{[21. light]}$$

$$d\phi = \pm\frac{b}{r^2}\left(1 - \frac{2M}{r}\right)dt \qquad \text{[22. light]}$$

The plus/minus signs in these two equations result from the square roots in their derivation. In each concrete application, the initial signs are chosen from the value and orientation of the launch angle θ_0 in Figure 4. [See *Caution* about ambiguities in angle in Sample Problem 3, page 5-20.]

Equations [21] and [22] apply to light moving under the influence of a spherically symmetric center of attraction. In order to plot the trajectory for the particular case given in this sample problem, substitute the value $b = 5.59M$ and integrate numerically by computer.

6 Effective Potential for Light

Orbits of light at a glance!

How to define effective potential?

Is there some way to set up an effective potential for light in order to visualize its trajectory simply and directly, the way we did in Chapter 4 for particles with mass? What is an effective potential, anyway? To define an effective potential for a particle with mass, we earlier made use of the equation for the radial motion (equations [30] and [31], pages 4-15 and 4-16):

$$\left(\frac{dr}{d\tau}\right)^2 = \left(\frac{E}{m}\right)^2 - \left(1 - \frac{2M}{r}\right)\left[1 + \frac{(L/m)^2}{r^2}\right]$$
$$= \left(\frac{E}{m}\right)^2 - \left(\frac{V(r)}{m}\right)^2 \qquad [23.\ m > 0]$$

Equation [23] gives us a recipe for the effective potential. On the left side of equation is a measure of the radial velocity of the particle. On the right side is the algebraic difference between two terms: The first term is a constant of the motion, independent of the position of the particle. The second term depends on the radius.

Try to find a similar equation for the motion of a light flash. The analogous measure of radial motion of such a pulse is given by equation [14]:

$$\left(\frac{dr}{dt}\right)^2 = \left(1 - \frac{2M}{r}\right)^2 - \left(1 - \frac{2M}{r}\right)^3 \frac{b^2}{r^2} \qquad [14.\ \text{light}]$$

Shell radial equation leads to effective potential.

Equation [14] does *not* meet the requirement that the first term on the right be a constant independent of the radial position of the particle. Therefore this equation as it stands *cannot* be used to define an effective potential. However, if we look instead at equation [17], we do see a constant first term on the right-hand side:

$$\left(\frac{dr_{\text{shell}}}{dt_{\text{shell}}}\right)^2 = 1 - \left(1 - \frac{2M}{r}\right)\frac{b^2}{r^2} \qquad [17.\ \text{light}]$$

In this equation, the "constant of the motion" is the impact parameter b. Put b into the first term on the right by dividing both sides of the equation by b^2:

$$\frac{1}{b^2}\left(\frac{dr_{\text{shell}}}{dt_{\text{shell}}}\right)^2 = \frac{1}{b^2} - \frac{\left(1 - \frac{2M}{r}\right)}{r^2} \qquad [24.\ \text{light}]$$

Effective potential for light

The left-hand side of this equation is a (rather strange!) measure of the radial velocity of the particle. The first term on the right-hand side depends, through b, on the choice of orbit but not on the Schwarzschild geometry. The second term on the right depends on the Schwarzschild geometry but not on the choice of orbit. This second term acts like the square of an effective potential:

$$\left(\begin{matrix}\text{effective potential}\\ \text{for a light flash}\end{matrix}\right)^2 = \frac{\left(1 - \frac{2M}{r}\right)}{r^2} \qquad [25.\ \text{light}]$$

Only one effective potential for light of ALL frequencies

The expression [25] for effective potential makes no reference to the energy of the light or its impact parameter b. Therefore it applies to light of all wavelengths. Only one effective potential is needed to analyze the motion of all light (including radio waves, radar pulses, and gamma rays)! A plot of this square of the effective potential is shown in Figure 5.

Using this effective potential we can simply and quickly predict the major features of light motion around a nonspinning spherically symmetric center of attraction.

Qualitative predictions are easy, even with strange measure of radial motion.

Equation [24] has one obvious drawback: The left-hand side expresses radial velocity in shell coordinates (with an extra coefficient $1/b^2$), rather than the accustomed Schwarzschild bookkeeper's coordinates r, ϕ, and t. The good news is that shell velocity is "real," the possible result of a local measurement. The bad news is that shell coordinates are only local coordinates, usable only over a small range of radial coordinates r. *Question:* Of what use are local shell measures of velocity? *Answer:* Equation [24] and the resulting graphical plots help us to gain a qualitative understanding of the motion of light even on a global level. They allow us to make statements such as the following: "Initially the light pulse moves to smaller radius." "At a particular radius r the radial component of velocity of the flash goes to zero, so the flash moves tangentially, perpendicular to the radial direction." "Finally the light pulse moves to larger radius again." Once we have such a qualitative understanding of the motion, we can use computations based on equations [21] and [22] to assemble a wider-reaching Schwarzschild bookkeeper's accounting of the trajectory expressed in coordinates r, ϕ, and t.

Different kinds of light trajectories for different values of b

Figure 5 shows the square of the effective potential for light. It has a maximum at $r = 3M$ and a value $1/(27\,M^2)$ at this maximum. Horizontal lines represent various possible values of $1/b^2$, where b is the impact parameter. According to equation [24], if the light beam has a value of $1/b^2$ greater than the peak of the effective potential (small enough impact parameter b), the light is captured by the black hole. In contrast, if $1/b^2$ has a value less than the peak of the effective potential (large enough impact parameter b), then the inward component of the light velocity goes to zero at the radius r for which the value of $1/b^2$ is equal to the effective potential. In this case the light subsequently moves outward again and flees the black hole. Figure 6 presents these results in another form.

Light orbits on a knife edge at $r = 3M$.

Finally, if $1/b^2$ is just equal to the peak of the effective potential—in other words, if $b = b_{critical} = (27)^{1/2}\,M = 5.20\,M$—then the light pulse stops its radial motion for some time at a coordinate radius $r = 3M$. But the tangential motion does not stop; the light moves for a while in a circular orbit. The light flash may stay at this radius for a fraction of an orbit or for many orbits, teetering on a knife edge before making a choice: return outward to a great distance or plunge on into the black hole. Which way it goes may seem random, because the choice is extremely sensitive to details of the way the light flash arrived in this orbit—similar to the question of which way a pencil balanced on its point will fall.

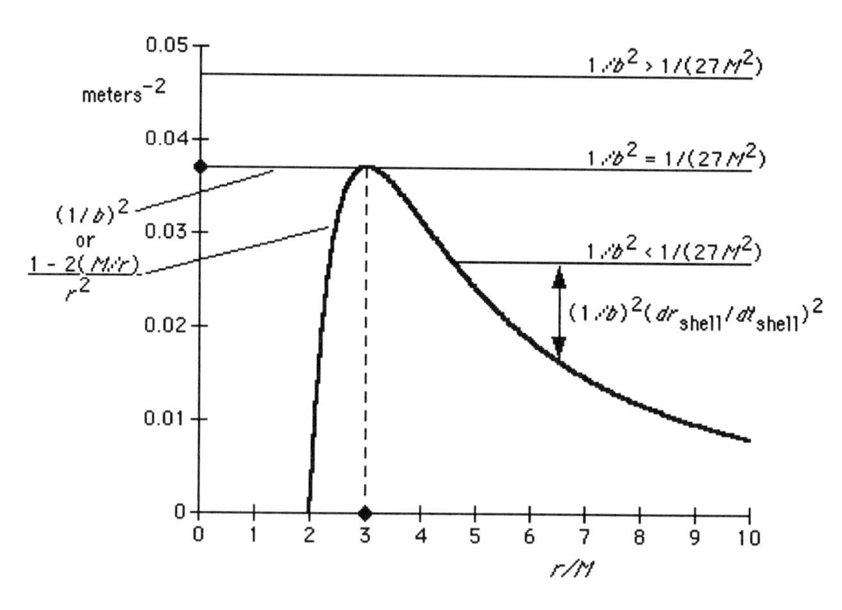

Figure 5 *Computer output: Square of the effective potential for light near a black hole. The same effective potential holds for light of all frequencies. There is no minimum in this potential, therefore no stable circular orbit for light. Trajectories can be described using the horizontal lines corresponding to different values of the quantity* $1/b^2$*. For a small value of the impact parameter b (large enough value of 1/b: top horizontal line), the light enters the black hole. For a large value of the impact parameter b (small enough value of 1/b: bottom horizontal line), the in-falling light reverses its radial component of motion and escapes to infinity. For the critical impact parameter* ($b_{critical} = 27^{1/2}M = 5.20M$*, line grazing the top of the effective potential) the light enters a circular knife-edge orbit of radius* $r = 3M$ *and may orbit the black hole for part of an orbit or for many turns before it escapes or plunges. Figure 6 shows schematically these three orbits themselves. (Note: Here we plot the* **square** *of the effective potential, whereas in the figures in Chapter 4 for particles with mass we plotted effective potential itself.)*

7 Schwarzschild Maps of the Motion of Light

The larger view that no observer observes!

A light flash moving under the influence of a spherically symmetric center of attraction of given mass M has an orbit whose size and shape, praise be, depends on only a single quantity, the impact parameter b.

Size and shape of orbit depends only on b.

The trajectory of a light flash near a black hole lends itself to a simple description using the effective potential. For example, Figure 6 is what we call a **Schwarzschild map** of the orbits of light for three sample values of the impact parameter b. The Schwarzschild map shows three light trajectories as a function of Schwarzschild bookkeeper coordinates r and ϕ. Figure 5 traces the radial motions along three such trajectories using the effective potential.

Describe orbits with "Schwarzschild map."

From these figures we can derive a qualitative description of the trajectory of any light pulse, no matter what the value of its impact parameter b. The more formal—and accurate—Schwarzschild map of the trajectory comes from integrating equations [14] and [15] themselves.

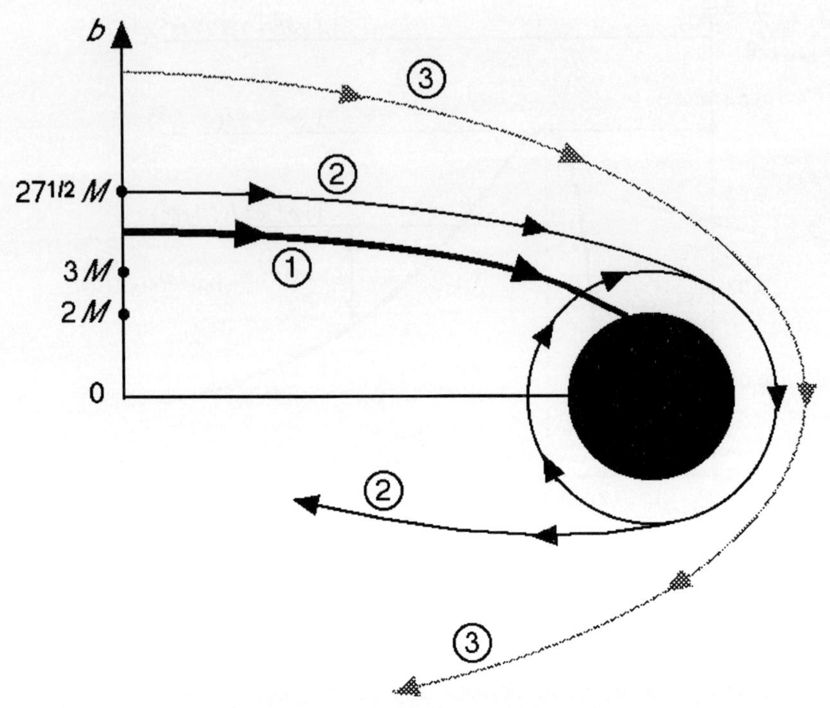

***Figure 6** **Schematic** Schwarzschild map of light trajectories around a black hole for the three values of the impact parameter b shown in Figure 5, page 5-13.*

1. *Light is captured for b less than the critical value $27^{1/2}$ M.*

2. *For critical impact parameter, light teeters on unstable circular orbit at r = 3M. Eventually the light will plunge into the black hole—or escape to infinity, as shown.*

3. *For larger impact parameters, the trajectory is deflected but light is not captured.*

SAMPLE PROBLEM 2 Escaping Light Flash?

Does the laser pulse described in Sample Problem 1, page 5-10, escape the black hole?

SOLUTION

That pulse had an impact parameter $b = 5.59\ M$. This value is greater than the critical impact parameter $b_{critical} = (27)^{1/2}\ M = 5.20\ M$. So that pulse escapes from the black hole.

The simplicity and power of the effective potential is witnessed by the brevity of this sample problem, which is the shortest in the book!

Schwarzschild map does not predict what shell observer sees.

Figures 5 and 6 do *not* tell us what we would see if we stood on a spherical shell near a black hole nor what color light we would perceive. Those figures focus on a Schwarzschild map, a plot artificially constructed from the accounting entries of the Schwarzschild bookkeeper, using coordinates r, ϕ, and t. The shell observer does not agree with the Schwarzschild bookkeeper about the direction of motion of these light beams. He does not even agree with the bookkeeper on the value of the speed of light! What we actually see as we stand on a shell is the subject of the following section. Here we explore Schwarzschild maps of some additional trajectories of light as plotted by the far-away bookkeeper.

As a specific example, think of light beams from different directions converging on a point at r-coordinate $r = 3M$, symbolized by a small open circle in Figure 7. Beams from a distant star located along the horizontal line move straight in from the right. Beams coming from stars at the left, directly behind the black hole, arrive at this point from both above and below at an angle of 90 degrees to the outward radial direction (radial component of velocity equal to zero where the $1/b^2$ horizontal line grazes the top of the effective potential, as shown in Figure 5). Beams from stars farther forward in angle arrive in directions less than 90 degrees to the outward radial direction, again from both above and below—and indeed from all transverse directions obtained by rotating Figure 7 around its horizontal axis.

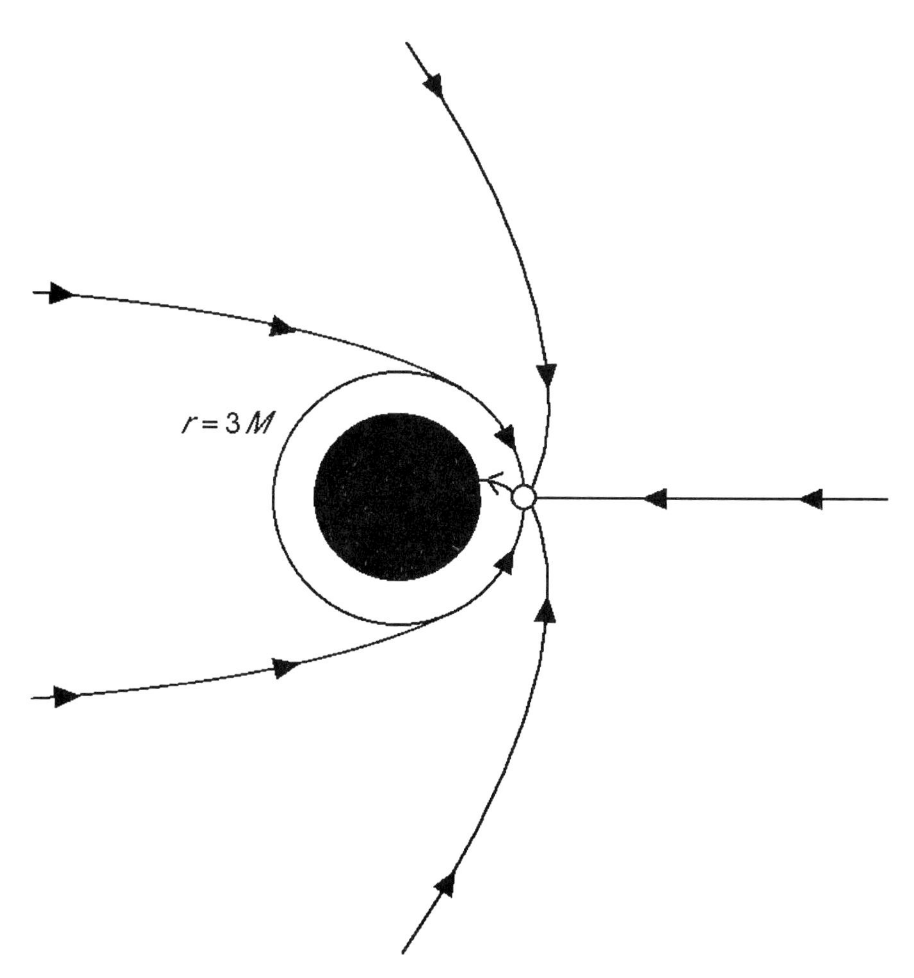

Figure 7 *Schematic Schwarzschild map of trajectories of light that converge on and are absorbed at a point (small open circle) on the shell at r = 3M.*

Shell observer's view of the black hole

What about light arriving from angles greater than 90 degrees from the radially outward direction? One sample light beam is shown leaving the given point at angle greater than 90 degrees and plunging into the black hole. No light can come the other way along the same trajectory because that light would have to come out of the black hole, which light—or anything else—cannot do! Hence a point at $r = 3M$ receives no light from angles greater than 90 degrees from the outward direction. In Section 9 we show that the viewer standing on the spherical shell at $r = 3M$ sees the edge of the black hole in this direction, at 90 degrees from the radially outward direction.

"Halo" around the black hole

The effective potential for light allows us to predict another effect unique to general relativity. Figures 5 and 6 show that the *r*-coordinate $r = 3M$ is the radius for the knife-edge orbit for a light beam approaching the black hole at the critical impact parameter $b_{\text{critical}} = (27)^{1/2} M$. But *some* light along the wave front from *every* star approaches the black hole at this critical impact parameter. Therefore every visible star contributes to a spherical shell of light circling the black hole at the radius $r = 3M$, each beam bound at least temporarily. These beams arrive at that radius from all sides at an angle of 90 degrees in the Schwarzschild map. In the following section we show that the viewer standing on this shell also sees these beams arriving from opposite directions at angles of ±90 degrees from the straight-outward direction. In other words, a viewer stationed at $r = 3M$ sees additional images of all the stars in the sky scattered on a narrow bright ring that extends all around him, transverse to the radially outward direction. This bright ring forms a "halo" around the image of the black hole. In Section 11 of this chapter we meet a similar image, called there an "awesome ring bisecting the sky."

The "light sphere"

These results apply to a reception point at $r = 3M$ on the so-called **light sphere**. For points on other spherical shells, the black hole will also be surrounded by a halo, but light from given stars will arrive at different angles than for $r = 3M$, and the view of the sky can be quite different. For example, Figure 8 describes light from a single star arriving from several directions at a fixed point on the shell, indicated by the small open circle in the figure. Looking around him, the shell observer sees multiple images of this single star in different directions.

8 Schwarzschild Map *vs.* Shell View

Different maps; different directions

Bookkeeper and shell observer disagree on direction of motion of flash.

The observer on a given shell and the Schwarzschild bookkeeper both track the path of a light flash between two events, A and B, that lie near the shell observer. The shell observer tracks this light flash moving past him at an angle θ_{shell} with respect to the radially outward direction (Figure 9). At what angle θ_{Schw} does the Schwarzschild bookkeeper record the light beam to be traveling? Because of the way we define the reduced circumference r (Section 4 of Chapter 2), the two observers agree on the tangential displacement component $rd\phi$ of motion as the light moves outward. However, they disagree about the radial separation between these shells.

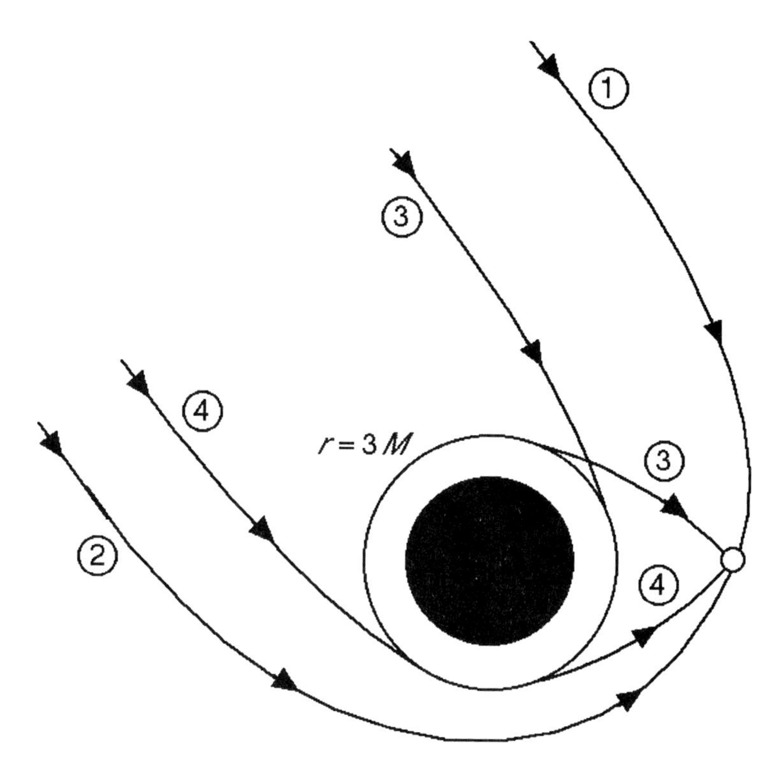

Figure 8 *Schematic (not computed!) Schwarzschild map showing the formation of multiple images of a single star. Light from a single distant star approaches the black hole along effectively parallel paths (shown here coming from the upper left). These beams can arrive at a given visual observer (open circle) along alternative trajectories around the black hole. Of all possible parallel light paths from a given star, four labeled with circled numbers are selected here for examination. Ray 1 is the most direct path arriving at the observer. A second ray from the distant star follows path 2, skirting the black hole on the opposite side from the first ray but closer to the black hole and therefore bent more. Incoming ray 3 circles the black hole once or many times clockwise near the radius 3M. Most of the light in this ray eventually falls into the black hole or escapes outward to infinity along various trajectories. But some small fraction of the light in ray 3 escapes along a trajectory that arrives at the observer. Likewise, incoming ray 4 circles the black hole counterclockwise and makes a fourth image of the distant star for the observer. In brief, the shell observer sees multiple images of the same star in several directions (see Section 8). Double images of distant galaxies corresponding to paths 1 and 2 in the map above have been observed from Earth. In each such case the attractive gravitational center between us and the imaged quasars is believed to be a low-luminosity galaxy or cluster of galaxies. (See the Einstein ring in Figure 14).*

The brightness of each of these different star images seen by the observer depends on the focussing properties of space near each trajectory. Do nearby rays converge or diverge along each path? Answering such questions is beyond the scope of this book.

The Schwarzschild bookkeeper reckons the angle to be given by the equation (from Figure 9):

$$\tan\theta_{\text{Schw}} = \frac{r\,d\phi}{dr} \qquad [26]$$

The shell observer claims that the radial separation between shells is not dr but rather dr_{shell}. So for the shell observer the tangent of the angle is

$$\tan\theta_{\text{shell}} = \frac{r\,d\phi}{dr_{\text{shell}}} \qquad [27]$$

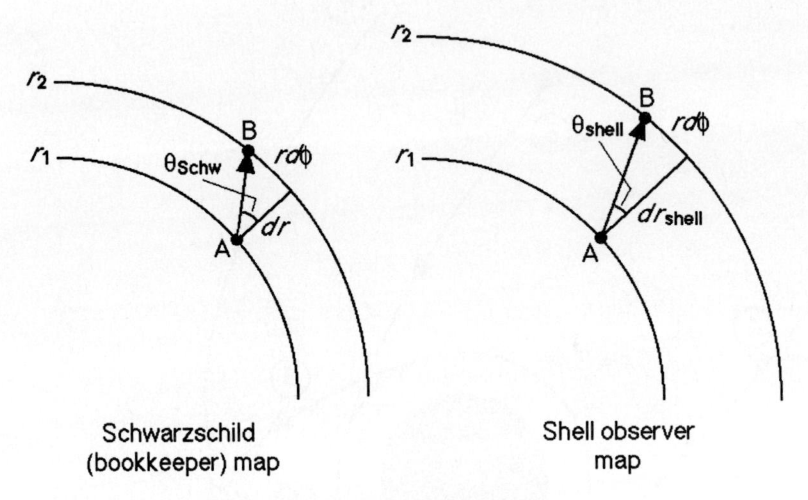

Figure 9 *Angle of light motion for bookkeeper and shell observer compared. Light flash moves from event* A *to event* B. *Each observer measures an angle of travel* θ *with respect to the radially outward direction. They agree that the tangential displacement is* $rd\phi$ *during this travel between adjacent shells. However, they disagree on the radial distance between these shells. The radial distance is greater as measured by the shell observer. As a result, the shell observer measures a smaller angle of motion* θ *with respect to the radially outward direction.*

But we know that the relationship between the two measures of radial distances is given by equation [D] in Selected Formulas at the end of the book:

$$dr_{shell} = \frac{dr}{\left(1 - \frac{2M}{r}\right)^{1/2}} \qquad [28]$$

Angles between light trajectories for bookkeeper and shell observer

Therefore the relation between the two tangents is

$$\tan\theta_{shell} = \left(1 - \frac{2M}{r}\right)^{1/2} \tan\theta_{Schw} \qquad [29]$$

Note that when the viewing angle is 90 degrees for the shell observer, the Schwarzschild angle is also 90 degrees, since the tangent of 90 degrees is infinite for both.

In the remainder of this chapter we deal primarily with the shell angle θ_{shell} of light propagation. Plotting the results on Schwarzschild maps such as Figures 7 and 8, however, requires the transformation of angles given by equation [29].

Relation between angles is general.

Note that the derivation of equation [29] does not depend on a light flash as the traveling object. It could be any material particle moving between some other events A and B. It could even be a fishing line stretched between the two points at which these events occur. All these cases lead to the same result, because a light beam or material particle that skims along a fishing line as observed in one reference frame must skim along the same fishing line as observed in all reference frames. Equation [29] is a general relationship between Schwarzschild and shell angles. Hence none of the equations [26] through [29] carries the label [light].

9 Outward View of the Stars

The shell observer finally looks around.

Using the results so far derived, we can now—finally!—stand up, look around, and view by eye both the black hole (Sample Problem 3) and the starry heavens around us. In the following we use the words *viewer* and *spectator* to describe the experience of someone who *sees*—who *looks* with his or her eye or takes photographs with a camera. As always, the word *observer* implies someone who does not look but rather *measures*—takes data on events using the local latticework of rods and clocks or summarizes these data, as does the Schwarzschild bookkeeper.

"Viewer" and "spectator": someone who sees by eye

There is a unique relation between the impact parameter b and the direction of any light beam seen by a shell viewer located at reduced circumference r, as shown in Figure 3 on page 5-9:

$$\sin\theta_{\text{shell}} = r\frac{d\phi}{dt_{\text{shell}}} = \left(1 - \frac{2M}{r}\right)^{1/2}\frac{b}{r} \qquad \text{[30. light]}$$

Here, θ_{shell} is the direction of the light beam observed by a shell viewer stationed at reduced circumference r, this angle measured with respect to the radial outward direction. Equation [30] tells us the direction of any light beam as viewed from the shell of any radius r, provided we know the impact parameter b of the beam. The light can move in either direction along the path (unless it crosses the horizon).

Star is seen in direction determined by impact parameter b.

Figure 11 shows a Schwarzschild map of a sample trajectory and emphasizes two angles, ϕ and θ_{Schw}. The angle ϕ indicates the direction in which the shell viewer would look to see the star if no center of attraction were present. The angle θ_{Schw} is the angle at which the light reaches the shell observer in the presence of the black hole *as calculated by the distant Schwarzschild bookkeeper* (not the shell viewer).

The box on page 5-22 displays the derivation of light trajectories that predicts at what angle θ_{shell} *the shell spectator* will see each star as he looks outward, compared with the angle ϕ at which he would look to see the star if the black hole were not present. Figure 13 shows the result of this numerical integration to find ϕ as a function of θ_{shell} for selected values of the shell radius r_0. Notice that these viewing angles are limited to the outward view of distant stars, star images that the shell viewer will see at angles less than 90 degrees from the radially outward direction.

SAMPLE PROBLEM 3 Shell View of the Black Hole

A viewer stands on a spherical shell at reduced circumference r. How much of the sky that he sees surrounding him is covered by the black hole?

SOLUTION

Start with equation [30] and think first of launching a light beam outward at shell angle θ_{shell} (measured from the radially outward direction). From Figures 5 and 6, we know that the critical impact parameter is $b_{critical} = (27)^{1/2}M$ for light that circles the black hole teetering on a knife edge before either plunging or escaping. So the critical angle for launching a beam that will barely escape is obtained by substituting this value of b into [30]:

$$\sin\theta_{\text{shell critical}} = \sqrt{27}\left(1 - \frac{2M}{r}\right)^{1/2}\frac{M}{r} \qquad [31.\ \text{light}]$$

Light launched at less than the critical angle with respect to the radially outward direction escapes from the black hole. Light launched at greater than the critical angle plunges into the black hole. Reverse these motions to discover what the shell observer sees. The shell viewer sees black when looking at angles greater than the critical angle of equation [31]. All the stars of the heavens are seen compressed within a cone of half-angle $\theta_{\text{shell critical}}$. Figure 10 presents "pie charts" whose black segments show the range of angles over which shell viewers at different r-coordinates see the black hole. In three space dimensions each pie-shaped segment becomes a cone, derived by rotating each pie chart about the horizontal axis through its center.

Caution: There is an ambiguity in the application of equations [30] and [31], because the sine function has a value for angle $180^\circ - \theta$ equal to its value for angle θ. Look at Figure 10 and think of two extreme cases. Case I: Very far from the black hole (very large r), equation [30] gives the sine equal to zero, implying a critical angle of 180 degrees, where sine is indeed equal to zero. If you are far from the black hole, you can fire the light flash in any direction—except directly backward 180 degrees at the black hole—and it will escape. Case II: Just outside the horizon the sine approaches zero again. In this case the critical angle is zero. There is no angle—except for a small range outward near zero degrees —at which you can fire the flash and have it escape. Even this angular spread is pinched off at the horizon. The dividing case occurs for $r = 3M$. At this radius, equation [31] tells us, the sine of the critical angle is unity and the critical angle 90 degrees, as shown in Figure 10. For $r > 3M$ find the value of θ (with respect to the *outward* direction) in the range $90^\circ < \theta_{\text{shell critical}} < 180^\circ$. In contrast, for $r < 3M$ find the value of $\theta_{\text{shell critical}}$ in the range $0 < \theta_{\text{shell critical}} < 90^\circ$.

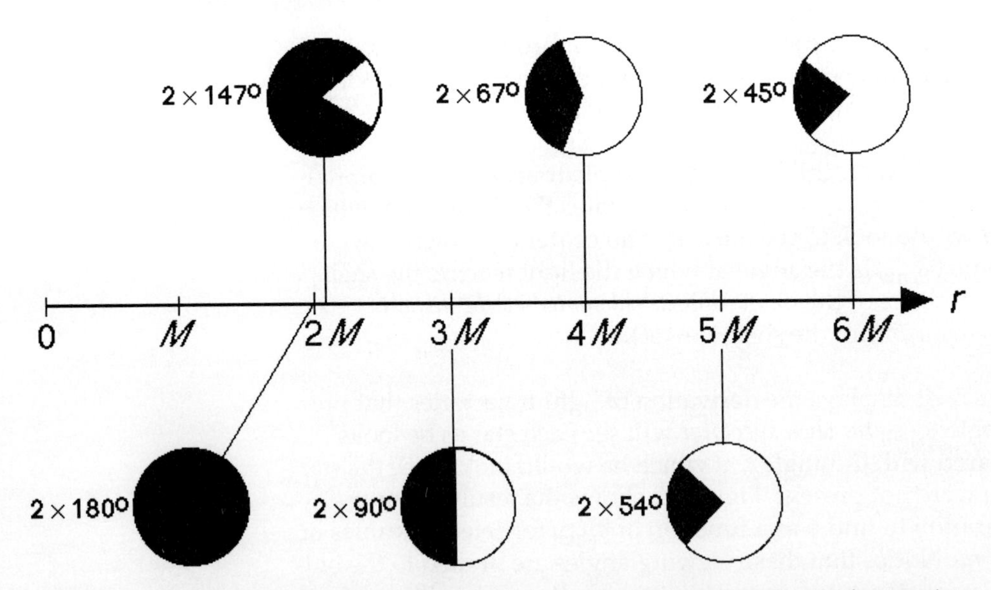

Figure 10 *"Pie charts" of the view from the shell. Each circle symbolizes the panoramic view from a point on a shell at that r-coordinate. Angles listed, such as 2 × 54°, are the span of angles of the black portions, representing the directions in which the shell viewer* sees *the black hole—or rather experiences the absence of light. White portions represent the directions into which rays are compressed arriving from all the stars of the visible heavens. Each pie chart can be made three-dimensional by rotating it about the horizontal axis through its center. We can draw no pie chart for a shell at radius less than 2M, because a shell cannot be built there.*

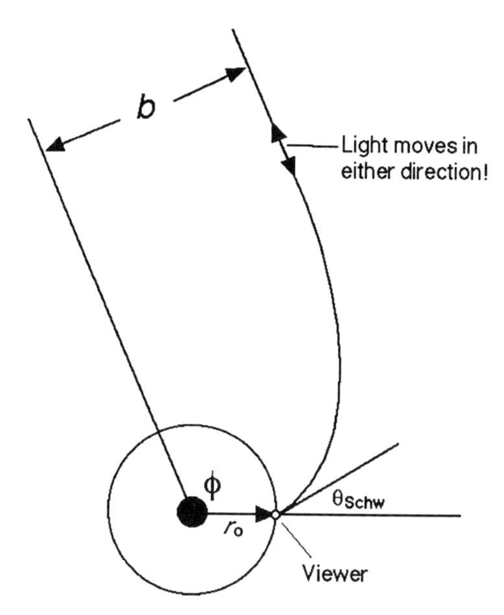

Figure 11 *Schematic Schwarzschild map of the trajectory of light that arrives from a distant star at the location of a viewer at rest on the shell at radius r_o. (It is an equally valid trajectory of a pulse or laser beam fired outward by this spectator.) The impact parameter b is determined by the angle θ_{shell} (with respect to the radially outward direction) at which the light beam intersects the shell. On this Schwarzschild map we use the angle θ_{Schw}, but equation [29] shows the relation between this angle and the angle θ_{shell} directly seen by the shell observer. Equation [30] then shows the relation between θ_{shell} and the impact parameter b. The angle ϕ tells us the direction in which the shell observer would look to see the distant star if there were no center of gravitational attraction nearby. The box on page 5-22 contains a full discussion of the angle ϕ.*

At the upper left of Figure 11 there are two parallel lines. Which of these lines points toward the star?

Both lines point in the same direction, a direction unique to that star. Think of light from the star as a wave. The star is so far away that the incoming wave is essentially flat and perpendicular to the parallel lines at the upper left of the figure, before the wave reaches the black hole. Light from one portion of the incoming wave moves along the straight line—the left one in Figure 11— that plunges radially and undeflected toward the center of the black hole. The shell observer shown in the figure does not see that portion of the incoming wave. Rather, he sees the portion that originates a distance *b* away from that straight trajectory, where *b* is the impact parameter.

Einstein ring

A star that is exactly behind the black hole with respect to any shell viewer will send its light equally around all sides of the black hole—by symmetry! The viewer on a shell near the horizon sees that star as a ring of light that surrounds the image of the black hole. This image is called an **Einstein ring** (see Figure 14, page 5-25). More generally, the black hole acts like a very bad lens, even producing multiple images of background stars. See Project D, Einstein Rings. *(Continued on page 5-24.)*

Looking Outward at the Stars

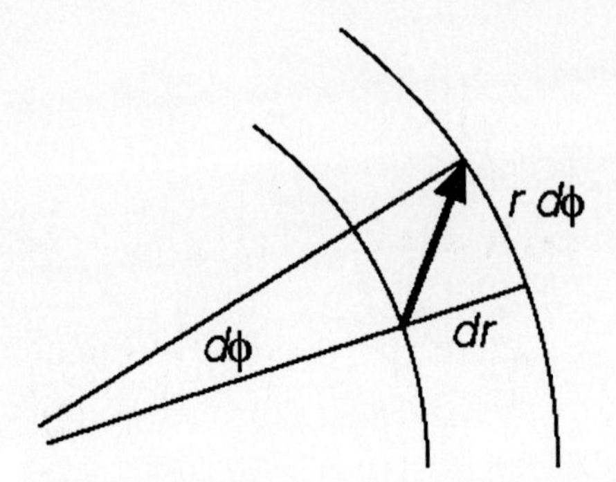

Figure 12 *Small segment of trajectory of light flash emitted by the shell spectator, in Schwarzschild coordinates.*

In what direction do we look to see a star when we stand on a shell centered on a black hole? How does that star direction compare with the direction in which we would see the star if no black hole were present? In this box we find how to calculate answers to these questions.

As long as light does not cross the horizon, it can move in either direction along its predicted trajectory. To simplify the following analysis, we assume that the shell viewer shines a laser beam outward, a narrow beam that ends up traveling along a line directed exactly toward a star (Figure 11). Then the launch direction of the beam will be the direction in which the shell occupant sees the star when light moves in the reverse direction, from star to observer.

For outward motion in the positive ϕ direction, equations [14] and [15], rewritten, read

$$\frac{dr}{dt} = \left(1-\frac{2M}{r}\right)\left[1-\left(1-\frac{2M}{r}\right)\frac{b^2}{r^2}\right]^{1/2} \qquad [32.\ \text{light}]$$

$$\frac{d\phi}{dt} = \left(1-\frac{2M}{r}\right)\frac{b}{r^2} \qquad [33.\ \text{light}]$$

Eliminate dt by dividing corresponding sides of [33] by those of [32] and multiplying the result through by dr:

$$d\phi = \frac{(b/r^2)dr}{\left[1-\left(1-\frac{2M}{r}\right)\frac{b^2}{r^2}\right]^{1/2}} \qquad [34.\ \text{light}]$$

Equation [34] tells us how much the direction of view ϕ changes for a radial change dr of the laser-beam trajectory (Figure 12). We sum (integrate) these changes from the shell observer's radius r_0 to infinite radius in order to find the final direction ϕ in which the laser beam moves. From Figure 3, page 5-9, we see that the denominator of this expression is equal to $\cos\theta_{shell}$, which goes to zero at $\theta_{shell} = 90$ degrees. Therefore we expect trouble in this particular computation when the observer looks tangentially, 90 degrees from radially outward. That is why we limit our analysis in this section to the outward view—viewing at angles less than 90 degrees from the radially outward direction.

To prepare for computation, simplify equation [34] by making the substitution $u = M/r$. Then

$$-\frac{M}{r^2}dr = du \qquad [35.\ \text{light}]$$

so the numerator on the right side of equation [34] becomes

$$(b/r^2)dr = -(b/M)du \qquad [36.\ \text{light}]$$

With this substitution, equation [34] can be written

$$d\phi = \frac{-du}{\left[\frac{M^2}{b^2} - u^2 + 2u^3\right]^{1/2}} \qquad [37.\ \text{light}]$$

The limits of integration are from $u = M/r_0$ to $u = 0$. To carry out this integration, we need the value of b, the impact parameter. Apply equation [30] to the radius r_0 of the shell occupied by the viewer:

$$\sin\theta_{o\ shell} = \left(1-\frac{2M}{r_o}\right)^{1/2}\frac{b}{r_o} \qquad [38.\ \text{light}]$$

Then solve for b/M:

$$\frac{b}{M} = \frac{r_o}{M}\left(1-\frac{2M}{r_o}\right)^{-1/2}\sin\theta_{o\ shell} \qquad [39.\ \text{light}]$$

The calculation goes as follows: Start with an angle of observation $\theta_{0\ shell}$ for a viewer on a shell at radius r_0. Use equation [39] to find the corresponding value of b/M. Substitute b/M into equation [37] and carry out a numerical integration to find the total angle ϕ between the shell spectator's radially outward direction and the "true" direction of the distant star. The angle ϕ tells the direction in which the spectator would see the star if there were no center of attraction under him. Figure 13 shows the result of this numerical integration for selected shell radii r_0.

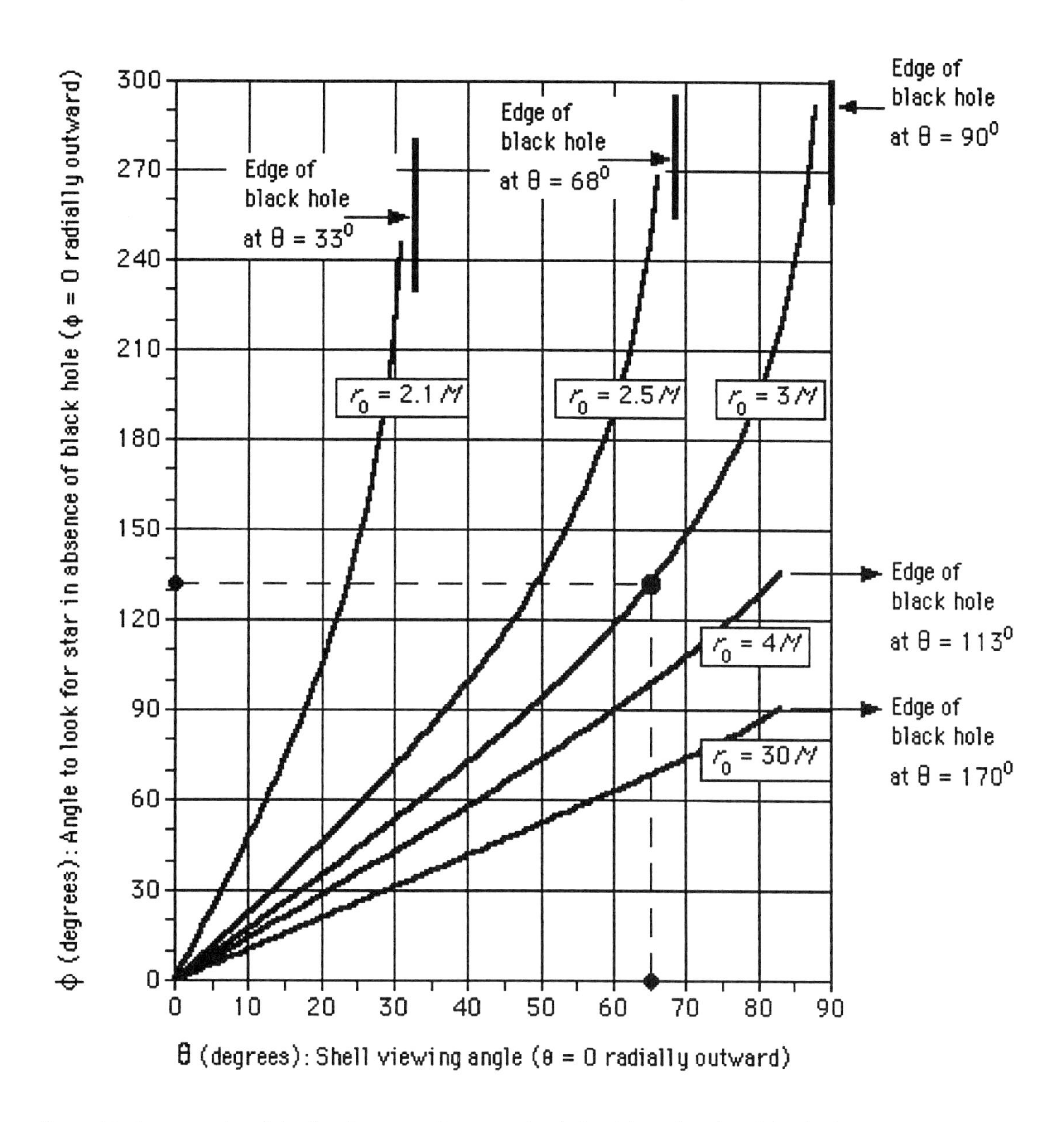

Figure 13 *Computer plot of the direction ϕ to a distant star (vertical scale) as a function of the viewing angle $\theta = \theta_{o\ shell}$ (horizontal scale) at which a spectator sees the star when perched on a shell at radius r_o. See Figures 10 and 11. Both angles are measured from the radially outward direction. Example (dashed lines): You stand on the shell at $r_o = 3M$ and see a star at 65 degrees from the radially outward direction (dot on horizontal axis). If there were no center of gravitational attraction behind you, this star would appear at 132 degrees from the radially outward direction (dot on vertical axis)—a bit behind you. For each value of r_o, the graph also shows the angle at which the edge of the black hole appears to this shell viewer.*

Light: blue shift

In addition to changing direction, light from a distant star also changes energy as seen by a shell inhabitant. Energy of starlight is upshifted (the so-called gravitational blue shift) as seen by the shell spectator. The formula for this energy change is equation [27], page 3-17.

Any astronomical body can act as a distorting lens, as Einstein himself recognized. However, Einstein doubted that we would ever see such an image, because it requires that the imaged star and the "lens star" line up almost perfectly. This alignment can be less perfect if both the distant object and the imaging object are galaxies. Figure 14 shows such an image, an **Einstein ring,** created by radio waves rather than light. The "ring" is actually the distorted image of a distant galaxy focused by an intermediate galaxy lying on a line between the imaged galaxy and our observation point, Earth. Figure 14 is the first Einstein ring ever observed; since then we have seen a number of them (and a large number of ring segments) at various wavelengths of electromagnetic radiation. For more on this subject see Project D, Einstein Rings.

10 Continuing Questions

So many questions, so little time!

This chapter presents the basic description of the motion of light in Schwarzschild spacetime and equations for predicting this motion. Further predictions are worked out in the exercises and in the following projects. Here are a few of the endless number of possible questions about the motion of light—and commentary on their answers.

Inward star view?

What does the star field look like when the shell spectator looks inward, toward the black hole? Section 9 describes the outward view of stars seen from a shell—stars that the shell spectator sees at angles less than 90 degrees from the radially outward direction. One can also ask for a description of the star field viewed inward, toward the black hole. The edge of the black disk appears at the so-called critical angle, given by equation [31] and shown in Figure 10. But trajectories of starlight are more difficult to calculate for the inward view. In addition to these images are those that result from trajectories that wrap once or many times around the black hole before reaching the viewer. (Recall paths 3 and 4 shown in Figure 8.) These trajectories traverse regions of great curvature so that the light may be focused or dispersed along the path. To predict the perceived intensity of star images for these wrapped paths, one must know the path of a pencil-thin beam and how curved spacetime focuses or disperses this nearly parallel beam. Detailed prediction about wrapped paths is too technical for treatment here. See the references.

View of nearby objects?

Where do we look to see *nearby* objects? Our outgoing laser beam finally heads toward a star—or we see the star by light that arrives in our vicinity from the same direction (Figure 11). For this view of *distant* stars, the shell viewing angle at which we see the star is locked uniquely to the direction toward that star in flat spacetime distant from the black hole, as embodied in Figure 13. (Technically the integration in the box on page 5-22 is

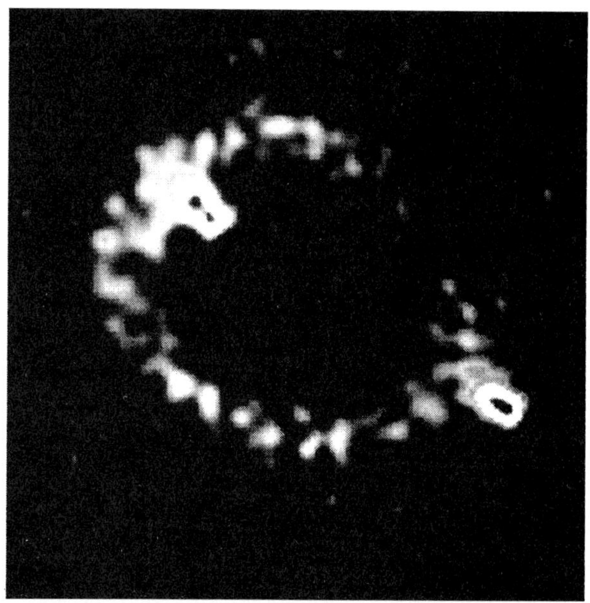

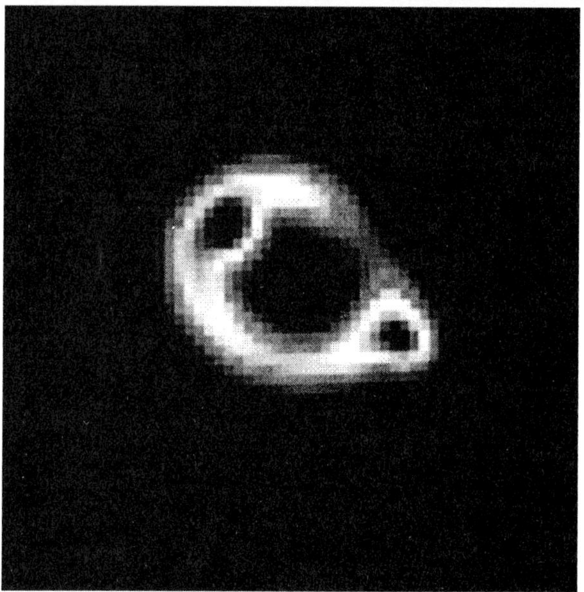

Figure 14 *The first Einstein Ring ever observed. It is the image of a distant radio galaxy detected not with visible light but rather at two different microwave frequencies at the Very Large Array (VLA) radio telescope in Socorro, New Mexico. The distorted view of the galaxy with a "hole in the middle" is due to microwaves deflected around all sides of a foreground galaxy located between the imaged radio galaxy and observers on Earth. The foreground lensing galaxy is a more ordinary galaxy that does not emit appreciably at radio frequencies. Though this foreground galaxy can be seen clearly in optical images as visible light, it is invisible to the VLA. Observations made at two microwave frequencies: 15 gigahertz = 15×10^9 cycles per second (upper image) and 5 gigahertz = 5×10^9 cycles per second (lower image). Data obtained using an array of 27 radio-telescope dishes located along the three legs of a Y, each leg approximately 20 kilometers long. (For credits, see the end of the chapter.)*

carried all the way to $r \longrightarrow \infty$.) The situation is quite different when we view an object lying on a nearby shell. Then the angle at which we see the object depends not only on the angle at which we would look to see it in the absence of the black hole but also on the r-value of the shell on which it lies. (Technically the integration in the box on page 5-22 would be carried out only to the r-value of the object at which we are looking.) In effect, we have to compute the trajectory from each such object separately in order to predict in what direction we must look to see it.

Orbiter view?

In what directions does the orbiting or plunging spectator see images of distant stars? How does the sky look if we orbit near a black hole or plunge radially toward its center? This question is fairly easy to answer for motion outside the horizon, provided one already knows the direction and frequency of light seen by occupants of the shell past which we are currently passing. The transformations of special relativity relate the shell observations to those of an observer moving past or through the shell. These transformations describe the **Doppler shift** (the change in frequency of light seen by a moving observer) and **aberration** (the alteration in direction of motion of light seen by that observer). Suppose we measure angles of observation with respect to the line of motion of the orbiter in the shell frame. Let ψ_{shell} be the direction with respect to that forward line of motion in which the shell spectator looks to see some object and E_{shell} be the shell-measured energy of a light flash. Then the orbiter passing close to the shell observer at speed v_{shell} sees the object in a direction $\psi_{orbiter}$ with respect to this forward direction and observes the light pulse to have energy $E_{orbiter}$ given by the equations

$$\cos\psi_{orbiter} = \frac{\cos\psi_{shell} + v_{shell}}{1 + v_{shell}\cos\psi_{shell}} \qquad \text{[40. light]}$$

$$E_{orbiter} = \frac{E_{shell}(1 + v_{shell}\cos\psi_{shell})}{(1 - v_{shell}^2)^{1/2}} \qquad \text{[41. light]}$$

(These equations derive from those in exercise 8-19 on page 263 of *Spacetime Physics*, Second Edition, taking into account the fact that the observer *looks* in a direction 180 degrees opposite to the direction in which the light *moves*. The result is to replace all cosines in the original equations by minus cosines, leading to equations [40] and [41].)

View from inside the horizon?

What does the sky look like to a spectator who is inside the horizon? Here we are in new territory, where our analysis that depends on shell observations is useless. Inside the horizon, radially inward motion is inevitable and no stationary shell exists. Yet light from the stars also passes inward through the horizon and, in fact, can be seen by plungers headed toward the center. Some of the background needed to analyze the motion of light inside the horizon is given in the earlier Project B, Inside the Black Hole, along with formulas for the final view as the plunger reaches the crunch-point at the center of the black hole.

Baked on the Shell?

As you stand on a spherical shell close to the horizon of a black hole, you will be crushed by an unsupportable local gravitational acceleration directed downward toward the center. If that is not enough, you will also be enveloped by an electromagnetic radiation field. William G. Unruh used quantum field theory to show that the temperature T of this radiation field in degrees Kelvin is given by the equation

$$T = \frac{h g_{\text{conv}}}{4\pi^2 k_B c} \qquad [42]$$

Here g_{conv} is the local acceleration of gravity in the conventional units meters/second^2, h is Planck's constant, c is the velocity of light (values inside the back cover), and k_B is the so-called **Boltzmann's constant,** which has the value 1.381 x 10^{-23} kilogram-meters^2/(second^2 degree Kelvin). The quantity $k_B T$ has the unit joules and gives an average value for the thermal energy this field can provide to local processes. (The same radiation field surrounds you when you accelerate at the rate g_{conv} in flat spacetime.)

In Exercise 9 of Chapter 3, you derived an expression for the local gravitational acceleration on a shell at radius r. This acceleration was expressed in the geometric unit meter^{-1}:

$$g_{\text{shell}} = \frac{g_{\text{conv}}}{c^2} = \frac{M}{r^2} \frac{1}{\left(1 - \frac{2M}{r}\right)^{1/2}} \qquad [43]$$

Substitute g_{conv} from [43] into [42] to obtain

$$T = \frac{hc}{4\pi^2 k_B} \frac{M}{r^2} \frac{1}{\left(1 - \frac{2M}{r}\right)^{1/2}} \qquad [44]$$

where M is in meters. Oddly, this temperature increases without limit as you approach the horizon at $r = 2M$. Therefore one would expect the radiation field near the horizon to shine brighter than any star when viewed by a distant observer. Why doesn't this happen? In a muted way it does happen. Remember that radiation is gravitationally red-shifted as it moves away from any center of gravitational attraction. From equation [C] in Selected Formulas at the end of this book we can show that every frequency is red-shifted by the factor $(1 - 2M/r)^{1/2}$, which cancels the corresponding factor in [44]. Let $r \longrightarrow 2M$ in the resulting equation. The distant viewer sees the radiation temperature

$$T_H = \frac{hc}{16\pi^2 k_B M} \qquad [45]$$

where M is in meters. The temperature T_H is called the **Hawking temperature** and characterizes the Hawking radiation from a black hole, described in the box on page 2-4. Notice that this temperature *increases* as the mass M of the black hole *decreases*. For a black hole whose mass is a few times that of our Sun, this temperature is extremely low, so from a distance such a black hole really looks *almost* black. See exercise 10.

The radiation field described by equations [42] through [44], although perfectly normal, leads to strange conclusions. Perhaps the strangest of all is that this radiation field is entirely undetected by a free-float plunging observer who passes the shell at radius r. The plunging traveler observes no such radiation field, while for the shell observer at the same radius the radiation is a surrounding presence. This apparent paradox cannot be resolved using the classical theory developed in this book; it requires quantum field theory. See the references.

How serious is the danger of being baked on a shell near the horizon of a black hole? In answer, compute the local acceleration of gravity for a shell on which the radiation field reaches a temperature equal to the freezing point of water, 273 degrees Kelvin. From equation [42] you can show that $g_{\text{conv}} = 6.7 \times 10^{22}$ meters/second^2, or almost 10^{22} times the acceleration of gravity on Earth's surface. Evidently we will be crushed by gravity long before we are baked by radiation!

11 The Plunging View

Floating to the center

Super cinema double feature

At the end of our journey through the world of space and time, momentum and energy, planets and black holes, we celebrate with a final parade of an all-star cast. Let's follow Richard Matzner, Tony Rothman, and Bill Unruh looking at the starry heavens as we free float straight down into a black hole so massive, so large, that even after crossing the horizon at the Schwarzschild radius we have four hours of existence ahead of us—the time of a super cinema double feature—to behold the whole marvelous ever-changing spectacle. Almost everything we have learned about relativity—both special and general—contributes to our appreciation of this mighty panorama. Section 9 of Project B previews some of these results.

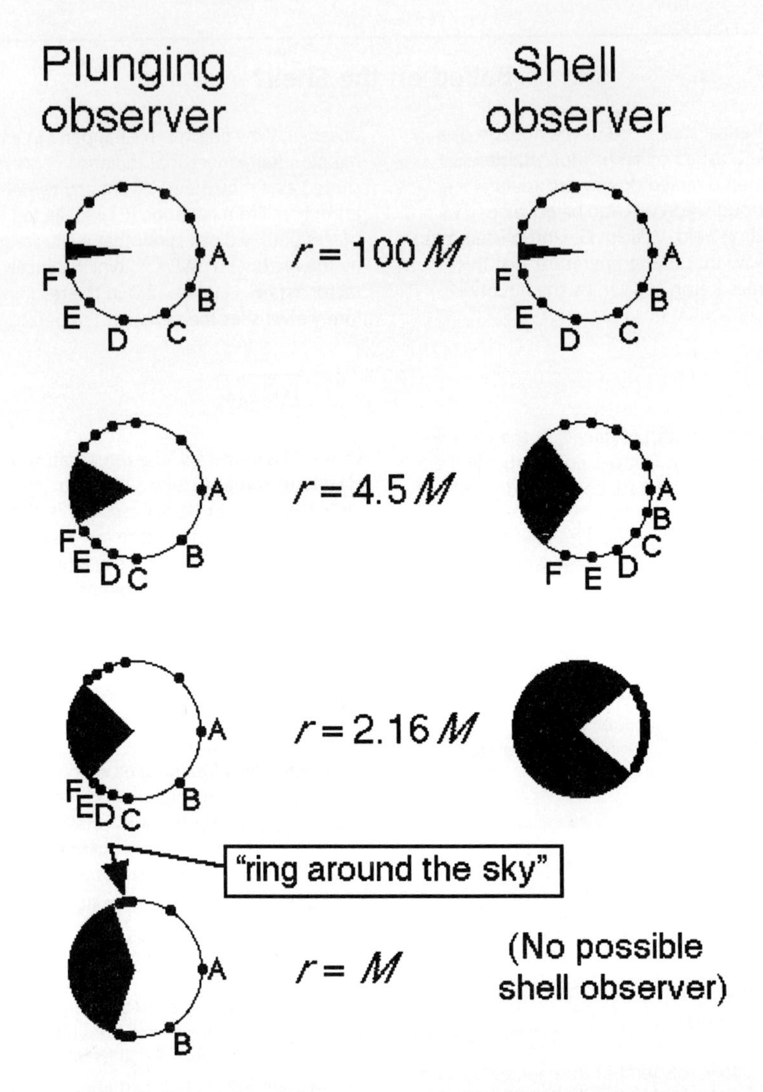

Figure 15 *Schematic drawing: View of black hole and stars seen by radial plunger and shell observer. The right column gives pie-chart views of the black hole from shells through which the traveler plunges. (See Figure 10 and equation [31].) Black sectors represent the angle spanned by the black hole as it is viewed at various distances from the hole. These radial distances are given in the center column. The left column gives the corresponding pie-chart view for the radially inward-plunging traveler. (Apply equation [40] to the critical angles for the shell viewer given in equation [31].) Letters A through F represent directions of stars evenly spaced around the sky for a viewer remote from the black hole. At r = 100 M, only a small section of the sky is dark, and only the stars whose directions lie nearest to that of the black hole show disturbed positions. At a stationary position just outside the horizon, the shell viewer sees most of the sky black. The star images are squeezed around behind him because of the bending of light around the black hole. In contrast, the plunging viewer passes through this shell at almost the speed of light. To her, only a 90-degree segment in front of her is dark and the stars are sweeping around in the sky toward the hole. Even inside the horizon, at r = M, the black hole fills less than half the sky for the plunging viewer. Notice that star positions are in general displaced forward more for the moving viewer than for the stationary spectator, showing that the changes in apparent star positions arise mainly from the aberration effect, due to the speed of the observer (equation [40]). In the final seconds of her journey the sky behind the plunging traveler is black, nearly empty of stars, and the sky ahead is black because of the black hole. Cleaving the forward half of the firmament from the backward half is a bright ring around the sky. This figure does not show secondary and higher-order star images from light that wraps once or many times around the black hole. (This figure is based on the work of M. Sikora, courtesy of M. Abramowicz.)*

The Final Plunge

—Adapted from Matzner, Rothman, and Unruh

With this background, let us now imagine a free-float journey into a billion-solar-mass black hole ($M = 10^9\ M_{Sun} = 1.5 \times 10^9$ kilometers = 1.6×10^{-4} light-years = about ten times the radius of Earth's orbit around Sun). The horizon radius of this hole—double the above figure—is about the size of our solar system. We begin our journey at one-tenth the velocity of light at a distance of $r = 20\,000\ M$ from the center of the black hole. We record each stage in the journey by giving both our wristwatch time remaining before we finally reach the singularity and our radius r.

30 years to the end

The beginning of the journey, 30 years before the end. At this point, the black hole is rather unimpressive. There is a small region (about 1 degree across—i.e., twice the size of Moon as seen from Earth) in which the star patterns look slightly distorted and within it a disk of total blackout. Careful examination of the stars shows that a few of those nearest the rim of the blacked-out region have second images on the opposite side of the rim. Had these images not been pointed out to us, we probably would have missed the black hole entirely.

10 days

Ten days before the end, at $r = 32\ M$. The image has grown immensely. There is now a pure dark patch ahead with a radius of about 10 degrees (approximately the size of a dinner plate held up at arm's length). The original star images that lay near the direction of the black hole have been pushed away from their original positions by about 15 degrees. Further, between the dark patch itself and these images lies a band of second images of each of these stars. Looking near the darkness with the aid of a telescope, we can even see faint second images of stars that actually lie behind us! This light has looped once around the black hole on its way to our eye.

4 hours

Four hours before the end. We are now at the horizon, and thus our shell speed is near that of light. Aberration effects are now extremely important. Anything we see from this moment on will be a secret taken to our grave, because we can no longer send any information out to our surviving colleagues. Although we are now "inside" the black hole, not all of the sky in front of us appears entirely dark. Because of our high speed, aberration causes light rays to arrive at our eyes at extreme angles. In fact, only the patch immediately in front of us is fully black, subtending an angle of 90 degrees—a substantial fraction of the forward sky.

Behind us we see the stars grow dim and spread out, moving around forward to meet the advancing edge of the black hole. This apparent star motion is again an aberration effect (Figure 15). But there is a more noticeable feature of the sky: We can now see second images of all the stars in the sky surrounding the black hole. These images are squeezed into a band about 5 degrees wide around the image of the black hole. These second images are now brighter than were the original stars because of the blueshifting of light falling into the hole. Surrounding the ring of second images are the still brighter first images. The rings encircle our direction of motion as the Arctic Circle surrounds the south-to-north axis of the Earth. The band of light caused by both the primary and secondary images now shines with a brightness ten times that of Earth's normal night sky.

4 minutes

Four minutes before oblivion; $r = M/7$ yet to cover. The black hole now subtends an angle of 150 degrees—almost the entire forward sky. Behind us stars are dimming and rushing forward. Only 20 percent of the stars are left in the sky behind us. In a 10-degree-wide band surrounding the outer edges of the black hole, not only second but also third and some fourth images are now visible. This band running around the sky now glows 1000 times brighter than the night sky viewed from Earth.

Final seconds

The final seconds. The sky everywhere except in that rapidly thinning band is dark. The luminous band—glowing ever brighter—runs completely around the sky perpendicular to our direction of motion. At 3 seconds before oblivion it shines brighter than Earth's Moon. New stars rapidly appear along the inner edge of the shrinking band as higher- and higher-order images become visible from light wrapped many times around the black hole. The stars of the universe seem to brighten and multiply as they are compressed into a thinner and thinner ring transverse to our direction of motion.

Ring bisecting the sky

Only in the last tenth of a second do the tidal forces become strong enough to end our journey and our view of that awesome ring bisecting the sky.

12 Summary

A light flash moves along a **null geodesic**, a locally straight worldline with zero proper time between any two events on this worldline (Figure 1). The same is true near a black hole. Every shell observer and every local free-float observer measures the speed of light to have the standard value unity. In contrast, the records of the Schwarzschild bookkeeper show light to move at a calculated speed less than unity near a black hole, with different radial and tangential speeds.

$$\frac{dr}{dt} = \pm\left(1 - \frac{2M}{r}\right) \qquad \text{[3. light, radial motion]}$$

$$r\frac{d\phi}{dt} = \pm\left(1 - \frac{2M}{r}\right)^{1/2} \qquad \text{[4. light, tangential motion]}$$

Near the horizon, the value of both components goes to zero. The reduced speed of light has been verified by Shapiro and his colleagues for radio waves passing close to Sun. (See Project E, Light Slowed Near Sun.)

A single constant of the motion, the impact parameter b, characterizes the trajectory of light around a center of gravitational attraction. For light that starts from a great distance, the value of this impact parameter is the perpendicular distance between the initial path of the light and the parallel path of a test particle that plunges radially straight into the black hole (Figure 2, page 5-6). For light that starts nearer to the black hole, the value of b must be derived from the tangential component of light velocity as measured by a local shell observer (Sample Problem 1, page 5-10).

A qualitative description of the motion of light derives easily from a plot of the constant quantity $1/b^2$ on the same graph as the effective potential for the motion of light (Figure 5, page 5-13).

$$\begin{pmatrix}\text{effective potential}\\ \text{for a photon}\end{pmatrix}^2 = \frac{1-\dfrac{2M}{r}}{r^2} \qquad \text{[25. light]}$$

A light beam with the critical impact parameter $b_{critical} = (27)^{1/2}M$ can circle in an unstable knife-edge circular orbit at $r = 3M$, the location of the peak of the effective potential. Light with impact parameter greater than the critical value reaches a minimum radius and then flees to infinite distance. Light with impact parameter smaller than the critical value plunges into the black hole. Experiment verifies the deflection of starlight by Sun (Project D, Einstein Rings).

The Schwarzschild map gives an overview of the orbit of light by plotting it in bookkeeper coordinates r and ϕ. No single observer measures this trajectory directly, but from this plot we can derive predictions about the visual panorama seen by a shell observer (Sections 8 and 9) and an observer in a free-float orbit (Section 10).

13 References and Credits

Initial quote: Emily Dickinson, poem 1129, from *The Complete Poems of Emily Dickinson*, edited by Thomas H. Johnson, Little Brown and Company, Boston, 1960. Reprinted with permission of Harvard University.

Figure 15 showing Einstein ring: Photo courtesy of Jacqueline N. Hewitt. The Very Large Array (VLA) is operated by the National Radio Astronomy Observatory under contract with the National Science Foundation. J. N. Hewitt, E. L. Turner, D. P. Schneider, B. F. Burke, G. I. Langston, and C. R. Lawrence, "Unusual Radio Source MG1131-0456: A Possible Einstein Ring," *Nature*, Volume 333, pages 537–540 (1988).

Figure 5 is adapted from Charles W. Misner, Kip S. Thorne, and John Archibald Wheeler, *Gravitation*, W. H. Freeman and Company, San Francisco (now New York), 1973, page 675.

For the view outward from near a black hole, see Joachim Schastok, Michael Soffel, and Hanus Ruder, "Stellar Sky as Seen from the Vicinity of a Black Hole," *American Journal of Physics*, Volume 55, Number 4, pages 336–341 (April 1987).

For an analysis of the intensity of star images deflected by a Schwarzschild black hole, see Hans C. Ohanian, "The Black Hole as Gravitational 'Lens'," *American Journal of Physics*, Volume 55, pages 428–432 (1987).

References for the box "Baked on the Shell?" on page 5-27: Historical background in Kip S. Thorne, Richard H. Price, and Douglas A. MacDonald *Black Holes: The Membrane Paradigm,* Yale University Press, 1986, pages 280–285 and 35–36; W. G. Unruh, "Notes on Black-Hole Evaporation", Physical Review D, Volume 14, Number 4, 15 August 1976, pages 870–892; William G. Unruh and Robert M. Wald, "What Happens When an Accelerating Observer Detects a Rindler Particle," Physical Review D, Volume 29, Number 6, 15 March 1984, pages 1047–1056.

Description of final plunge (Section 11) is adapted from Richard Matzner, Tony Rothman, and Bill Unruh, "Grand Illusions: Further Conversations on the Edge of Spacetime," in *Frontiers of Modern Physics: New Perspectives on Cosmology, Relativity, Black Holes and Extraterrestrial Intelligence,* edited by Tony Rothman, Dover Publications, Inc., New York, 1985, pages 69–73.

Robert Wilson pointed out the ambiguity in shell angles for viewing the edge of the black hole, described in the box on page 5-20. Eric Sheldon and Joel Therrien gave us the idea for the exercise Shadow of a Black Hole. Mary Nickles gave us the idea for the exercise Measuring Your Distance from a Black Hole.

Chapter 5 Exercises

Note: With the background of Chapters 4 and 5, you may want to go back and examine Sections 8 through 10 of Project B, Inside the Black Hole.

THOUGHT/RESEARCH QUESTIONS

Thought/research questions are open-ended queries that may or may not have answers and can or cannot be answered qualitatively after thought. Research occurs when qualitative answers are followed by derivation of formulas and numbers, so the answers could, in principle, be tested by experiment. A typical strategy for thought/research questions is to replace a complicated question about a real situation with a simple question about an idealized situation. With luck and hard work, the simplified answer then leads back step by step to a more realistic analysis.

1. View from the Light Sphere

You stand on a shell at $r = 3M$, on what is called the *light sphere*. You look in the tangential direction, that is, perpendicular to the radial direction. What do you *see?* Do you see the back of your own head? What do you see when you turn your head right and left, scanning many tangential directions? Do you see stars? If so, which stars? What do you see if you look inward from the tangential direction? outward from the tangential direction?

2. Shadow of a Black Hole?

According to legend, a vampire can be recognized by the fact that he (or she!) has no reflection in a mirror and casts no shadow. Does a black hole cast a shadow? Does a black hole cast a shadow at some distances but not at others? What set of experiments might answer these questions? What results do you predict for these experiments?

3. Measuring Your Distance from a Black Hole

You are the pilot of a spaceship approaching or cruising near a black hole. You would like to know how close you are to it. What experiment can you do to find out? Assume that the mass M of the black hole is known. Can you use laser or radar signals or the view of the stars to tell you what your reduced circumference r is with respect to the black hole? Do you need the presence of other spaceships, at rest or in orbit, in order determine your distance?

4. Crossing the Horizon

Pete Brown disagrees with the statement in an earlier chapter, "No special event occurs as we fall through the Schwarzschild horizon" (page 3-19). He says, "Suppose you go feet first through the horizon. Since your feet hit the horizon before your eyes, then your feet should disappear for a short time. When your eyes pass across the horizon, you can then see what is inside, including your feet. So tie your sneakers tightly or you will lose them in the dark!" Is Pete right? Suppose another spaceship free floats ahead of you across the horizon. Will you lose sight of the leading spaceship after it crosses the horizon but before you do?

EXERCISES

5. Effective Potential for Light

Start with expression [25], page 5-11, for the square of the effective potential for light:

$$\begin{pmatrix}\text{effective}\\ \text{photon}\\ \text{potential}\end{pmatrix}^2 = \frac{1-\dfrac{2M}{r}}{r^2} \qquad [25]$$

Take the derivative of this function with respect to r and determine the value of r for which the effective potential is a maximum. (The maximum of a squared function occurs at the same r-value as the maximum of a positive function itself.) Determine also the value of the effective potential at this maximum. Compare your results with those quoted in the text. *Optional:* Take the second derivative of the expression for squared effective potential. From the sign of this second derivative at the position of the zero for the first derivative, verify that the effective potential has a *maximum* value at this position.

6. Firing a Laser Pulse Outward

The shell commander on the shell at r-coordinate r fires a laser pulse of energy E_{shell} radially outward.

A. What will be the energy of the laser pulse when it reaches a great distance? Answer this question using the following outline or some other method. (Do not assume the answer that is asserted for equation [27], page 3-17.)

Redder color, lower frequency, smaller quantum energy characterizes the pulse received by the remote observer compared with the pulse launched by the shell observer at radius r. The amount of this *gravitational red shift* can be spelled out from the equation that connects the quantum description of a photon with the classical description of light as a wave of frequency f and period T:

$$\begin{pmatrix}\text{quantized}\\ \text{energy E}\\ \text{of photon}\end{pmatrix} = \begin{pmatrix}\text{Planck's}\\ \text{constant h}\end{pmatrix} \times \begin{pmatrix}\text{classical}\\ \text{frequency f of}\\ \text{corresponding}\\ \text{wave}\end{pmatrix} \qquad [46]$$

$$E = hf = h\left(\frac{1}{\text{time for one cycle}}\right) = \frac{h}{T} \qquad [47]$$

Call T_{shell} the period of the light as measured by a shell observer and E_{shell} the corresponding photon energy. And let the corresponding symbols E and T_{far} refer to measurements by an observer far from the black hole. We know that equation [C] in the Selected Formulas at the back of the book relates the period of light T_{shell} measured by the shell observer to the period T_{far} measured by the remote observer:

$$T_{shell} = \left(1 - \frac{2M}{r}\right)^{1/2} T_{far} \qquad [D]$$

From these equations, show that

$$E = \left(1 - \frac{2M}{r}\right)^{1/2} E_{shell} \qquad [48]$$

This equation applies to *anything* moving near a black hole, one more example of the general equation [27], page 3-17.

B. Suppose the laser pulse is shot not straight out but at some angle to the outward direction. Nevertheless, it escapes from the black hole and arrives at the distant observer. Will equation [48] of part A still describe the energy E of this pulse as measured by the distant observer?

C. Instead of moving outward, the laser pulse is fired radially *inward* so that it crosses the horizon. By how much, ΔM, does the black hole increase in mass?

Note: We rarely use the word *photon* in this book, because the photon is a quantum object with strange properties. Typically, to detect a photon is to deflect or destroy it, so that in a deep sense the trajectory of a photon is unobservable. General relativity is a classical (nonquantum) theory, so we usually talk about classical light beams or classical light flashes.

7. The Horizon as a One-Way Barrier

A light flash is launched from the shell at r-coordinate r_1. Its energy at launch is $E(r_1)_{shell}$ as measured by observers on the launching shell. Now the flash moves radially inward or outward and is received at a shell of different reduced circumference r_2.

A. What energy $E(r_2)_{shell}$ does the light flash have as measured by the observers on the shell at r_2? *Hint:* Run equation [48] backward and forward, assuming constant energy E measured at infinity.

B. Take the limit of your expression derived in part A as $r_1 \longrightarrow 2M$ to show that a light flash launched from the horizon cannot be detected at an r-coordinate even 1 millimeter above the horizon. This outcome illustrates the general-relativity result that nothing, not even light, can cross the horizon in an outward direction. The Newtonian analysis, in contrast, predicts that a particle moving outward from $r_1 = 2M$ with speed c can just make it to an infinite distance, and a particle launched from $r_1 = 2M$ with speed less than c nevertheless climbs to some maximum radius before falling back. (The speed of light c is not a natural speed limit in the Newtonian analysis.) These predictions of the Newtonian analysis are incorrect.

C. Now assume that the light flash is launched *inward* from r-coordinate r_1 just outside the horizon. What is the shell-measured energy of this light flash as it crosses the horizon? Does this result make sense to you?

8. Energy Production by a Quasar?

Note: Some results of exercises in Chapter 4, as indicated, are used in solving the following exercise.

A quasar (contraction of the name *quasi-stellar object*) is an astronomical object that pours out a prodigious amount of energy. Because they are so bright, quasars are the most distant visible objects, some of them as much as 14×10^9 light-years distant. A single bright quasar can give off energy at a rate greater than that of all the stars in our galaxy, though most quasars emit energy at a more modest rate than that. What is the source of this energy? We do not know. The energy emission rate can sometimes change significantly in a short time, implying that the emitting structure is small. (Otherwise the limiting speed of light would prevent different parts of the structure from "cooperating" to change the emission rate.) Evidence is accumulating that the energy comes from stars torn apart and spiraling into a black hole, material that is heated to extreme temperatures in the process and emits radiation copiously prior to disappearing across the horizon of the black hole. In essence, some fraction of the energy at infinity of the in-falling material is converted into light.

Most theories of quasar energy production assume that the black hole involved is rotating rapidly. The rotating black hole is the subject of Project F, The Spinning Black Hole. Here we analyze a simpler (and less realistic) model of energy production around a nonrotating black hole.

To show the order of magnitude of gravitational energy that is available, consider the following simplified encounter: A stone of mass m is in the stable orbit of smallest possible radius around a black hole. A second stone of equal mass m, initially at rest at a great distance, plunges radially toward the black hole and collides with the stone in the circular orbit. Assume that the entire kinetic energy of the pair, as observed by a local shell observer, is converted into light that travels outward. (Do not worry about conservation of momentum for the shell observer.)

A. What is the total *kinetic* energy of the two stones measured by the shell observer? Assume that all of this kinetic energy is converted into a light flash directed radially outward.

B. What is the total energy of the resulting light flash received by a remote observer?

C. What fraction of the total rest energy $m + m = 2m$ has been turned into light at infinity? Compare this with an energy-conversion fraction of 0.1% or less for nuclear reactions on Earth.

9. Plunger Wink-Out Time

You are stationed a great distance from a black hole. A beacon moves directly away from you, plunging radially into a black hole of mass M as the beacon emits radially outward a light signal of constant proper frequency f_o as observed in its own free-float frame. How long does it take for this beacon to "go black" or "wink out" as far as you, the distant observer, are concerned?

This exercise is a mini project. You may stop reading at this point and answer the question posed in the preceding paragraph, interpreting as you go what the question might mean. Or you may read on to receive a skeletal structure for answering the question according to one of several possible interpretations.

What does it mean for a plunging beacon to "go black" or "wink out" for a distant observer? One interpretation is that as the beacon moves inward emitting a proper frequency f_o, the signal received by the distant clock drops from a frequency equal to $0.9\,f_o$ to a frequency $0.1\,f_o$. We adopt this interpretation in what follows and ask how long is the far-away time between the reception of these two frequency-shifted signals.

What is the relation between the proper frequency f_o of the signal emitted by the descending beacon and the frequency f_{far} of the signal received by the distant observer? We answer this question in two steps: (1) What is the relation between the frequency f_{shell} observed by the shell observer at radius r and the frequency f_{far} observed by the far-away bookkeeper? (2) What is the relation between the proper frequency f_o of the falling beacon and the frequency f_{shell} of that beacon detected by the shell observer at radius r?

A. A shell observer detects a signal of frequency f_{shell} received from the descending beacon and rebroadcasts the signal (or simply lets it pass) to the remote bookkeeper, who receives it as frequency f_{far}. What is the relation between f_{shell} and f_{far}? Start with the equation for a wave $f = 1/T$, where T is the period of the wave. Use equation [C] in Selected Formulas at the back of the book to show that

$$f_{\text{shell}} = \frac{f_{\text{far}}}{\left(1 - \frac{2M}{r}\right)^{1/2}} \qquad [49]$$

What frequency f_{shell} does the shell observer detect for the signal broadcast with proper frequency f_o from the passing beacon? Here we must take into account the Doppler shift of the beacon just after it has passed the shell observer but is still effectively at the shell radius r. The shell observer looks down (radially inward) and receives the Doppler-shifted signal from the beacon according to the special relativity result (Taylor and Wheeler, *Spacetime Physics*, page 114)

$$f_{\text{shell}} = \left(\frac{1 - v_{\text{shell}}}{1 + v_{\text{shell}}}\right)^{1/2} f_o \qquad [50]$$

We have not yet determined how fast the beacon is moving when it passes the shell observer. Assume that it falls from rest at a great distance from the black hole. Then the square of the speed measured by the shell observer comes from equation [24], page 3-15:

$$v^2_{\text{shell}} = \frac{2M}{r} \qquad [51]$$

B. Substitute [51] into [49] and equate the expressions for f_{shell} in the modified equation [49] with the same expression in equation [50]. Show that for the conditions we have assumed,

$$\frac{f_{\text{far}}}{f_o} = 1 - |v_{\text{shell}}| = 1 - \left(\frac{2M}{r}\right)^{1/2} \qquad [52]$$

C. Find the radius r_1 the beacon is passing when it emits a signal whose frequency is observed to be $f_{\text{far}} = 0.9 f_o$ when received by the distant observer. Our answer is $r_1 = 200M$. Is our answer correct? If not, find the correct answer to three significant digits.

D. Find the radius r_2 the beacon is passing when it emits a signal whose frequency is observed to be $f_{\text{far}} = 0.1 f_o$ when received by the distant observer. Our answer is that r_2 has a value a little greater than $2M$; find the answer to three significant digits.

Now we know the two radii r_1 and r_2 at which the beacon emits signals that are measured by the far-away observer to have frequencies $0.9f_o$ and $0.1f_o$, respectively. Next we need to determine how long a time the far-away observer has to wait between the reception of these two signals. This will be the "wink-out time" for the beacon. This wink-out time is made up of two parts: (1) the far-away time that it takes for the beacon to fall from r_1 to r_2 and (2) the far-away time it takes for the light to climb back up from r_2 to r_1. *Note:* We do not need to take into account the time it takes light to climb from the outer radius r_1 to the position of the far-away observer. Whatever value this time has, it is the *same* time for the initial flash from r_1 as for the second flash that passes r_1 on its way from the inner radius r_2. We want to know the time *difference* for the distant observer between the arrival of these two flashes.

E. Find the time determined by the far-away observer for the beacon to fall from r_1 to r_2, these two radii defined in parts C and D. Start with equation [21], page 3-13:

$$\frac{dr}{dt} = -\left(1 - \frac{2M}{r}\right)\left(\frac{2M}{r}\right)^{1/2} \qquad [53]$$

The integration of dt to find the far-away time lapse is not an easy one. You will have your own favorite procedure. We made the substitution $z = r/(2M)$ followed by the substitution $u^2 = z$. The result is

$$dt = \frac{4Mu^4 du}{1 - u^2} \qquad [54]$$

Our table of indefinite integrals contains the formula

$$\int \frac{u^4 du}{1 - u^2} = -\frac{u^3}{3} - u + \frac{1}{2}\ln\left|\frac{u+1}{u-1}\right| \qquad [55]$$

From this analysis, find an expression for the time for the beacon to fall from r_1 to r_2 given by the results of parts C and D. Our result to two significant digits is $1400M$; find the result to at least three significant digits.

F. Find the value of the far-away time required for the light from the beacon at r_2 to rise to the higher radius r_1. Start with the positive solution to equation [3], page 5-3:

$$\frac{dr}{dt} = 1 - \frac{2M}{r} \qquad [56]$$

This integration is much easier, though it still requires a table of integrals. Our result for the time for light to climb from r_2 to r_1 is, to one significant digit, $200M$; find the result to three significant digits.

G. Find the value of the elapsed far-away time Δt_{far} between the arrival of the signal shifted to $0.9 f_o$ and the arrival of the signal shifted to $0.1 f_o$.

H. What is the wink-out time Δt_{far} for a beacon descending into a black hole of mass M equal to ten times that of Sun? Express your answer in seconds.

10. Temperature of a Black Hole

A. Use equation [45] to find the temperature, when viewed from a great distance, of a black hole of mass five times the mass of Sun.

B. What is the mass of a black hole whose temperature, viewed from a great distance, is 1800 degrees Kelvin (the melting temperature of iron)? Express your answer as a fraction or multiple of the mass of Earth.

Project D Einstein Rings

- *Why does light change direction as it passes our Sun?*
- *What causes Einstein rings? How can I see them?*
- *How can we observe dark objects in our galaxy?*
- *What accounts for the weirdly distorted image on the cover of this book?*

Project D

Einstein Rings

Einstein was discussing some problems with me in his study when he suddenly interrupted his explanation and handed me a cable from the windowsill with the words, "This may interest you." It was the news from Eddington confirming the deviation of light rays near the sun that had been observed during the eclipse. I exclaimed enthusiastically, "How wonderful, this is almost what you calculated." He was quite unperturbed. "I knew that the theory was correct. Did you doubt it?" When I said, "Of course not, but what would you have said if there had not been such a confirmation?" he retorted, "Then I would have to be sorry for dear God. The theory is correct."

— Ilse Rosenthal-Schneider

1 "Did you doubt it?"

Arthur Eddington's verification of the deflection of starlight by Sun in 1919 made Albert Einstein an instant celebrity. In this project we reproduce Einstein's prediction (though not by his method) and apply it to important modern astronomical observational techniques. Einstein's results are correct for deflection of starlight by most spherical astronomical objects and for light passing a Schwarzschild black hole at a large radial distance such that $r >> 2M$. This is called the **weak field approximation**. Section 7 uses the weak field approximation to describe **Einstein rings** such as those displayed in Figure 14, page 5-25. The lens-like concentration of light can also increase the amount of light received from a distant star, an effect called **microlensing**, described in Section 8. In Section 9 we use Einstein rings to account for some features of the image of Saturn on the cover of this book.

Light passing close to a black hole does not satisfy the weak field approximation but is radically deflected and can even go into temporary orbit (Chapter 5). Such radical deflection accounts for the "diamond necklace" in the center of the upper image on the cover, as described in Section 10. (For even wilder behavior of light see Project F, The Spinning Black Hole.)

2 Newtonian Deflection of Light

Before carrying out the analysis of general relativity, begin with an approximate Newtonian prediction of light deflection under the assumption that light accelerates downward in a gravitational field (Figure 1). We make the approximations (1) light is deflected only during its passage across the diameter of Sun, and (2) during this deflection the acceleration is perpendicular to the direction of motion of light and equal in magnitude to the acceleration at the surface of Sun.

QUERY 1	**Gravitational acceleration at the surface of Sun.** Using physical constants from inside the back cover, calculate the acceleration of gravity at the surface of Sun, according to Newton, in conventional units meters/second^2.

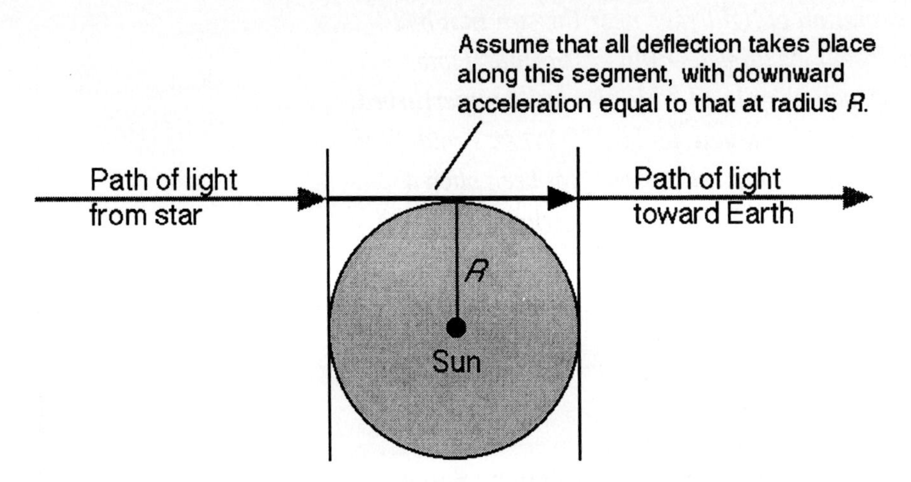

Figure 1 *Newtonian analysis of light deflection by Sun. Assume that light, along with material particles, undergoes gravitational acceleration. The change in transverse velocity, according to Newton, is reckoned using the acceleration at the surface of Sun acting during the time the light crosses the diameter of Sun (calculus proof!). The deflection, less than two seconds of arc, is too small to be visible in this or later figures, which therefore will show this deflection greatly exaggerated.*

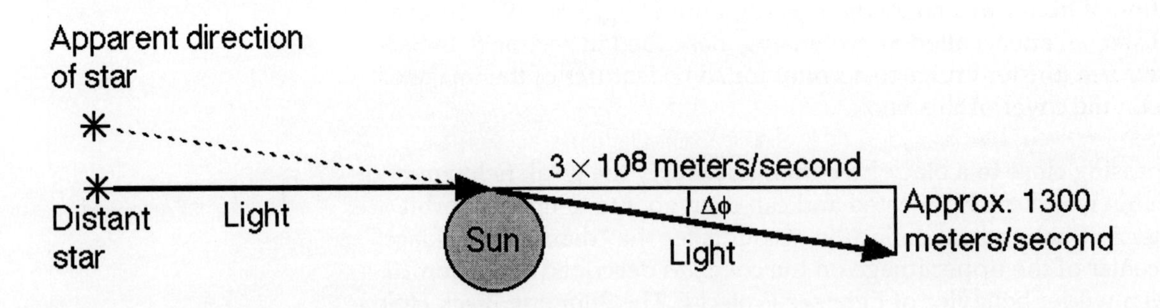

Figure 2 *Schematic diagram (greatly exaggerated) of the deflection of starlight in the Newtonian analysis outlined in Figure 1.*

QUERY 2	**Time for light to cross Sun.** Calculate the length of time in seconds it takes light to move a distance equal to the diameter of Sun.
QUERY 3	**Change in "sunward" motion of light.** Assume that light grazing Sun experiences a constant acceleration perpendicular to its path equal to that calculated in Query 1 for a time calculated in Query 2 (and no acceleration anywhere else along its path). Then the light picks up a "sunward" component of velocity equal to approximately 1300 meters/second. See Figure 2. Find the result to four-digit accuracy.
QUERY 4	**"Newtonian" angle of deflection.** Let $\Delta\phi$ be the angle of deflection of light in radians, as shown in Figure 2. Because of the small angle of deflection, assume that $\tan(\Delta\phi) \approx \Delta\phi$. From the results of the preceding queries, this deflection has the approximate value $\Delta\phi \approx 4 \times 10^{-6}$ radian. Find the result to three-digit accuracy.
QUERY 5	**Seconds of arc.** The deflection of starlight by Sun is usually expressed in seconds of arc. There are 60 minutes of arc in one degree and 60 seconds of arc in one minute of arc. The deflection $\Delta\phi$ of light by Sun is approximately 1 second of arc—according to this Newtonian analysis. Find the result to three-digit accuracy.

From the Newtonian result (too small by a factor of 2, as we shall see), we draw the provisional conclusion that the deflection is very small. This assumption is important in the general relativistic derivation that follows, and must be verified again for the results of that derivation.

3 Setting Up the General-Relativistic Analysis

Now we move on to the general-relativistic prediction of light deflection. A pulse of light approaches, passes, and recedes from Sun, its position tracked by the azimuthal angle ϕ (Figure 3). If there is no deflection, the angle ϕ sweeps through π radians as the pulse moves from distant approach to distant recession. We want to find the *additional* angle $\Delta\phi$ caused by gravitational deflection (Figure 4).

How much, $d\phi$, does the tracking angle ϕ change for each small change in radius dr (Figure 3)? The answer is embodied in the expression $d\phi/dr$, derivable from equations [15] and [14], page 5-8:

$$\left(\frac{d\phi}{dt}\right)^2 = \frac{b^2}{r^4}\left(1 - \frac{2M}{r}\right)^2 \qquad [1]$$

$$\left(\frac{dr}{dt}\right)^2 = \left(1 - \frac{2M}{r}\right)^2 - \left(1 - \frac{2M}{r}\right)^3 \frac{b^2}{r^2} \qquad [2]$$

Here b is the impact parameter (defined in Figure 2, page 5-6).

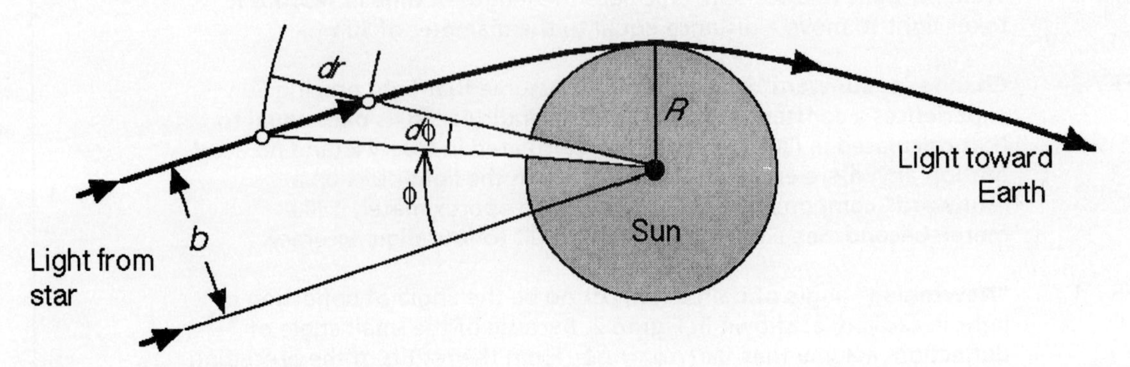

Figure 3 *Measuring the change dϕ in azimuthal angle ϕ as a light pulse changes radius dr. The deflection is greatly exaggerated. If there is no deflection, the angle ϕ will sum to π as r goes from the distant star to R and out to distant Earth. To predict the actual deflection, we need a relation between dr and dϕ. Equation [4] gives this relation. The distance b is the impact parameter.*

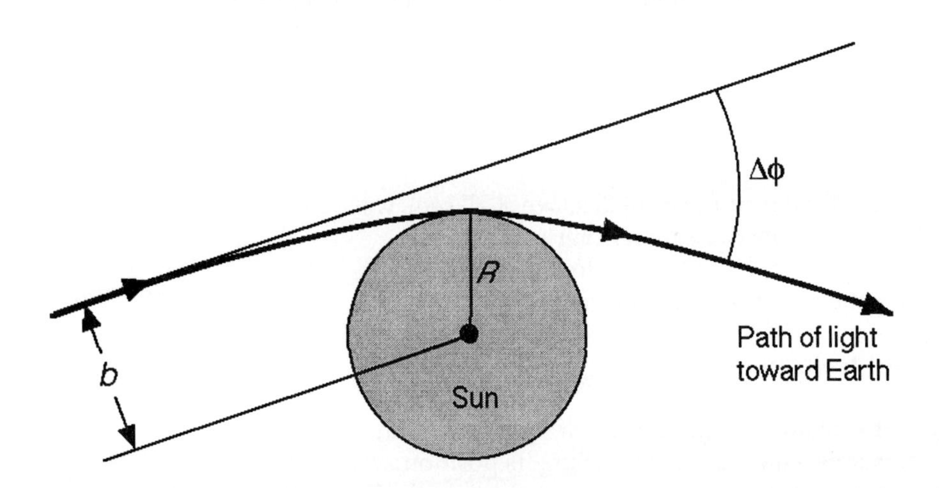

Figure 4 *Gravitational deflection angle Δϕ of starlight by Sun—greatly exaggerated.*

QUERY 6 **Angle ϕ changes as r changes.** Divide corresponding sides of equations [1] and [2] and simplify to show that

$$\left(\frac{d\phi}{dr}\right)^2 = \frac{b^2/r^4}{1-\left(1-\frac{2M}{r}\right)\frac{b^2}{r^2}} \qquad [3]$$

Show that this can be modified to yield

$$d\phi = \frac{dr}{r^2\left[\frac{1}{b^2}-\frac{1}{r^2}\left(1-\frac{2M}{r}\right)\right]^{1/2}} \qquad [4]$$

Now, in principle we simply integrate equation [4] from $r = \infty$ to $r = R$, the radius of closest approach. The total deflection will be twice this result. When $M \longrightarrow 0$ (no mass at the center), the result is $\phi = \pi$ radians. However, with mass M present the result will be *a little more* than π. This "little more," namely, $\Delta\phi$ of Figure 4, is what we are trying to determine; we expect its value to be approximately 10^{-5} radians.

The right side of equation [4] does not appear in an integral table, so we make a substitution that can lead, with approximations below, to an expression that can be integrated.

QUERY 7 **Prepare for integration.** First make the substitution $u = R/r$ (a standard substitution even in Newtonian orbital mechanics). Then $dr = -r^2du/R$. We will integrate the right side of the resulting expression outward from $r = R$ to $r = \infty$ (from $u = 1$ to $u = 0$). Show that equation [4] becomes, after some manipulation

$$d\phi = \frac{-du}{\left[\frac{R^2}{b^2} - u^2 + 2\frac{M}{R}u^3\right]^{1/2}} \qquad [5]$$

Before we can carry out this integration, we must take care of one detail, namely, that the impact parameter b is a function of both M and R. Why? Because the value of b is not arbitrary; we *choose* b so that the light ray just grazes the surface of Sun (Figures 3 and 4). What relation connects b, R, and M? Return to equation [24], page 5-11, namely,

$$\frac{1}{b^2}\left(\frac{dr_{\text{shell}}}{dt_{\text{shell}}}\right)^2 = \frac{1}{b^2} - \frac{1}{r^2}\left(1 - \frac{2M}{r}\right) \qquad [6]$$

The light moves tangentially as it skims the surface of Sun, so for $r = R$, we have zero radial motion and the left side of [6] is equal to zero.

QUERY 8 **Eliminating the impact parameter *b*.** Substitute $r = R$ into equation [6] and set the left side equal to zero. Manipulate the result to show that

$$\frac{R^2}{b^2} = 1 - \frac{2M}{R} \qquad [7]$$

Substitute equation [7] into [5]:

$$d\phi = \frac{-du}{\left[1 - u^2 - \frac{2M}{R}(1 - u^3)\right]^{1/2}} = \frac{-(1 - u^2)^{-1/2}du}{\left[1 - \frac{2M}{R}\frac{(1 - u^3)}{(1 - u^2)}\right]^{1/2}} \qquad [8]$$

4 Approximations

So far there have been no approximations. Now recall that M/R is extremely small. The value of M, the mass of Sun, is 1477 meters and the value of R, the radius of Sun, is approximately 7×10^8 meters. Therefore $M/R \approx 2 \times 10^{-6}$. The small value of this quantity encourages us to use our by now standard approximation:

$$(1 + d)^n \approx 1 + nd \qquad \text{provided} \qquad |d| \ll 1 \qquad \text{and} \qquad |nd| \ll 1 \qquad [9]$$

When this equation is applied to equation [8], then d in [9] corresponds to the second term under the square root in the denominator on the right hand side of equation [8].

QUERY 9 **Approximation for small *M/R*.** Apply approximation [9] to equation [8] and show that the result can be written

$$d\phi = -(1 - u^2)^{-1/2}\left[1 + \frac{M}{R}\frac{(1 - u^3)}{(1 - u^2)}\right]du$$

$$= \frac{-du}{(1 - u^2)^{1/2}} - \frac{M}{R}\frac{du}{(1 - u^2)^{3/2}} + \frac{M}{R}\frac{u^3 du}{(1 - u^2)^{3/2}} \qquad [10]$$

5 Results for General Relativity

Now it's time to integrate! We integrate from $r = R$ to $r = \infty$ or use the limits $u \equiv R/r = 1$ to $u = 0$. In either case, multiply the result by 2 to account for both "legs" of the trajectory in Figures 3 and 4. From a table of integrals, the first integral on the right side (multiplied by 2) gives the change in angle when there is no Sun present:

$$\int_1^0 \frac{-2du}{(1 - u^2)^{1/2}} = [-2 \arcsin u]\Big|_1^0 = 2 \arcsin 1 = 2\frac{\pi}{2} = \pi \qquad [11]$$

We want the small *additional* angle of deflection $\Delta\phi$ beyond the undeflected value of π. So, referring to Figures 3 and 4, set

$$\phi_{total} = \pi + \Delta\phi \qquad [12]$$

The value of $\Delta\phi$ comes from integration of the second and third terms in [10] (again, multiplied by 2). Evaluate these two integrals together using a table of integrals.

$$\Delta\phi = -\frac{2M}{R}\left[\int_1^0 \frac{du}{(1-u^2)^{3/2}} - \int_1^0 \frac{u^3\,du}{(1-u^2)^{3/2}}\right]$$

$$= -\frac{2M}{R}\left[\frac{u}{(1-u^2)^{1/2}} - (1-u^2)^{1/2} - \frac{1}{(1-u^2)^{1/2}}\right]\Bigg|_1^0 \qquad [13]$$

$$= -\frac{2M}{R}\left[\frac{u-1}{(1-u^2)^{1/2}} - (1-u^2)^{1/2}\right]\Bigg|_1^0$$

$$= \frac{4M}{R}$$

The result comes from evaluation at the upper limit: $u = 0$. There is a slight problem at the lower limit, since the denominator $(1 - u^2)^{1/2}$ of the first term approaches zero at the lower limit $u = 1$. However, notice that the first term can be written

$$\frac{u-1}{(1-u^2)^{1/2}} = -\frac{1-u}{(1-u)^{1/2}(1+u)^{1/2}} = -\left(\frac{1-u}{1+u}\right)^{1/2} \qquad [14]$$

which clearly goes to zero as u goes to 1. Therefore the lower limit in equation [13] yields zero in the result. The upper limit yields $4M/R$ which is the deflection of starlight by Sun in radians. The result of all our calculations is the simple little equation

$$\Delta\phi \approx \frac{4M}{R} \qquad \text{[15. weak field approximation]}$$

Note: If you carry through the equations for the Newtonian analysis in Queries 1 through 4, you will find the result to be exactly half that in [15].

QUERY 10 **Deflection of starlight in radians.** Substitute values for M and R for Sun and calculate the value for the deflection $\Delta\phi$ in radians.

QUERY 11 **Deflection of starlight in seconds of arc.** Convert your answer to Query 10 to units of seconds of arc. To one significant figure, the result is 2 seconds of arc. Find the result to three significant digits.

The answer to Query 11 is the prediction of general relativity. Although we have made an approximation, this approximation is very good for Sun. Equation [15] actually shows the first term in a series expansion whose next term will be proportional to $(M/R)^2$, yielding a contribution approximately 2×10^{-6} times that of the term on the right side of [15]. This added correction in predicted angle is far smaller than the accuracy of observation and so can be ignored for small deflections of light.

For most spherical astronomical objects except black holes the distance of closest approach R will be very much greater than the mass M of the object, measured in units of length. To analyze light passing close to these objects we can validly apply the results of this section, the weak field approximation [15]. Motion of light passing close to a black hole is analyzed in Chapter 5 and in Section 10 of this project.

QUERY 12 **Deflection of light by a neutron star.** How great is the deflection of light that skims past the surface of a neutron star? Assume that a nonspinning neutron star has a mass 1.4 times the mass of Sun and a radius of 10 kilometers. What is the deflection of light that skims tangentially past its surface, according to equation [15]? Express your result in degrees. Do you trust this conclusion? That is, does this case satisfy approximations used in deriving equation [15]?

6 Comparison with Observation

How well do observations verify the prediction of light deflection by Sun given in equation [15]? Table 1 shows measurements using visible light. *Note:* Observed stars were at various angles from the eclipsed Sun. Figures in the fourth column are values recomputed for the deflection of a light ray that just grazes Sun's surface.

Table 1 Deflection of starlight by Sun; values deduced in various eclipses

Eclipse Date	Location	Number of stars	Deflection (seconds of arc)	References (end of project)
May 20, 1919	Sobral Principe	7 5	1.98 ± 0.16 1.61 ± 0.40	a
September 1, 1922	Australia Australia Australia Australia	11–14 18 62–85 145	1.77 ± 0.40 1.42 to 2.16 1.72 ± 0.15 1.82 ± 0.20	b c d e
May 9, 1929	Sumatra	17–18	2.24 ± 0.10	f
June 19, 1936	USSR Japan	16–29 4–7	2.73 ± 0.31 1.28 to 2.13	g h
May 20, 1947	Brazil	51	2.01 ± 0.27	i
February 25, 1952	Sudan	9–11	1.70 ± 0.10	j

Much more consistent and accurate results come from radio astronomy—using radio waves instead of visible light. Each October the Sun moves across the image of the quasar labeled 3C279 seen from Earth. Radio astronomers use this **occultation** to measure the change in angle of the signal as the source approaches Sun, crosses the edge of Sun, and moves behind Sun. This change in angle is measured with respect to the undeflected signal from another quasar, labeled 3C273, that is about 10 degrees away from Sun as seen from Earth. The accuracy of angle measurement is increased by employing an experimental technique called **very long baseline interferometry** (VLBI) that uses widely separated dish antennas. A recent observation by D. E. Lebach and collaborators employed one dish—actually a pair of dishes—in Big Pine, California, and another pair of dishes in Westford, Massachusetts. Their observational results correspond to a gravitational deflection 0.9996 ± 0.0008 times that predicted by general relativity. This result straddles the value 1.00000, for which agreement between theory and experiment would be perfect. See the references at the end of this project.

7 Computing Einstein Rings

The first Einstein ring ever observed is shown in Figure 14 on page 5-25. We see an Einstein ring when light from a distant source is deflected toward us around all sides of an intermediate gravitating object (Figure 5). The Einstein ring appears only when the source, the intermediate object, and the observer lie along the same straight line.

The results of preceding sections allow us to predict the approximate angular diameter of the Einstein ring for a point source. The geometric construction is shown in Figure 6. In computing deflection by Sun we wanted to use Sun's radius R. Equation [7] allowed us to eliminate the impact parameter b. The light that enters our eye to form the Einstein ring of a distant star, however, may not be the light that grazes the surface of the intermediate dark object. But equation [7] tells us that for $r >> 2M$ we have simply $R \approx b$. Then equation [15] becomes

$$\Delta\phi \approx \frac{4M}{b} \qquad [16]$$

We use additional approximations in the analysis of Figure 6. A more or less complete list follows. These approximations are generally justified by the fact that the distances of source and observer from the intermediate object (defined in Figure 6) are literally astronomical compared with the distance from the intermediate object at which deflections take place.

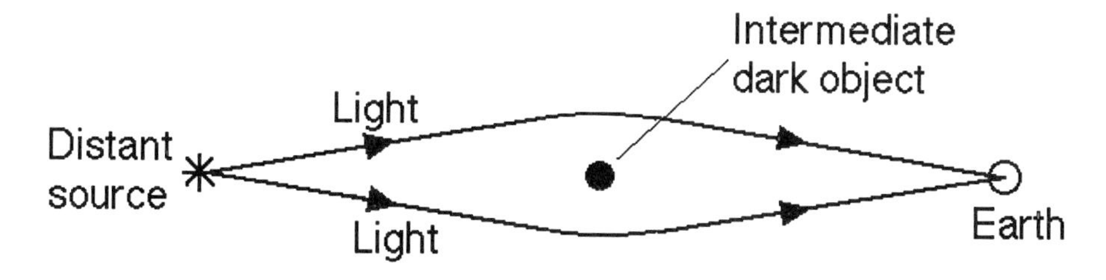

Figure 5 *Schematic diagram of the formation of an Einstein ring. Not to scale—and the intermediate object need not be halfway between the distant star and Earth, as shown here.*

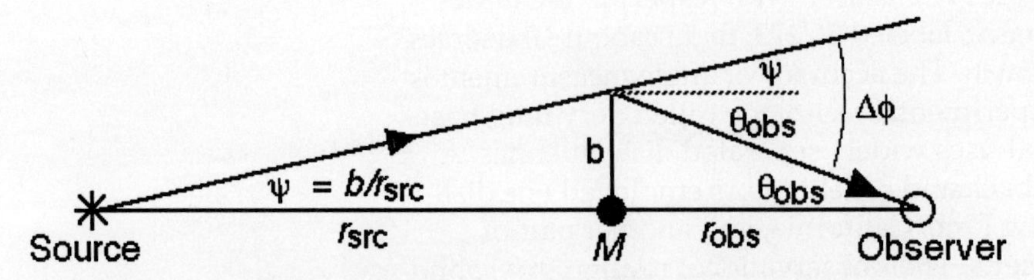

Figure 6. *Schematic diagram, not to scale, for the derivation of the angle of observation θ_{obs} for the Einstein ring at the eye of the observer. This figure defines the quantities r_{src} and r_{obs}.*

1. Assume that $r_{src} >> 2M$ and $r_{obs} >> 2M$ and $r_{src} >> b$ and $r_{obs} >> b$.

2. Use Euclidean geometry.

3. Assume that deflection occurs at a single point near the intermediate object. This is reasonable, since the distance from M to the source and the distance from M to the observer are both literally astronomical compared with the distance along the trajectory over which deflection takes place.

4. Measure the impact parameter b vertically, as shown in Figure 6, rather than perpendicular to the incoming ray of light as b is usually defined.

5. The sine or tangent of a small angle is approximately equal to the value of that angle in radians.

Apply approximations 1 and 5 to equation [38] on page 5-22, where the angle θ_{shell} is measured with respect to the radially *outward* direction. The angles called θ in Figure 6 are defined with respect to the radially *inward* direction, but the two have the same sine. (Recall the *Caution* in Sample Problem 3 on page 5-20.) Call the result θ_{obs} for "observation angle" and the radial distance from the intermediate object r_{obs}. Then from Figure 6,

$$\theta_{obs} = \frac{b}{r_{obs}} \tag{17}$$

Use the subscript "src" for "source" and write down the relation between angles shown at the upper right of Figure 6. The result is

$$\Delta\phi = \psi + \theta_{obs} = \frac{b}{r_{src}} + \theta_{obs} \tag{18}$$

QUERY 13 **Einstein ring angle.** Substitute equation [16] into [18] and use equation [17] to eliminate *b*. Show that the result is

$$\theta_{obs} = \theta_{Ein} = \left[\frac{4Mr_{src}}{r_{obs}(r_{obs}+r_{src})}\right]^{1/2} \qquad [19]$$

where the subscript "Ein" refers to the Einstein ring.

8 Microlensing

The visible stars in rotating galaxies do not have enough total mass to hold these galaxies together; these galaxies should fly apart as they rotate. But these galaxies do not fly apart. This and other lines of evidence convince observers that galaxies contain much more mass than can be counted in the visible stars they contain. This largely unknown presumed presence earns the name **dark matter.** Individual dark matter objects have been whimsically named **MAssive Compact Halo Objects,** abbreviated **MACHOs,** which presumably occupy the halo surrounding galaxies.

By what means could we detect this dark matter in, say, our own galaxy? In 1986 Bohdan Paczynski proposed as a practical procedure a method that had long been known only as a theoretical possibility. The method depends on the deflection of starlight by a dark object. When a dark object lines up between Earth and a distant visible star, the dark object focuses light that would otherwise miss Earth (Figure 5). The result is an increase in the amount of light received from the distant star, even when the equipment is not able to resolve the structure of the Einstein ring. This increase in focused light is called **microlensing.** Microlensing can be used to study objects in our galaxy that do not emit sufficient light to be detected by other means. Paczynski's proposal started a whole new field of observation, and dozens of microlensing events have been detected.

The great difficulty with microlensing is that, at any given time, only one star in 2 million or so will have its image augmented due to microlensing. As a result, Paczynski predicted, microlensing would be a rare event requiring the monitoring of many stars simultaneously.

To have the best chance to observe microlensing, one wants a background rich in visible stars to increase the chance that one of them will lie behind some dark object in the foreground. There are several such star-rich backgrounds for Earth observations: the galactic bulge in the center of our galaxy and the so-called Magellanic Clouds, two satellite galaxies of our Milky Way Galaxy visible in the southern hemisphere. (The name *Magellanic* comes from observation of these clouds by Ferdinand Magellan's crew during the first circumnavigation of the globe, 1519–1522 A.D.)

When a microlensing event occurs, the amount of light received on Earth from the distant star grows and fades over a period of days or weeks and at peak may be many times the normal flux of light from that star.

Microlensing is so recent a technique that path breaking developments occur from month to month. Preliminary results appear to show that the dark matter observed by microlensing does not account for the missing mass of our galaxy—the mass in addition to that of the visible stars needed to keep the galaxy from flying apart. See the references at the end of this project.

QUERY 14 **Identifying microlensing events.** Discussion questions: How can we distinguish between the increased amount of light from a star due to microlensing and greater light flux due to some internal mechanism of the star? The following inquiries are meant to help answer this question.

Will the increase in light due to microlensing be the same for all colors of the spectrum of light from the star? Suppose a natural increase in light is due to greater burning rate, leading to a higher temperature of the star. Do you expect that this process will result in a changed spectrum of light from the star?

Is a microlensing event likely to occur more than once for the same star? Might an increase in the flux of light due to internal processes occur more than once in a given star?

Do you expect that the time profile of a microlensing event—the curve of light intensity vs. time—can be predicted by astronomers? Is it likely that this time profile will be the same as that for an increase in light due to internal processes?

9 The Einstein Donut

Images on the cover of this book illustrate some gravitational effects of a black hole on the visual appearance of background objects.

The undistorted lower image of Saturn was taken with an infrared camera on the Hubble Space Telescope. In addition to Saturn's familiar rings, you see a dot at the lower left of the image, which is Saturn's moon Dione. Another moon, Tethys, forms a bright spot at the upper right edge of Saturn's round disk. (We brightened the images of these two satellites to make them more obvious.)

Now place the center of a black hole on the direct line of sight between us and Saturn so that we the observers are at a radial distance (reduced circumference) $r = r_{obs} = 10M$ on one side of the black hole. Saturn is at a *much* greater distance on the other side. We assume that Saturn, black hole, and the viewer are all relatively at rest, and that there is no change in the structure of Saturn (or us!) due to the curvature of spacetime induced by the black hole. The upper image on the cover results. What do we see?

First, Saturn itself is distorted into a donut surrounding a (nearly) black disk, to which we give the name **Einstein donut**. The donut shape results when the source is an apparent disk instead of a point. In what follows we will use a generalization of the Einstein ring equation [19] to explore major features of the donut. But we cannot legitimately calculate numerical values from the result, because its derivation depends on the assumption that $r_{obs} >> 2M$, whereas we placed the observer at $r_{obs} = 10M$ to create the upper cover image. This upper image is much more dramatic than the one seen by an observer who is at a much greater distance.

Second, we see duplicate distorted images of Saturn's rings. One image lies along the top of the outer surface of the donut, while the other image, reversed right-for-left and top-for-bottom, hugs the inside bottom of the donut.

Third, gravitational deflection smears the image of Saturn's moon Dione and creates a second, faint image just inside the Einstein donut on the upper right.

Fourth, a circular "diamond necklace" of dots shines faintly inside the Einstein donut.

QUERY 15 **Donut size.** The upper and lower images on the cover are displayed to the same scale. Explain in one sentence why the average diameter of the donut in the upper image is larger than the diameter of Saturn shown in the lower image.

Figure 7 is a generalization of Figure 6 for a source point off the axis between the observer and the center of the black hole. Here θ_o is the angle at which the source point would be viewed if the intermediate deflecting mass were removed.

Using Figure 7 and its caption we can repeat the derivation that led to equation [19], but for this new case. This derivation depends on all five assumptions (approximations) given on page D-10. The present derivation is a bit complicated and therefore optional. The result is:

$$\theta_{obs} = \theta_o + \frac{4Mr_{src}}{\theta_{obs} r_{obs}(r_{obs} + r_{src})} = \theta_o + \frac{\theta_{Ein}^2}{\theta_{obs}} \qquad [20]$$

where, recall, the Einstein ring angle θ_{Ein} (given by equation [19]) is the angle of observation in the special case that the source, deflecting mass, and observer all lie along the same straight line.

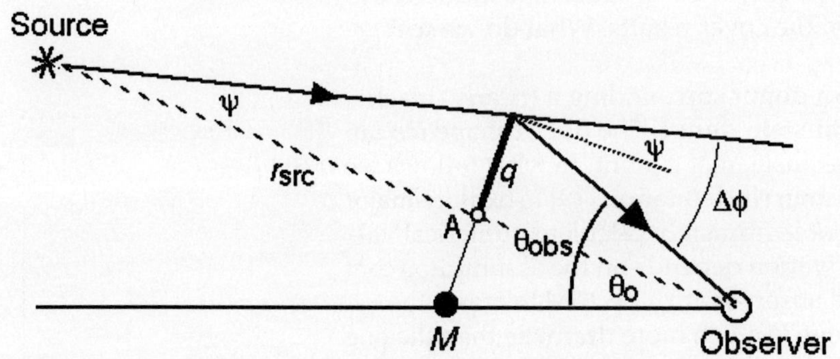

Figure 7 *Construction figure (not to scale) for the derivation of the observation angle* θ_{obs} *(equation [20]) for a source off the axis that runs between the observer and the deflecting mass. The angle* θ_o *is the observation angle in the absence of the deflecting mass M. The length q and construction angle* ψ *are used in the derivation of equation [20]. Some labels have been omitted for clarity. For example, the distance from the observer to point A is (to a good approximation) the observer radius* r_{obs}*. The small unmarked angle below the symbol* ψ *on the upper right has the value* $\theta_{obs} - \theta_o$ *.*

QUERY 16 **Observation angle for off-axis source.** Equation [20] is a quadratic equation. Show that the solution is

$$\theta_{obs} = \frac{\theta_o \pm (\theta_o^2 + 4\theta_{Ein}^2)^{1/2}}{2} \qquad [21]$$

What is the value of θ_{obs} when $\theta_o = 0$? What is the physical meaning of $\theta_o = 0$? What is the physical meaning of the plus or minus in front of your result for $\theta_o = 0$?

QUERY 17 **Off-axis source: First example.** Consider the special case $\theta_o = \theta_{o1} = 5^{1/2}\theta_{Ein}$. Using equation [21], find the two resulting values of θ_{obs1} . Why is one value negative? What does this negative value mean physically?

QUERY 18 **Off-axis source: Second example.** Find the two values of θ_{obs2} for a second undeflected angle $\theta_{o2} = 12^{1/2}\theta_{Ein}$. Compare your results to those obtained in Query 17. Is θ_{obs2} greater or less than θ_{obs1}? (Careful!)

We need to keep reminding ourselves that equation [21] does *not* predict numerically correct details of the donut structure in the upper cover image, because $r_{obs} = 10M$ for that figure, violating our assumption that $r_{obs} >> 2M$ that led to [21]. Nevertheless, *qualitative* features of the upper image make sense when analyzed using the results of Queries 16 through 18. (The "diamond necklace" circling the center of the upper image results from a far greater deflection of light near the black hole and cannot be described at all by the approximate equation [21]. Section 10 discusses the diamond necklace.)

QUERY 19 **Features of Einstein donut on the cover.** Refer to points labeled A, B, C in Figure 8, which shows a top view of Saturn, its ring, and a distant observer. For each of the points A, B, and C the angle θ_{obs} will have double values. When examining these values, distinguish between angles at which the observer looks to the LEFT of center, as indicated in Figure 8, and angles at which the observer looks to the RIGHT. Use your results for Queries 16–18 to answer the following qualitative questions:

1. For each of the points A, B, C as seen on the LEFT, is the magnitude of θ_{obs} greater than or less than the magnitude of θ_o, the angle of observation in the absence of the deflecting mass M?

2. For each of the points A, B, C as seen on the RIGHT, is the magnitude of θ_{obs} greater than or less than the magnitude of θ_o?

3. Reading from left to right, what is the *order* of points A, B, C as seen on the LEFT side of the center of Saturn?

4. Reading from left to right, what is the *order* of points A, B, C as seen on the RIGHT side of the center of Saturn?

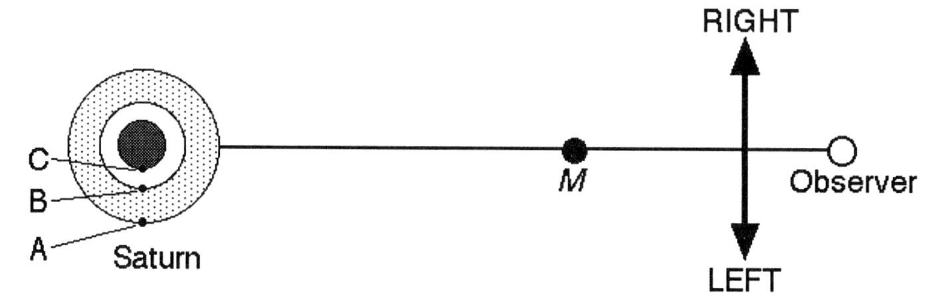

Figure 8 *Top view of Saturn and its ring as if looking down from above on Saturn displayed in the lower image on the cover. Not to scale. Query 19 asks questions about the angles at which the observer at the right sees points A, B, and C when the black hole M is in place.*

QUERY 20 **Describing the cover.** Write a paragraph that describes for the layperson the two images on the cover of this book. Do not attempt to explain the physics behind these transformations and distortions. Instead, tell in some detail where portions from the lower image appear on the upper image. Use as an example the shadow on the rings to the left of Saturn's disk in the undistorted lower image. How many times (and where) does this shadow appear in the upper image? Show the cover and your resulting paragraph to a roommate or friend and revise your description in light of his or her reaction.

10 The Diamond Necklace

Circling around inside the black middle of the Einstein donut on the cover is a faint ring, a "diamond necklace" of dots. This necklace shines with light that, on its way from Saturn to our eye, circles once around the black hole just outside the photon sphere at $r = 3M$ (Figure 9). The necklace is very thin and faint because the large spacetime curvature near the black hole fans out the light rays like a demagnifying lens. (Gravitational "lenses" obviously create horrible distortions!) In principle there are infinitely many concentric circular necklace images, resulting from light circling the black hole 1, 2, 3, 4 . . . times from the source to our eye. The greater the number of rotations, the closer the radius of the orbit is to that of the light sphere at $3M$. But the images for greater number of revolutions are so strongly defocused that only the first one is visible. Light deflections leading to this necklace are so extreme that the approximate Einstein ring analysis of Section 7 is useless.

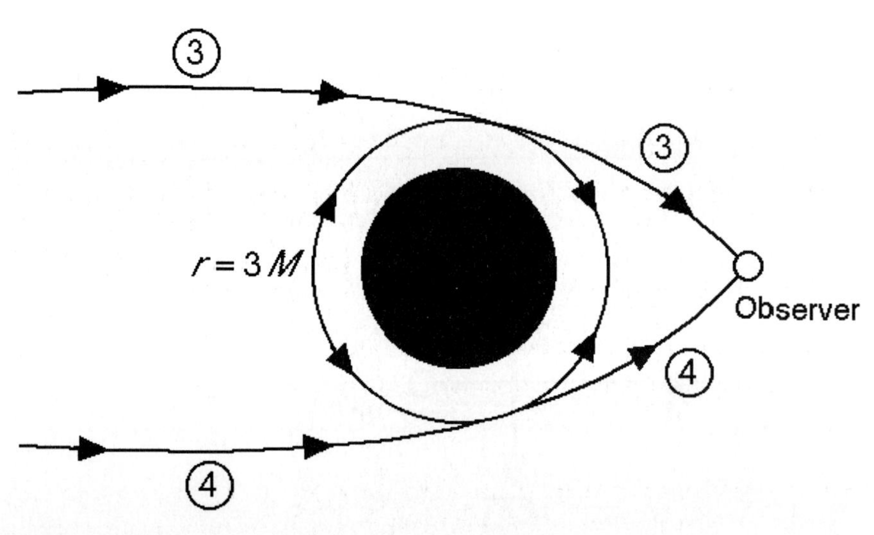

Figure 9 *Schematic Schwarzschild map showing sample trajectories of the light that forms the diamond necklace in the center of the Einstein donut on the cover. The circular portions of these paths are actually just outside the radius of the photon sphere at r = 3M.*

QUERY 21 **Viewing angle of the necklace.** The upper image on the cover is made for an observer at rest on a shell of radius (reduced circumference) $r = 10M$. Substitute this value into equation [31], page 5-20, and find the observation angle at which the shell observer sees the necklace. Express your answer in degrees. This analysis assumes that the light circles the black hole at $r = 3M$, whereas the r-value of this circle is slightly more than $3M$. (Note that your angle is measured with respect to the radially *inward* direction; see the *Caution* at the end of the box on page 5-20.)

What would the upper image on the cover look like if light were not deflected by "gravity" (not affected by curved spacetime)? How can we answer this question, since both of our theories, the general-relativistic and the Newtonian, predict deflection of light by gravity? Instead we ask another question. Replace the black hole with a thin spherical shell of so little mass that gravitational effects can be ignored. Now we can validly ask how large this spherical shell looks to an observer. Even then we need to be careful in specifying how the size of the new spherical shell compares with that of the original black hole.

QUERY 22 **Observation angle for a black spherical shell.** The black hole between Saturn and the observer is replaced with a thin metal spherical shell painted black. This spherical shell has a negligible mass and is given a radius (measured by Euclidean methods) with the same numerical value as the reduced circumference of the original black hole's horizon. An observer views the sphere at a distance from the center equal to five times its radius.

A. *Guess*: Will the viewer see the radius of this sphere to be less than or more than the radius of the diamond necklace with the black hole in place? Write down your guess.

B. Now calculate the angle between the center of the disk and its edge as seen by the observer. Compare this angle with that between the center and the necklace for the original black hole. Was your guess in part A correct?

C. From your result in B, calculate the observed radius of the black spherical shell as a fraction of the radius of the observed diamond necklace in the *upper* image of Saturn on the cover. Cut a disk of this size out of black paper and tape it temporarily at the center of the *lower* image of Saturn on the cover. This is as close as we can come to showing the resulting image of Saturn "if light were not deflected by gravity."

11 References and Acknowledgment

Edmund Bertschinger and Miriam L. Castellano provided much of the material and text for Sections 7, 9, and 10, which analyze their images on the cover.

Initial quote: Ilse Rosenthal-Schneider in *Some Strangeness in the Proportion*, edited by Harry Woolf, Addison-Wesley, Reading, MA, 1980, page 523.

Other treatments of the deflection of starlight by Sun: Robert M. Wald, *General Relativity*, University of Chicago Press, 1984, pages 143–146, and Steven Weinberg, *Gravitation and Cosmology*, John Wiley and Sons, New York, 1972, pages 188–191.

For analysis of large deflections of light near a Schwarzschild black hole and analysis of intensity of this light, see Hans C. Ohanian, "The Black Hole as Gravitational 'Lens'," *American Journal of Physics*, Volume 55, pages 428–432 (1987).

For large deflection of light, see W. M. Stucky "The Schwarzschild Black Hole as a Gravitational Mirror," *American Journal of Physics*, Volume 61, pages 448– 456 (1993).

Table 1 was adapted with permission from Steven Weinberg, *Gravitation and Cosmology*, John Wiley and Sons, New York, 1972, pages 193–194. The references shown in the table are as follows:

a. F. W. Dyson, A. S. Eddington, and C. Davidson, *Philosophical Transactions of the Royal Society*, Volume 220A, page 291 (1920); *Memoirs of the Royal Astronomical Society*, Volume 62, page 291 (1920).

b. G. F. Dodwell and C. R. Davidson, *Monthly Notices of the Royal Astronomical Society*, Volume 84, page 50 (1924).

c. C. A. Chant and R. K Young, *Publications of the Dominion Astronomical Observatory*, Volume 2, page 275 (1924).

d. W. W. Campbell and R. Trumpler, *Lick Observatory Bulletins*, Volume 11, page 41 (1923); *Publications of the Astronomical Society of the Pacific*, Volume 35, page 158 (1923).

e. W. W. Campbell and R. Trumpler, *Lick Observatory Bulletins*, Volume 13, page 130 (1928).

f. E. F. Freundlich, H. von Klueber, and A. von Brunn, *Abhandlungen der Preussischen Akademie der Wissenschaften*, No. 1, 1931; *Zeitschrift für Astrophysik*, Volume 3, page 171 (1931).

g. A. A. Mikhailov, *Doklady Akademii nauk SSSR: Comptes rendues de l'Académie des sciences de l'URSS*, Volume 29, page 189 (1940).

h. T. Matukuma, A. Onuki, S. Yosida, and Y. Iwana, *Japanese Journal of Astronomy and Astrophysics*, Volume 18, page 51 (1940).

i. G. van Biesbroeck, *Astronomical Journal*, Volume 55, pages 49 and 247 (1949).

j. G. van Biesbroeck, *Astronomical Journal*, Volume 58, page 87 (1953).

Results of the VLBI radio-astronomy determination of deflection is from D. E. Lebach, B. E. Corey, I. I. Shapiro, M. I. Ratner, J. C. Webber, A. E. E. Rogers, J. L. Davis, and T. A. Herring, "Measurement of the Solar Gravitational Deflection of Radio Waves Using Very-Long-Baseline

Interferometery," *Physical Review Letters,* Volume 75, Number 8, pages 1439–1442 (1995).

Einstein rings, See A. Einstein, "Lens-like action of a star by the deviation of light in the gravitational field," *Science,* Volume 84, pages 506–507 (1936).

Image of Saturn, lower image on the cover, is from the Hubble Space Telescope, taken by Erich Karkoschka, University of Arizona, and the United States National Aeronautics and Space Administration. Upper image on the cover is by Edmund Bertschinger and Miriam L. Castellano.

Microlensing paper describing an update of his original proposal: B. Paczynski, "Gravitational Microlensing of the Galactic Bulge Stars," *The Astrophysical Journal Letters,* Volume 371, pages L63–L67 (1991).

Report on two early projects to look for microlensing events: C. Alcock, R. A. Allsman, D. Alves, R. Ansari, E. Aubourg, T. S. Axelrod, P. Bareyre, J.-Ph. Beaulieu, A. C. Becker, D. P. Bennett, S. Brehin, F. Cavalier, S. Char, K. H. Cook, R. Ferlet, J. Fernandez, K. C. Freeman, K. Griest, Ph. Grison, M. Gros, C. Gry, J. Guibert, M. Lachieze-Rey, B. Laurent, M. J. Lehner, E. Lesquoy, C. Magneville, S. L. Marshall, E. Maurice, A. Milsztajn, D. Minniti, M. Moniez, O. Moreau, L. Prevor, F. Queinnec, P. J. Quinn, C. Renault, J. Rich, M. Spiro, C. W. Stubbs, W. Sutherland, A. Tomaney, T. Vandehei, A. Vidal-Madjar, L. Vigroux, and S. Zylberajch, "EROS and MACHO Combined Limits on Planetary-Mass Dark Matter in the Galactic Halo," *Astrophysical Journal Letters,* Volume 499, pages L9–L12 (1998).

A current microlensing project, called OGLE-II, is hitting its stride in 1999. Use this title as a keyword for searching the World Wide Web.

Project E Light Slowed Near Sun

- *What happened to "Light always moves with the same speed"?*
- *Who says light slows down near Sun?*
- *How much does light slow down near Sun?*
- *Does observation verify the predicted value of the slow-down?*

Project E

Light Slowed Near Sun

I sometimes ask myself how it came about that I was the one to develop the theory of relativity. The reason, I think, is that a normal adult never stops to think about problems of space and time. These are things which he [or she] has thought about as a child. But my intellectual development was retarded, as a result of which I began to wonder about space and time only when I had already grown up.

—Albert Einstein

1 Introduction

The Schwarzschild bookkeeper records a "smaller speed of light" than do shell or free-float observers. Near the black hole, bookkeeper light speed is decisively less than unity. That is the prediction of our analysis in equations [3], [4], and [16], pages 5-3 and 5-7. Does this prediction have a physical meaning at all? Is there any way to measure this "slowing down of the speed of light" as reckoned by remote observers?

"Yes" and "yes" were the answers predicted by Irwin Shapiro and demonstrated by him and coworkers. They showed that the total time of transit of a light flash along a trajectory in space that brushes past our Sun will be longer than that predicted for flat spacetime. Then they measured the time delay of a round-trip radio signal—an electromagnetic wave of lower frequency than light but traveling at the speed of light—first between Earth and Mercury, later between Earth and Venus, finally between Earth and Mars when these planets were situated on opposite sides of Sun (Figure 1 shown for Mars). As the Earth observer sees the image of Venus approaching the edge of Sun, for example, the time for a round-trip for the radio signal increases by approximately 190 microseconds (see Figure 3) due to the slowing of light near Sun. Later results for Mars, good to one part in a thousand, are much more accurate than those for Venus.

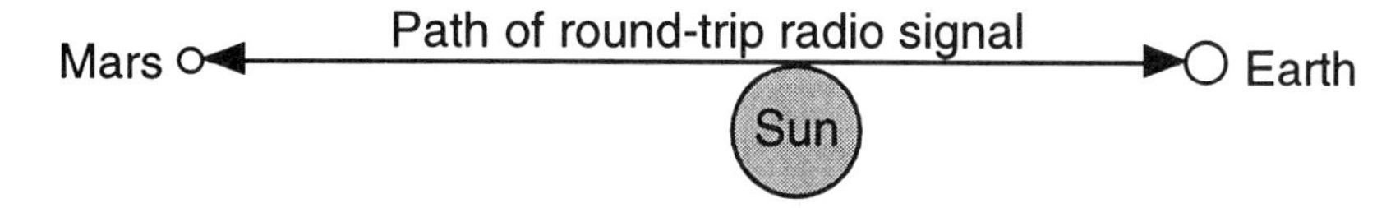

Figure 1 *Diagram, not to scale, of the path of radio signals that graze the surface of Sun on their round-trip between tracking stations on Earth, right, and a Viking lander on the surface of Mars, left. Note: The path bends near Sun, but by an angle less than 2 seconds of arc, far too small to show on this or later diagrams.*

How can anyone possibly detect the minuscule 190-microsecond delay in the very much longer round-trip time light requires to go from Earth past Sun to Venus, then back again to Earth along the return path? This round-trip requires nearly half an hour. This difficulty is compounded because Earth and Mars are in different orbits, so the distance between them is constantly changing!

This difficulty is one of many overcome by Shapiro and his coworkers. (Another is the uncertainty of the speed of their radio waves passing through the cloud of ionized atoms near the surface of Sun—called Sun's **corona**.) When Venus is not near Sun as viewed from Earth, time-delay effects are small. In principle one can track the round-trip time for the signal as Venus approaches, passes, and leaves the vicinity of Sun as viewed from Earth. The gradually changing Earth-Venus distance leads to a smooth curve of time delay, on which the additional delay due to slowed propagation appears as a spike, shown for Venus in Figure 3. In practice Shapiro and his colleagues employed a much more sophisticated analysis that made use of all relevant data simultaneously.

Can we predict the value of approximately 190 microseconds (Venus) or the more accurately determined 250 microseconds (Mars) for the round-trip delay of light or a radio pulse that skims past the surface of Sun? Yes, if we use an approximation, described and spelled out in what follows.

2 Approximating the Light Path

The actual "skimming" path of light approaches Sun obliquely along the thin line in Figure 2. Along the actual path the light moves neither along a radius nor tangentially (except tangentially at the grazing point). Therefore we cannot use our expressions for radial and tangential velocity, equations [3] and [4] on page 5-3. Even the expression for the skimming path, equation [16] page 5-7, is awkward to integrate. Instead we obtain an estimate of the delay time by assuming a slightly different path for the light, a path that comes in radially from Earth, circles halfway around Sun in a tangential semicircular path, then moves out radially straight outward to Venus or Mars (thick curve in Figure 2). This alternative path provides a crude model of the actual path, but yields an estimate much closer to a precise calculation than we might have expected. Now for the details.

3 Radial Path Segments

Begin with Segment 1 in Figure 2. Equation [3], page 5-3 says that for radial motion of light (plus or minus sign omitted):

$$\frac{dr}{dt} = 1 - \frac{2M}{r} \qquad [1]$$

(We also omit the label $m = 0$ on equations that describe the motion of light, since *all* equations in this project describe the motion of light.) From equation [1] we can find the total far-away time t required for the light to travel from the radius R of Sun to the distant radial position r_{Eorbit} of Earth. (It does not matter which way we make the integration: Assume that the light moves outward.) Then the total time for Segment 1 comes from integrating the expression

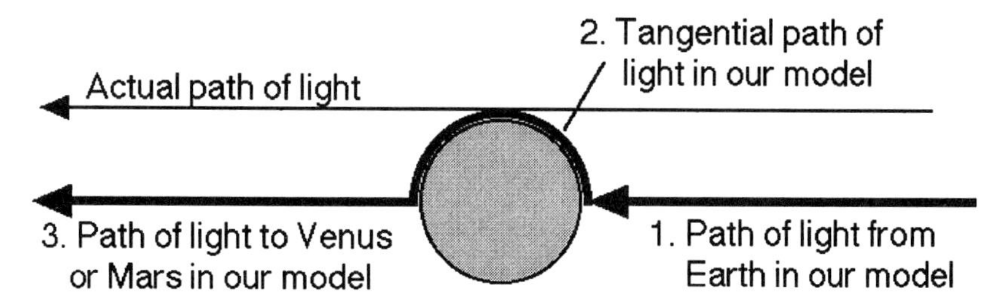

Figure 2 *Our model of time delay of light or radio signal grazing the surface of Sun on its way from Earth to Venus (or Mars). The actual path is along the thin line. Our model replaces this actual path with three segments, numbered above. Segment 1: Radial path inward from Earth to surface of Sun. Segment 2: Tangential path that skims the surface of Sun in a semicircle. Segment 3: Radial path outward from surface of Sun to Venus (or Mars).*

$$dt = \frac{dr}{1 - \frac{2M}{r}} \qquad [2]$$

Now, the fraction $2M/r$ is very small. Its largest value occurs at the surface of Sun. In round numbers, $2M \approx 3000$ meters and $R \approx 7 \times 10^8$ meters. So the maximum value of $2M/r$ is less than 4×10^{-6}, which is very much less than unity. Therefore we can use our favorite simple approximation (the first of several times in this project):

$$(1 + d)^n \approx 1 + nd \qquad \text{provided that} \qquad |d| \ll 1 \quad \text{and} \quad |nd| \ll 1 \qquad [3]$$

This approximation works for positive and negative d and for positive, negative, or fractional powers n.

QUERY 1 **Earth-Sun time delay.** Apply the approximation [3] to equation [2] and integrate both sides of the resulting equation from R, the radius of Sun, to r_{Eorbit}, the radial distance of Earth from the center of Sun. Show that the result has two terms, one of which is simply the distance from Sun's surface to Earth, equal to the time in meters for light to travel this distance at the conventional speed unity. The second term is the expected delay for light predicted by general relativity. Show that this delay Δt_1 along Segment 1 has the form:

$$\Delta t_1 = 2M \ln\left(\frac{r_{\text{Eorbit}}}{R}\right) \qquad [4]$$

QUERY 2 **Value of Earth-Sun time delay.** Substitute values from inside the back cover to show that the time delay Δt_1 along the first segment has the approximate value 53 microseconds.

QUERY 3 **Sun-Venus time delay.** Apply the same analysis to Segment 3, from Sun to Venus. (Recall that direction of radial motion does not matter; light moves more slowly in both directions.) The average radius of the orbit of Venus is 1.08×10^{11} meters. What is the time delay for this segment of the path for Venus?

QUERY 4 **Sun-Mars time delay.** Use the value $r_{Morbit} = 2.28 \times 10^{11}$ meters for the average radius of Mars' orbit around Sun. Show that the segment from Sun to Mars has a time delay of approximately 57 microseconds.

4 Tangential Path Segment

For Segment 2, the portion of our model trajectory that follows a semicircle around Sun in Figure 2, we assume that the speed is the tangential speed, equation [4], page 5-3 (plus or minus sign omitted):

$$\frac{ds}{dt} = r\frac{d\phi}{dt} = \left(1 - \frac{2M}{r}\right)^{1/2} \qquad [5]$$

We assume that $r = R$ for the entire Segment 2 in Figure 2. Then the time t_2 for the light to traverse Segment 2 is obtained by solving equation [5] for dt and summing it along the semicircular path of length πR.

$$t_2 = \frac{\pi R}{\left(1 - \frac{2M}{R}\right)^{1/2}} \qquad [6]$$

QUERY 5 **Time around Sun.** Use the approximation in [3] again to obtain the transit time for Segment 2. As before, show that result has two parts, one of which is simply the expected time for light to move in a semicircle around Sun at speed unity. The second part is the delay Δt_2 predicted by general relativity. Show that this delay is approximately 15 microseconds.

5 Comparing Prediction and Observation

QUERY 6 **Total round-trip time delay for Venus.** Add up the time delays for the three segments of the one-way trip between Earth and Venus. Multiply by 2 for the round-trip. Figure 3 shows results for Venus when, as seen from Earth, Venus approached, but did not reach, the edge of Sun. *Should* your result for the edge of Sun be more or less than the maximum in Shapiro's plot? Is it?

QUERY 7 **Total round-trip time delay for Mars.** Now add up the time delays for the three segments of the one-way trip between Earth and Mars. Multiply by 2 to describe the round-trip, and show that the total round-trip delay is approximately 250 microseconds.

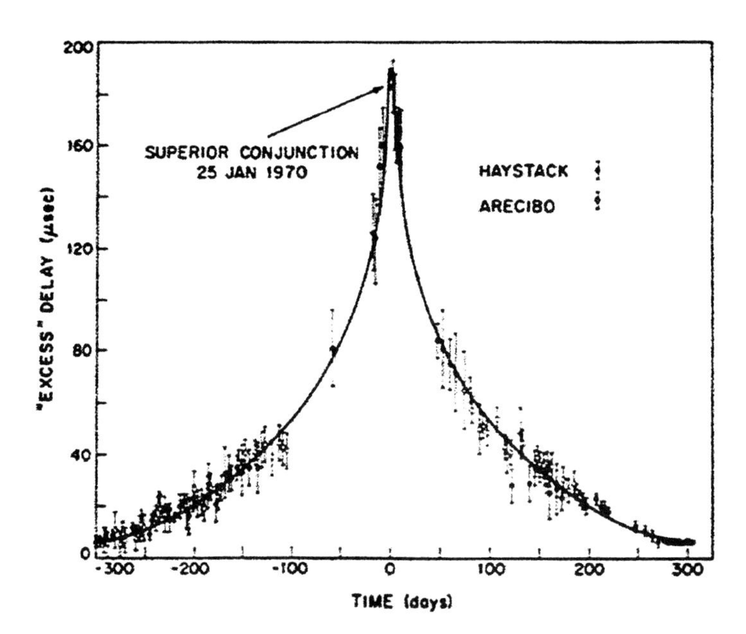

Figure 3 *Total time delay of round-trip radio signal to Venus as measured by Irwin Shapiro and colleagues in January 1970. As seen from Earth, Venus approached, but did not reach, the edge of Sun. In this version of the experiment the signal was reflected from Venus. The solid curve is the prediction of general relativity. The dots and vertical error bars represent experimental determinations. See the reference at end of this project.*

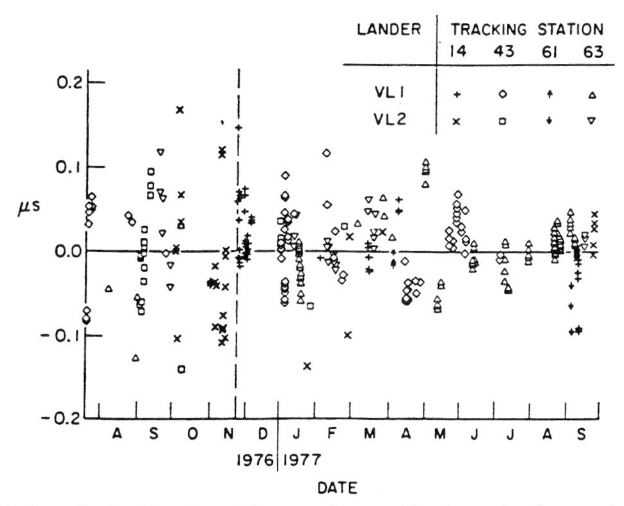

Figure 4 *Final results for Mars time delay experiments. Shapiro and colleagues increased the amplitude of the return signal greatly by using a Viking lander on the surface of Mars. The radio signal from Earth triggered a return signal from the Viking lander. This figure shows the differences (technical term:* ***residuals****) between the observed data and the predictions of general relativity. The dashed vertical line marks the time of closest approach of Mars to Sun on November 25, 1976. This figure looks much less dramatic than Figure 3 but "squeezes" general relativity much harder. Note that the vertical scale is marked in tenths of a microsecond. See the reference at end of this project.*

How lucky we are! We calculated 250 microseconds for the Earth-Mars round-trip delay, a value very close to the observed result. Our calculated result is also very close to the more precise value calculated by Shapiro using (1) the straight path that skims past Sun and (2) the Earth-Sun-Mars distances correct at the time of the experiment.

Figures 3 and 4 present observed results of Shapiro and colleagues for Venus and Mars respectively.

QUERY 8 **Distance to Moon.** Astronauts have left several corner reflectors on the surface of Moon. These are shaped like the corner of a room and have the property that they reverse the path of any laser pulse incident on them. As described by Clifford Will (Chapter 7 in the reference at the end of this project), a round-trip laser pulse sent from Earth can be timed to measure the Earth-Moon distance between laser emitter and Moon reflector. Initially the uncertainty was about 15 centimeters; more recently the uncertainty has been reduced to the one-centimeter range. Does the Shapiro time delay have to be taken into account in doing this measurement? Will the distance measurement be wrong if no account is taken of the reduced speed of light in the curved spacetime near Earth and Moon? If so, what will be the approximate error in the measurement? In traveling to Moon, remember, the pulse moves outward through Earth's gravitational field and inward through Moon's gravitational field. Directions are reversed on the return trip, but the time delay adds for each direction. *Hints:* (1) Adapt equation [4] of the text separately for Earth and Moon. (2) The average distance between the centers of Earth and Moon is given inside the back cover. (3) The "equilibrium point" where (as Newton would say) the gravitational attraction is equal and opposite to Earth and Moon is at 0.90 the distance from Earth's center to Moon's center. (4) The ratio of masses is $M_{Moon}/M_{Earth} = 0.0123$. (5) The ratio of Moon to Earth radii is $r_{Moon}/r_{Earth} = 0.273$.

QUERY 9 **Delayed replay.** *Discussion question.* You are in a stable circular free-float orbit around a black hole. You send yourself a message coded on a laser signal, firing your laser pulse inward so that its trajectory lies across your orbit, approaching nearer the black hole than you do. Can you use the light-retardation effect in such a way that you receive your own signal when you arrive somewhere else in your orbit? If not, why not? If so, what limits are there on the amount of delay you can experience in receiving your message? *Hint:* Look at Figure 6 on page 5-14.

6 References and Acknowledgments

Initial quote, *The Quotable Einstein,* edited by Alice Calaprice, Princeton University Press, 1996, page 182, and *Albert Einstein* by Albrecht Fölsing, Penguin Books, New York, 1997, page 13.

Figure 3. "Fourth Test of General Relativity: New Radar Result," Irwin I. Shapiro, Michael E. Ash, Richard P. Ingills, William B. Smith, Donald B. Campbell, Rolf B. Dyce, Raymond F. Jurgens, and Gordon H. Pettengill, *Physical Review Letters,* Volume 26, Number 18, pages 1132–1135 (3 May 1971).

Figure 4. R. D. Reasenberg, I. I. Shapiro, P. E. MacNeil, R. B. Goldstein, J. C. Breidenthal, J. P. Brenkle, D. L. Cain, T. M. Kaufman, T. A. Komarek, and A. I. Zygielbaum, "Viking Relativity Experiment: Verification of Signal Retardation by Solar Gravity," *Astrophysical Journal Letters,* Volume 234, pages L219–L221 (15 December 1979).

The fascinating story of experiments verifying the time delay of radio signals passing near Sun is told by Clifford M. Will in his book *Was Einstein Right? Putting General Relativity to the Test,* Second Edition, Basic Books/ Perseus Group, New York, 1993, Chapter 6.

Irwin Shapiro read and commented on this project.

Query 8 was suggested by Glen Govertsen. Query 9 was suggested by Charles Holbrow.

Project F The Spinning Black Hole

- *How fast can a black hole spin?*
- *How can I observe space dragging around a spinning black hole?*
- *Does a rapidly spinning black hole keep me from falling through the horizon?*
- *How many Sun masses per year does a quasar convert to light?*

PROJECT F

The Spinning Black Hole

Black holes are macroscopic objects with masses varying from a few solar masses to millions of solar masses. To the extent they may be considered as stationary and isolated, to that extent, they are all, every single one of them, described exactly by the Kerr solution. This is the only instance we have of an exact description of a macroscopic object. Macroscopic objects, as we see them all around us, are governed by a variety of forces, derived from a variety of approximations to a variety of physical theories. In contrast, the only elements in the construction of black holes are our basic concepts of space and time. They are, thus, almost by definition, the most perfect macroscopic objects there are in the universe. And since the general theory of relativity provides a single unique two-parameter family of solutions for their description, they are the simplest objects as well.

—S. Chandrasekhar

1 Introduction

In this project we explore some of the properties of spacetime near a spinning black hole. Analogous properties describe spacetime external to the surface of the spinning Earth, Sun, or other spinning uncharged heavenly body. For a black hole these properties are truly remarkable. Near enough to a spinning black hole—even outside its horizon—you cannot resist being swept along tangentially in the direction of rotation. You can have a negative total energy. From outside the horizon you can, in principle, harness the rotational energy of the black hole.

Do spinning black holes exist? The primary question is: Do black holes exist? If the answer is yes, then spinning black holes are inevitable, since astronomical bodies most often rotate. As evidence, consider the most compact stellar object short of a black hole, the neutron star. Detection of radio and X-ray pulses from some spinning neutron stars (called **pulsars**) tells us that many neutron stars rotate, some of them very rapidly. These are impressive structures, with more mass than our Sun, some of them spinning once every few milliseconds. Conclusion: If black holes exist, then spinning black holes exist.

General relativity predicts that when an isolated spinning star collapses to a black hole, gravitational radiation quickly (in a few seconds of far-away time!) smooths any irregularities in rotation. Thereafter the metric exterior to the horizon of the spinning black hole will be the Kerr metric used in this project.

However, the typical spinning black hole is not isolated; it is surrounded by other matter that is attracted to it. The inward-swirling mass of this **accretion disk** may affect spacetime in its vicinity, distorting the metric away from that of the isolated spinning black hole that we analyze here.

2 Angular Momentum of the Black Hole

An isolated spinning uncharged black hole is completely specified by just two quantities: its mass M and its angular momentum. In Chapter 4 (page 4-3) we defined the angular momentum per unit mass for a particle orbiting a nonspinning black hole as $L/m = r^2 d\phi/d\tau$. In this expression, the angle ϕ has no units and proper time τ has the unit meter. Therefore L/m has the unit meter. To avoid confusion, the angular momentum of a spinning black hole of mass M is given the symbol J and its angular momentum per unit mass is written J/M. The ratio J/M appears so often in the analysis that it is given its own symbol: $a = J/M$. We call the constant "a" the **angular momentum parameter.** Just as the angular momentum L/m of a stone orbiting a non-rotating black hole has the unit meter, so does the angular momentum parameter $a = J/M$ have the unit meter. In what follows it is usually sufficient to treat the angular momentum parameter a as a positive scalar quantity.

Newman and others found the metric for a spinning black hole with net electric charge (see equation [51] and references at the end of this project). The most general steady-state black hole has mass, angular momentum, and electric charge. However, we have no evidence that astrononiical bodies carry sufficient net electric charge (which would ordinarily be rapidly neutralized) to affect the metric. If actual black holes are uncharged, then the Kerr metric describes the most general stable isolated black hole likely to exist in nature.

3 The Kerr Metric in the Equatorial Plane

For simplicity we are going to study spacetime and particle motion in the **equatorial plane** of a symmetric spinning black hole of angular momentum J and mass M. The equatorial plane is the plane through the center of the spinning black hole and perpendicular to the spin axis.

Here is the **Kerr metric** in the equatorial plane, expressed in what are called **Boyer-Lindquist coordinates**. The angular momentum parameter a appears in a few unaccustomed places.

$$d\tau^2 = \left(1 - \frac{2M}{r}\right)dt^2 + \frac{4Ma}{r}dt\,d\phi - \frac{dr^2}{1 - \frac{2M}{r} + \frac{a^2}{r^2}} - \left(1 + \frac{a^2}{r^2} + \frac{2Ma^2}{r^3}\right)r^2 d\phi^2 \quad [1]$$

For the *nonrotating* black hole examined in Chapters 2 through 5, the Schwarzschild metric describing spacetime on a plane is the same for *any* plane that cuts through the center of the black hole, since the Schwarzschild black hole is spherically symmetric. The situation is quite different for the spinning Kerr black hole; the metric [1] is correct *only* for the plane passing through the center of the black hole and perpendicular to its axis of rotation. We choose the equatorial plane because it leads to the simplest and most interesting results.

The time t in equation [1] is the "far-away time" registered on clocks far from the center of attraction, just as for the Schwarzschild metric. In contrast, for $a > 0$ the Boyer-Lindquist r-coordinate does *not* have the simple geometrical meaning that it had for the Schwarzschild metric. More on the meaning of r in Sections 4 and 9. The metric [1] provides a *complete* description of spacetime in the equatorial plane outside the horizon of a spinning uncharged black hole. No additional information is needed to answer every possible question about its (nonquantum) properties and (with the Principle of Extremal Aging) about orbits of free particles and light pulses in the equatorial plane.

You say that the Kerr metric provides a complete nonquantum *description of the spinning black hole. Why this reservation? What more do we need to know to apply general relativity to quantum phenomena?*

In answer, listen to Stephen Hawking as he discusses the "singularity" of spacetime at the beginning of the Universe. A similar comment applies to the singularity inside any black hole.

The general theory of relativity is what is called a classical theory. That is, it does not take into account the fact that particles do not have precisely defined positions and velocities but are "smeared out" over a small region by the uncertainty principle of quantum mechanics that does not allow us to measure simultaneously both the position and the velocity. This does not matter in normal situations, because the radius of curvature of space-time is very large compared to the uncertainty in the position of a particle. However, the singularity theorems indicate that space-time will be highly distorted, with a small radius of curvature at the beginning of the present expansion phase of the universe [or at the center of a black hole]. In this situation, the uncertainty principle will be very important. Thus, general relativity brings about its own downfall by predicting singularities. In order to discuss the beginning of the universe [or the center of a black hole], we need a theory that combines general relativity with quantum mechanics.

—Stephen Hawking

Suggestion: As you go along, check the units of all equations, the equations in the project and also your own derived equations. An equation can be wrong if the units are right, but the equation cannot be right if the units are wrong!

Do Spinning Black Holes Power Quasars?

In contrast to dead solitary black holes, the most powerful steady source of energy we know or conceive or see in all the universe may be powered by a spinning black hole of many millions of solar masses, gulping down enormous amounts of matter swirling around it. Maarten Schmidt, working at the Palomar Mountain Observatory in 1956, was the first to uncover evidence for these **quasi-stellar objects**, or **quasars**, starlike sources of light located not billions of kilometers but billions of light-years away. Despite being far smaller than any galaxy, the typical quasar manages to put out more than a hundred times as much energy as our entire Milky Way with its hundred billion stars. Quasars—unsurpassed in brilliance and remoteness—can justly be called lighthouses of the heavens.

Observation and theory have come together to explain in broad outline how a quasar operates. A spinning black hole of some hundreds of millions of solar masses, itself perhaps built by accretion, accretes more mass from its surroundings. The incoming gas, and stars converted to gas, does not fall in directly, any more than the water rushes directly down the bathtub drain when the plug is pulled. This gas, as it goes round and round, slowly makes its way inward to regions of ever-stronger gravity. In the process it is compressed and heated and finally breaks up into positive ions and electrons, which emit copious amounts of radiation at many wavelengths. The in-falling matter brings with it some weak magnetic fields, which are also compressed and powerfully strengthened. These magnetic fields link the swirling electrons and ions into a gigantic accretion disk. Matter little by little makes its way to the inner boundary of this accretion disk and then, in a great swoop, falls across the horizon into the black hole. During that last swoop, hold on the particle is relinquished. Therefore, the chance is lost to extract as energy the full 100 percent of the mass of each in-falling bit of matter. However, magnetic fields do hold on to the ions effectively enough and long enough to extract, as radiant energy, several percent of the mass. In contrast, neither nuclear fission nor nuclear fusion is able to obtain a conversion efficiency of more than a fraction of 1 percent. Of all methods to convert bulk matter into energy, no one has ever seen evidence for a more effective process than accretion into a spinning black hole, and no one has ever been able to come up with a more feasible scheme to explain the action of quasars. See Section 11 for more details.

QUERY 1 **Equatorial-plane Kerr metric in the limit of zero angular momentum.** Show that for zero angular momentum ($a = J/M = 0$), the Kerr metric, equation [1], reduces to the Schwarzschild metric (equation [A] in Selected Formulas at the end of this book).

QUERY 2 **Motion stays in plane.** Make an argument from symmetry that a free object that begins to orbit a spinning black hole in the equatorial plane will stay in the equatorial plane.

The Kerr metric has four central new features that distinguish it from the Schwarzschild metric.

The first new feature of the Kerr metric is a new *r*-value for the horizon. In the Schwarzschild metric, the coefficient of dr^2 is $1/(1 - 2M/r)$. This coefficient increases without limit at the Schwarzschild horizon, $r_H = 2M$. For the Kerr metric, in contrast, the horizon—the point of no return—has an r-value that depends on the value of the angular momentum parameter a. (*Note:* A true proof that a horizon exists requires the demonstration that worldlines can run through it only in the inward direction, not outward. See Project B, pages B-14–15. Our choice of the horizon at the place where the coefficient of dr^2 blows up is an intuitive, if correct, choice.)

QUERY 3 **Radial coordinate of the horizon.** Show that for the spinning black hole, the coefficient of dr^2 increases without limit at the r-value:

$$r_H = M \pm (M^2 - a^2)^{1/2} \qquad [2]$$

Look first at the case with the plus sign. What value does r_H have when $a = 0$? For a spinning black hole, is the value of r_H greater or less than the corresponding r-value for the Schwarzschild horizon?

Unless stated otherwise, when we say "the horizon" we refer to equation [2] with the plus sign.

Research note: Choosing the minus sign in equation [2] leads to a second horizon that is *inside* the outer, plus-sign horizon. This inner horizon is called the **Cauchy horizon**. Theoretical research shows that spacetime is stable (correctly described by the Kerr metric) immediately inside the outer horizon and most of the way down to the inner (Cauchy) horizon. However, near the Cauchy horizon, spacetime becomes unstable and therefore is *not* described by the Kerr metric. At the Cauchy horizon is located the so-called *mass-inflation singularity* described in the box on page B-5. The presence of the mass-inflation singularity at the Cauchy horizon bodes ill for a diver wishing to experience in person the region between the outer horizon and the center of a rotating black hole. It is delightful to read in a serious theoretical research paper a sentence such as the following: "Such . . . results strongly suggest (though they do not prove) that inside a black hole formed in a generic collapse, an observer falling toward the inner [Cauchy] horizon should be engulfed in a wall of (classically) infinite density immediately after seeing the entire future history of the outer universe pass before his eyes in a flash." (Poisson and Israel)

4 The Kerr Metric for Extreme Angular Momentum

In this project we want to uncover the central features of the spinning black hole with minimum formalism. The equations become simpler for the case of a black hole that is spinning at the maximum possible rate.

QUERY 4 **Maximum value of the angular momentum.** How "live" can a black hole be? That is, how large is it possible to make its angular momentum parameter $a = J/M$? Show that the largest value of the angular momentum parameter, a, consistent with a real value of r_H is $a = M$. This maximum value of the angular momentum parameter a is equivalent to angular momentum $J = M^2$. What happens to the (inner) Cauchy horizon in this case?

A black hole spinning at the maximum rate derived in Query 4 is called an **extreme Kerr black hole**. How fast are existing black holes likely to spin; how "live" are they likely to be? Listen to Misner, Thorne, and Wheeler (page 885): "Most objects (massive stars; galactic nuclei; . . .) that can collapse to form black holes have so much angular momentum that the holes they produce should be 'very live' (the angular momentum parameter $a = J/M$ nearly equal to M; J nearly equal to M^2)."

QUERY 5 **Maximum angular momentum of Sun?** A recent estimate of the angular momentum of Sun is 1.91×10^{41} kilogram meters^2 per second (see the references). What is the value of the angular momentum parameter $a = J/M$ for Sun, in meters? (Hint: Divide the numerical value above by M_{kg}, the mass of Sun in kilograms, to obtain an intermediate result in units of meter^2/second. What conversion factor do you then use to obtain the result in meters?) What fraction a/M is this of the maximum possible value permitted by the Kerr metric?

The metric for the equatorial plane of the extreme-spin black hole results if we set $a = M$ in equation [1], which then becomes

$$d\tau^2 = \left(1 - \frac{2M}{r}\right)dt^2 + \frac{4M^2}{r}dt d\phi - \frac{dr^2}{\left(1 - \frac{M}{r}\right)^2} - R^2 d\phi^2 \qquad \text{[3. extreme Kerr]}$$

Note how the denominator of the dr^2 term in the Kerr metric differs in two ways from the dr^2 term in the Schwarzschild metric: here the denominator is squared and also contains M/r instead of $2M/r$.

Equation [3] has been simplified by defining

$$R^2 \equiv r^2 + M^2 + \frac{2M^3}{r} \qquad \text{[4. extreme Kerr]}$$

The form $R^2 d\phi^2$ of the last term on the right side of equation [3] tells us that R is the **reduced circumference** for extreme Kerr spacetime. That is, the value of R is determined by measuring the circumference of a stationary ring in the equatorial plane concentric to the black hole and dividing this circumference by 2π. This means that r is *not* the reduced circumference but has a value derived from equation [4]. Finding an explicit expression for r in terms of R requires us to solve an equation in the third power of r, which leads to an algebraic mess. Rather than solving such an equation, we carry along expressions containing both R and r. Note from equation [4] that R is not equal to r even for large values of r, although the *percentage* difference between R and r does decrease as r increases.

QUERY 6 **Limiting values of R.** What is r_H, the value of r at the horizon? What is R_H, the value of R at the horizon? Find the *approximate* range of r-values for which the value of R differs from the value of r by less than one part in a million.

QUERY 7	**More general R_a.** Consider the more general case of arbitrary angular momentum parameter a given in equation [1]. What is the expression for R^2 (call it R_a^2) in this case? What is the value of R_a in the limiting case of the nonspinning black hole?

Now move beyond the new r-value for the horizon—the first new feature of the Kerr metric—to the **second new feature of the Kerr metric,** which is the presence of the product $dtd\phi$ of two different spacetime coordinates, called a **cross product**. The cross product implies that coordinates ϕ and t are intimately related. In the following section we show that the Kerr metric predicts **frame dragging**. What does "frame dragging" mean? Near any center of attraction, radial rocket thrust is required to keep a stationary observer at a fixed radius. Near a spinning black hole, an additional *tangential* rocket thrust is required to prevent orbiting, that is to keep the fixed stars in steady position overhead. One might say that spacetime is swept around by the rotating black hole: spacetime itself on the move!

Unless otherwise noted, everything that follows applies to the equatorial plane around an extreme Kerr black hole.

5 The Static Limit

The third new feature of the Kerr metric is the presence of a so-called **static limit**. The horizon of a rotating black hole lies at an r-value *less* than $2M$ (equation [2] with the plus sign). The horizon is where the metric coefficient of dr^2 blows up. In contrast, for the equatorial plane, the coefficient of dt^2, namely, $(1 - 2M/r)$, goes to zero at $r = 2M$, just as it does in the Schwarzschild metric for a nonrotating black hole. The r-value $r = 2M$ in the equatorial plane at which the coefficient of the dt^2 term goes to zero is called the **static limit**. A comparison of equations [3] and [1] shows that the expression for the static limit in the equatorial plane is the same whatever the value of the angular momentum parameter a, namely

$$r_S = 2M \qquad [5]$$

The static limit gets its name from the prediction that for radii smaller than r_S (but greater than that of the horizon r_H) an observer cannot remain at rest, cannot stay *static*. The space between the static limit and the horizon is called the **ergosphere**. Inside the ergosphere you are inexorably dragged along in the direction of rotation of the black hole. No matter how powerful your rockets, you cannot stand at one fixed angle ϕ. For you the fixed stars cannot remain at rest overhead. In principle, a small amount of frame dragging is detectable near any spinning astronomical object. An experimental Earth satellite (Gravity Probe B), now under construction at Stanford University, will measure the extremely small frame-dragging effects predicted near the spinning Earth. Inside the static limit of a rotat-

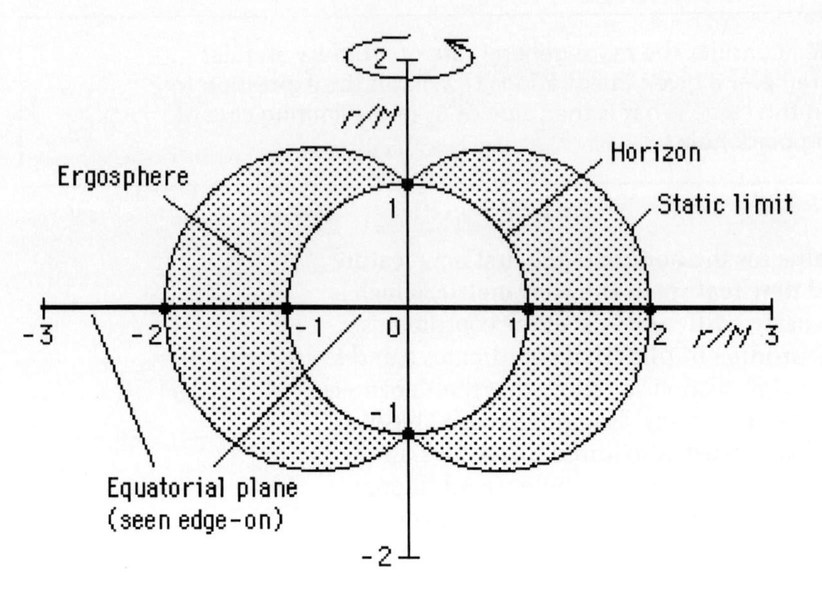

Figure 1 *Computer plot of the cross-section of an extreme black hole showing the static limit and horizon using the Kerr bookkeeper (Boyer-Lindquist) coordinate r (not R). From inside the horizon no object can escape, even one traveling at the speed of light. Between the horizon and the static limit lies the* ***ergosphere,*** *shaded in the figure. Within this ergosphere everything—even light—is swept along by the rotation of the black hole. Inside the ergosphere, too, a stone can have a negative total energy (Section 10).*

ing black hole, in contrast, the frame dragging is irresistible, as will be described on the following page.

The Kerr metric for three space dimensions—not discussed in this book—reveals that the horizon has a constant *r*-value in all directions (is a sphere) while the static limit has cusps at the poles. Figure 1 shows this result. This figure is drawn in Kerr bookkeeper (Boyer-Lindquist) coordinates, which present only one possible way to view these structures. Other coordinate systems, representing a more "intrinsic" geometry, stretch the horizon in the horizontal direction, giving it the approximate shape of a hamburger bun.

QUERY 8 **Reduced circumference of the static limit.** For the extreme black hole, find an expression for R_S, the reduced circumference of the static limit, in the equatorial plane.

QUERY 9 **Viewing the spinning black hole from above.** Draw a cross-section of the extreme black hole in the equatorial plane. That is, show what it would look like to display the static limit and horizon in bookkeeper coordinates on a plane cut through the horizontal axis of Figure 1, as if looking downward along the vertical axis in that figure. Label the static limit, horizon, and ergosphere and put in expressions for their radii.

Now look more closely at the nature of the static limit in the equatorial plane. Examine the Kerr metric for the case of light moving initially in the ϕ direction ($dr = 0$). (Only the initial motion in the equatorial plane will be in this tangential direction; later the beam may be deflected radially inward or outward.) Because this is light, the proper time is zero between adjacent events on its path: $d\tau = 0$. Make these substitutions in the metric [3], divide through by dt^2, and rearrange to obtain

$$R^2\left(\frac{d\phi}{dt}\right)^2 - \frac{4M^2}{r}\left(\frac{d\phi}{dt}\right) - \left(1 - \frac{2M}{r}\right) = 0 \qquad \text{[6. light]}$$

Equation [6] is quadratic in the angular velocity $d\phi/dt$.

QUERY 10 **Tangential motion of light.** Solve equation [6] for $d\phi/dt$. Show that the result has two possible values:

$$\frac{d\phi}{dt} = \frac{2M^2}{rR^2} \pm \frac{2M^2}{rR^2}\left[1 + \frac{r^2R^2}{4M^4}\left(1 - \frac{2M}{r}\right)\right]^{1/2} \qquad \text{[7. light]}$$

Look closely at this expression at the static limit, namely, where $r = 2M$ and $R^2 = 6M^2$. The two solutions are

$$\frac{d\phi}{dt} = 0 \qquad \text{and} \qquad \frac{d\phi}{dt} = \frac{4M^2}{rR^2} = \frac{1}{3M} \qquad \text{[8. light]}$$

To paraphrase Schutz (see references), the second solution in [8] represents light sent off in the same direction as the hole is rotating. The first solution says that the other light flash—the one sent "backward"—does not move at all as recorded by the far-away bookkeeper. The dragging of orbits has become so strong that this light cannot move in the direction opposite to the rotation! Clearly, any material particle, which must move slower than light, will therefore have to rotate with the hole, even if it has an angular momentum arbitrarily large in the sense opposite to that of hole rotation.

QUERY 11 **Light dragging in the ergosphere.** Show that inside the ergosphere (r such that $r_H < r < r_S$), light launched in *either* tangential direction in the equatorial plane moves in the direction of rotation of the black hole as recorded by the far-away bookkeeper. That is, show that the initial tangential angular velocity $d\phi/dt$ is always positive.

The static limit creates a difficulty of principle in measuring the reduced circumference R, defined by equation [4] on page F-6. According to that definition, one measures R by laying off the total distance—the circumference—around a stationary ring in the equatorial plane concentric to the black hole and dividing the circumference by 2π to find the value of R. But inside the static limit no such ring can remain stationary; it is inevitably swept along in a tangential direction, no matter how powerful the rockets we use to try to keep it stationary. Thus, for the present, we have no practical definition for R inside the static limit. We will overcome this difficulty in principle in Section 9.

For completeness, we mention here the fourth new feature of the Kerr metric, which is analyzed further in Sections 10 and 11.

The fourth new feature of the Kerr metric is available energy. No net energy can be extracted from a nonspinning black hole (except for the quantum "Hawking radiation," page 2-4, which is entirely negligible for star-mass black holes). For this reason, the nonspinning black hole carries the name *dead*. In contrast, energy of rotation is available from a spinning black hole, which therefore deserves its name *live*. See Section 12.

6 Radial and Tangential Motion of Light

QUERY 12 **Radial motion of light.** For light ($d\tau = 0$) moving in the radial direction ($d\phi = 0$), show from the metric that

$$\frac{dr}{dt} = \pm\left(1-\frac{M}{r}\right)\left(1-\frac{2M}{r}\right)^{1/2} \qquad \text{[9. light, } d\phi = 0\text{]}$$

Show that this radial speed goes to zero at the static limit and is imaginary (therefore unreal) inside the ergosphere. *Meaning:* No purely radial motion is possible inside the ergosphere. See Figure 2.

For light ($d\tau = 0$) moving in the tangential direction ($dr = 0$), we call the tangential velocity $Rd\phi/dt$ as recorded by the Kerr bookkeeper. From equation [7], this tangential velocity is given by

$$R\frac{d\phi}{dt} = \frac{2M^2}{rR} \pm \frac{2M^2}{rR}\left[1+\frac{r^2R^2}{4M^4}\left(1-\frac{2M}{r}\right)\right]^{1/2} \qquad \text{[10. light, } dr = 0\text{]}$$

The second term on the right side of [10] can be simplified by substituting for R^2 in the numerator from equation [4]. (Trust us or work it out for yourself!) Equation [10] becomes

$$R\frac{d\phi}{dt} = \frac{2M^2}{rR} \pm \frac{r-M}{R} \qquad \text{[11. light, } dr = 0\text{]}$$

QUERY 13 **Light dragging at the horizon.** What happens to the light dragging at the horizon (r_H given by equation [2] with the plus sign and $a = M$, and R_H derived in Query 6)? Show that at the horizon the initial tangential rotation $d\phi/dt$ for light has a single value whichever way the pulse is launched. Show that the bookkeeper velocity $Rd\phi/dt$ for this light at the horizon has the value shown in Figure 2.

The radial and tangential velocities of light in equations [9] and [11] are *bookkeeper velocities,* reckoned by the Kerr bookkeeper using the coordinates r and ϕ and the far-away time t. Nobody measures the Kerr bookkeeper velocities directly, just as nobody measured directly bookkeeper velocities near a non-spinning black hole (Chapters 3 through 5).

Figure 2 shows the radial and tangential bookkeeper velocities of light for the extreme Kerr metric. Note again that these plots show the *initial* velocity of a light flash launched in the various directions. After launch, a radially moving light flash may be dragged sideways or a tangentially moving flash may be deflected inward.

QUERY 14 **Locked-in motion?** (Optional) Kip Thorne says, "I guarantee that, if you send a robot probe down near the horizon of a spinning hole, blast as it may it will never be able to move forward or backward [in either tangential direction] at any speed other than the hole's own spin speed. . . ." What evidence do equation [11] and Figure 2 give for this conclusion? What is "the hole's own spin speed"? (See Kip S. Thorne, *Black Holes and Time Warps*, W. W. Norton & Co., New York, 1994, page 57.)

7 Wholesale Results, Extreme Kerr Black Hole

Now suppose that you have never heard of the Kerr metric and someone presents you with the "anonymous" metric [3] (which we know to be the metric for the extreme Kerr black hole) plus the definition of R:

$$d\tau^2 = \left(1 - \frac{2M}{r}\right)dt^2 + \frac{4M^2}{r}dtd\phi - \frac{dr^2}{\left(1 - \frac{M}{r}\right)^2} - R^2 d\phi^2 \qquad [3]$$

$$R^2 \equiv r^2 + M^2 + \frac{2M^3}{r} \qquad [4]$$

You say to yourself, "This equation is just a crazy kind of mixed-up Schwarzschild-like metric, with a nutty denominator for the dr^2 term, a cross-term in $dtd\phi$, and R^2 instead of r^2 as a coefficient for $d\phi^2$. Still, it's a metric. So let's try deriving expressions for angular momentum, energy, and so forth for a particle moving in a region described by this metric in analogy to similar derivations for the Schwarzschild metric." So saying,

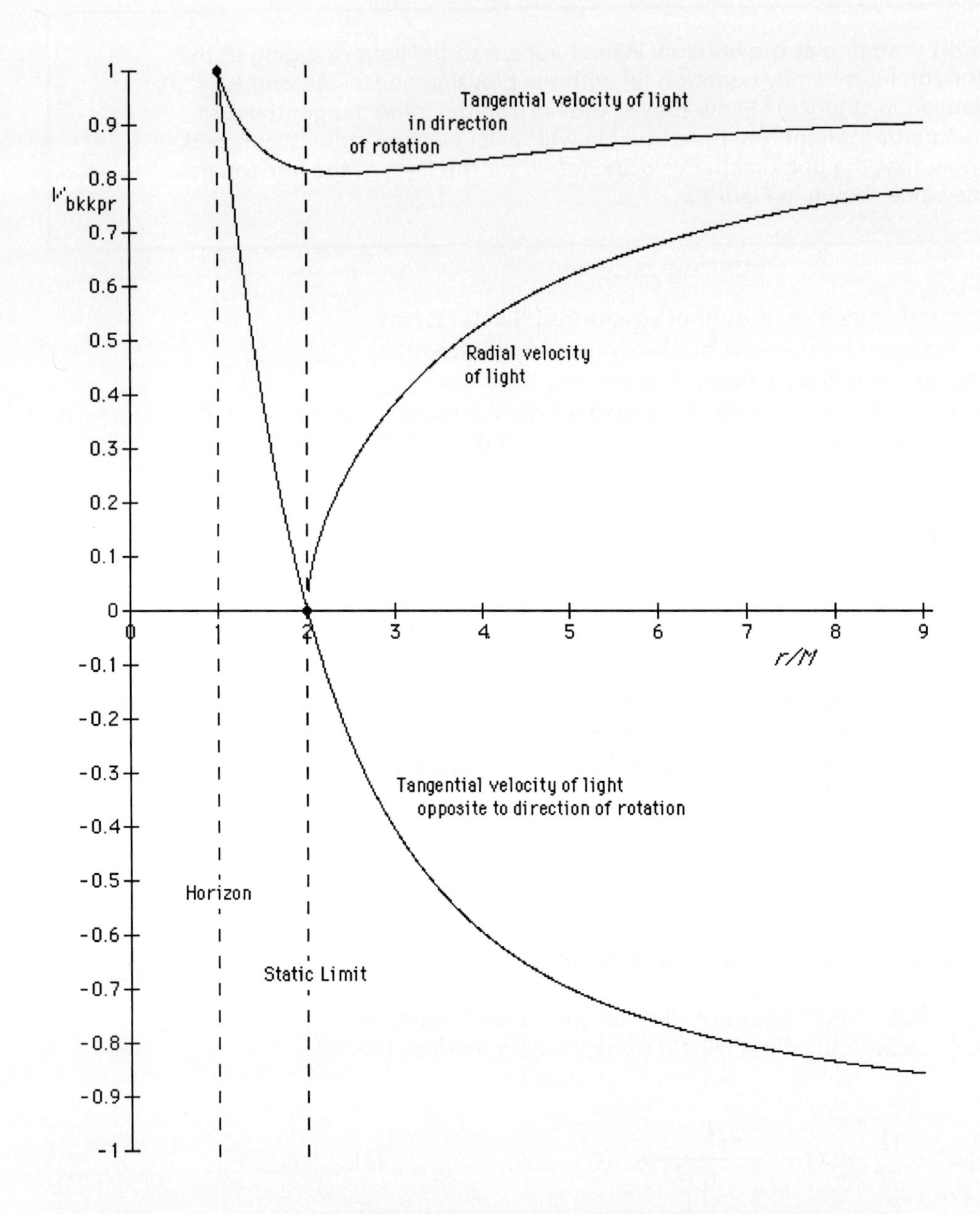

Figure 2 *Computer plot of bookkeeper radial and tangential velocities of light near an extreme Kerr black hole (a = J/M = M). Note that as r/M becomes large, the different bookkeeper velocities all approach plus or minus unity. Note also that purely radial motion of light is not possible inside the static limit. Important: These are* **initial** *velocities of light just after launch in the given direction. After launch, the light will generally change direction. For the case of a nonrotating black hole, see Figures 6 and 7, pages B-18–19.*

Table 1 Comparison of results of nonspinning and extreme-spin black holes

Quantity	Nonspinning Schwarzschild black hole	Extreme-spin Kerr black hole ("shell" = stationary ring outside static limit)
Define r and R	Reduced circumference = $r \equiv \frac{\text{(circumference of shell)}}{2\pi}$ [12]	Reduced circumference R given by: $R^2 \equiv r^2 + M^2 + \frac{2M^3}{r}$ [13]
Shell time vs. far-away time: (gravitational red shift)	$dt_{\text{shell}} = \left(1 - \frac{2M}{r}\right)^{1/2} dt$ [14]	$dt_{\text{shell}} = \left(1 - \frac{2M}{r}\right)^{1/2} dt$ [15. stationary]
dr_{shell} vs. dr	$dr_{\text{shell}} = \left(1 - \frac{2M}{r}\right)^{-1/2} dr$ [16]	$dr_{\text{shell}} = \left(1 - \frac{M}{r}\right)^{-1} dr$ [17. stationary]
Energy (constant of the motion)	$\frac{E}{m} = \left(1 - \frac{2M}{r}\right)\frac{dt}{d\tau}$ [18]	$\frac{E}{m} = \left(1 - \frac{2M}{r}\right)\frac{dt}{d\tau} + \frac{2M^2}{r}\frac{d\phi}{d\tau}$ [19]
Angular momentum (constant of the motion)	$\frac{L}{m} = r^2\frac{d\phi}{d\tau}$ [20]	$\frac{L}{m} = R^2\frac{d\phi}{d\tau} - \frac{2M^2}{r}\frac{dt}{d\tau}$ [21]

you use the Principle of Extremal Aging and other methods of Chapters 2 through 5 to derive expressions similar to results in those chapters and enter them in the last column of Table 1.

Notes: (1) We limit ourselves to the equatorial plane. (2) Outside the static limit we can still set up stationary spherical shells (which we have limited to stationary *rings* in the equatorial plane), but we must use continual tangential rocket blasts to keep these rings from rotating in the tangential direction.

QUERY 15 **Energy and angular momentum as constants of the motion.** Derive Table 1, entries [19] and [21] for energy and angular momentum of a free object moving in the equatorial plane of an extreme Kerr black hole.

8 Plunging: The "Straight-In Spiral"

For the nonrotating black hole the simplest motion was radial plunge (Chapter 3). What is the simplest motion near a spinning black hole? By analogy, let us examine motion starting from infinity and proceeding with zero angular momentum.

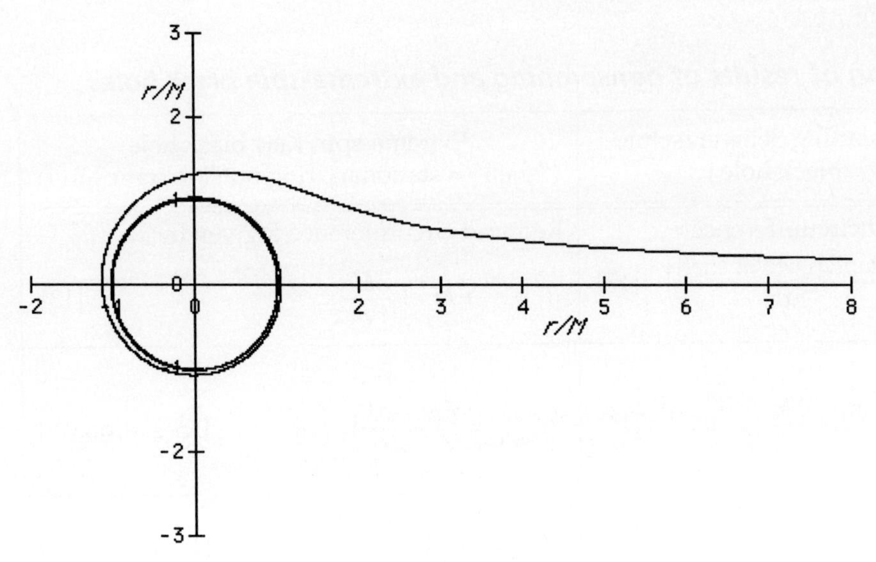

Figure 3 *Computer plot: Kerr map (Kerr bookkeeper plot) of the trajectory in space of a stone dropped from rest far from a black hole (therefore with zero angular momentum). According to the far-away bookkeeper, the stone spirals in to the horizon at r = M and circulates there forever.*

QUERY 16 **No angular momentum. But angular motion!** Set angular momentum [21] equal to zero and verify the following equation:

$$\frac{d\phi}{dt} = \frac{2M^2}{rR^2} \qquad [22.\ L = 0]$$

Equation [22] gives the remarkable result that a particle with zero angular momentum nevertheless circulates around the black hole! This result is evidence for our interpretation that the black hole drags nearby spacetime around with it. Figure 3 shows the trajectory of an inward plunger with zero angular momentum, as calculated in what follows.

Let's see if we can set up the equations to follow a stone that starts at rest far from a rotating black hole and moves inward with zero angular momentum. At remote distance, in flat spacetime, the stone has energy $E/m = 1$. It keeps the same energy as it falls inward. From equation [19] in Table 1,

$$\frac{E}{m} = 1 = \left(1 - \frac{2M}{r}\right)\frac{dt}{d\tau} + \frac{2M^2}{r}\frac{d\phi}{d\tau} \qquad [23]$$

Equations [22] and [23] are two equations in the four unknowns dr, dt, $d\tau$, and $d\phi$. A third equation is the metric [3] for the extreme-spin black hole. With these three independent equations, we can eliminate three of the four unknowns to find a relation between any two remaining differentials. We choose the quantities dr and $d\phi$, because we want to draw the trajectory, the Kerr map. Don't bother doing the algebra—it is a mess. After substituting equation [4] for R^2 into the result, one obtains the relation between dr and $d\phi$:

$$\begin{aligned} dr &= \frac{r-M}{r}\left[\frac{r^5}{2M^3} - \frac{r^4}{M^2} + \frac{r^3}{M} - r^2 + \frac{Mr}{2}\right]^{1/2} d\phi \\ &= \frac{(r-M)^2}{r}\left[\frac{r^3}{2M^3} + \frac{r}{2M}\right]^{1/2} d\phi \end{aligned} \qquad [24.\ L = 0]$$

The computer has no difficulty integrating and plotting this equation, as shown in Figure 3. Since we used the Kerr bookkeeper angular velocity [22], the resulting picture is that of the Kerr bookkeeper. For her, the zero-angular-momentum stone spirals around the black hole and settles down in a tight circular path at $r = M$, there to circle forever.

QUERY 17	**Final circle according to the bookkeeper.** Verify that dr goes to zero (that is, r does not change) once this stone reaches the horizon.

Remember that for the nonspinning black hole an object plunging inward slows down as it approaches the horizon, according to the records of the Schwarzschild bookkeeper. For both spinning and nonspinning black holes, the in-falling stone with $L = 0$ never crosses the horizon when clocked in far-away time.

QUERY 18	**Bookkeeper speed in the "final circle."** At the horizon, what is the numerical value of the tangential speed $Rd\phi/dt$ of the stone dropped from rest at infinity, as measured by the Kerr bookkeeper?

The observer who has fallen from rest at infinity has quite a different perception of the trip inward! For her there is no pause at the horizon; she has a quick, smooth trip to the center (assuming that the Kerr metric holds all the way to the center!). An algebra orgy similar to the previous one gives a relation between dr and $d\tau$, where $d\tau$ is the wristwatch time increment of the in-faller:

$$\begin{aligned} \left(\frac{dr}{d\tau}\right)^2 &= \frac{2Mr^3 - 4M^2r^2 + 4M^3r - 4M^4 + \dfrac{2M^5}{r}}{r^2(r-M)^2} \\ &= \frac{2M}{r}\left(1 + \frac{M^2}{r^2}\right) \end{aligned} \qquad [25.\ L = 0]$$

Figure 4 compares the magnitude of the square root of this expression with the magnitude of the velocity of the stone dropped from rest at a great distance in the Schwarzschild case (equation [32], page 3-22):

$$\frac{dr}{d\tau} = -\left(\frac{2M}{r}\right)^{1/2} \qquad \text{[26. } L = 0 \text{ Schwarzschild]}$$

Both equations [25] and [26] show radial components of speed greater than unity in the region of small radius. The resulting speed is even more impressive when one adds the tangential motion forced on the diver descending into the spinning black hole (Figure 2 and 3). Does such motion violate the "cosmic speed limit" of unity for light? A similar question is debated for the Schwarzschild black hole in Section 3 of Project B, Inside the Black Hole, pages B-6–12.

Research note: When applied inside the horizon, equation [25] assumes that the Kerr metric correctly describes spacetime all the way to the center of the extreme Kerr black hole. This may not be the case. See the box Egg-beater Spacetime? on page B-5.

9 Ring Riders

Equation [22] in Section 8 describes the angular rotation rate ω of an in-falling stone that has zero angular momentum:

$$\frac{d\phi}{dt} \equiv \omega = \frac{2M^2}{rR^2} \qquad \text{[27. } L = 0\text{]}$$

In some way, ω in this equation describes the angular rate at which space is "swept along" by the nearby spinning black hole. What happens if we "go with the flow," moving tangentially at angular rate ω given by this equation? Will we cease to feel a tangential force? What happens to us at the static limit?

To pursue these ideas, we envision a set of nested rings in the equatorial plane and concentric to the black hole (Figure 5). Each of these rings revolves at an angular rate given by equation [27] as reckoned by the Kerr bookkeeper. Rings at different values of r rotate at different angular rates.

The result of this construction is a set of observers in the equatorial plane whom we call **ring riders**. A ring rider is an observer who stands at rest on one of the zero angular momentum rotating rings. In times past, ring riders were known as **locally nonrotating observers**, but now the customary name is **zero angular momentum observers** or **ZAMOs**. Each ring rider, like each shell observer in Schwarzschild geometry, is subject to a gravitational acceleration directed toward the center of the black hole, but experiences no frame dragging in the tangential direction (because he rides along with the rotating ring). In both cases the radially inward gravitational acceleration becomes infinite at the horizon, destroying any possible circumferential ring structures at or inside the horizon. According

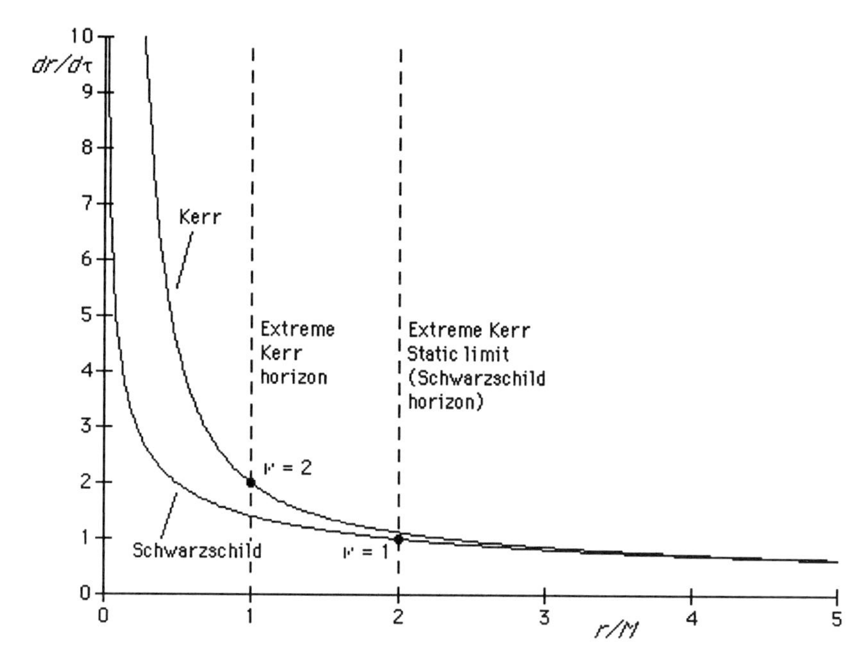

Figure 4 *Comparison of radial components of plunge velocities experienced by an in-faller who drops from rest (so with $L = 0$) at a great distance from Schwarzschild and extreme Kerr black holes.*

Figure 5 *Kerr map (perspective plot) of rings surrounding a rotating black hole. The rings rotate in the same direction as the black hole but at angular rates that differ from ring to ring.*

to ring rider measurements, light has speed unity, the same speed in both tangential directions, as we shall see.

QUERY 19 **Ring slippage.** Will the inner rings rotate with larger or smaller angular velocity than the rings farther out? Justify your choice.

QUERY 20 **Ring speed according to the bookkeeper.** What are the units of ω in equation [27]? What is the numerical value of the bookkeeper speed $R\omega$ for each of the rings $r = 100M$, $r = 10M$, $r = 2M$, and $r = M$? Express the answers as a fraction of the speed of light.

QUERY 21 **Does rain fall vertically?** Present an argument that a stone dropped from rest starting at a great radial distance falls vertically past the rider on every ring. *Guess:* Is the same true if the stone is *flung* radially inward from a great distance? *Guess:* What about light?

Can we write a simplified metric for the ring rider? Probably not for events separated radially because of shearing, the slippage between adjacent rings. So limit attention to events separated tangentially along the ring. According to the remote observer, each ring revolves with angular velocity ω given by equation [27]. Define an azimuthal angle increment $d\phi_{ring}$ that the ring rider measures along the ring with respect to some zero mark on the ring. Let an object move along the ring. Then the Kerr bookkeeper will observe different angular rates of change in her ϕ coordinate and along the ring:

$$\frac{d\phi}{dt} = \frac{d\phi_{ring}}{dt} + \omega \qquad [28]$$

The positive direction of both $d\phi$ and $d\phi_{ring}$ is in the direction of rotation of the black hole.

Now think of two events separated by the angle $d\phi_{ring}$ along the ring and at far-away time separation dt. Then the angular separation $d\phi$ between these two events for the far-away observer is, from [27] and [28],

$$d\phi = d\phi_{ring} + \frac{2M^2}{rR^2}dt \qquad [29.\ dr = 0]$$

The metric [3] with the same limitation to motion along the ring ($dr = 0$) is

$$d\tau^2 = \left(1 - \frac{2M}{r}\right)dt^2 + \frac{4M^2}{r}dtd\phi - R^2 d\phi^2 \qquad [30.\ dr = 0]$$

QUERY 22 **New metric for the ring.** Substitute equation [29] into [30]. Show first that the coefficient of the cross-term in $dtd\phi_{ring}$ is equal to zero. Second, collect terms in dt^2 and $d\phi_{ring}{}^2$ to show that the resulting metric is given by equation [31] for motion along the ring. *Hint:* Group over a common denominator r^2R^2, then substitute in the numerator for R^2 (equation [4]):

$$d\tau^2 = \frac{r^2}{R^2}\left(1 - \frac{M}{r}\right)^2 dt^2 - R^2 d\phi_{ring}^2 \qquad [31.\ dr = 0]$$

QUERY 23 **Time on the ring rider clock.** A ring rider is at rest on the ring. Show that the time dt_{ring} between ticks on his clock and the time dt between ticks on the far-away clock are related by the equation

$$dt_{ring} = \frac{r - M}{R}dt \qquad [32.\ dr = d\phi_{ring} = 0]$$

Show that, with this substitution, the metric for $dr = 0$ becomes

$$d\tau^2 = dt_{ring}^2 - R^2 d\phi_{ring}^2 \qquad [33.\ dr = 0]$$

In brief, for nearby events along the ring, the metric [33] looks like that of flat spacetime. But spacetime is *not* flat on a rotating ring near a spinning black hole; the coordinates that appear in [33] do not form a coordinate system useful globally as the Kerr bookkeeper coordinate system is. Equation [33] describes a *local frame*, useful only in analyzing events and experiments that are strictly limited in space and time and for which the local "gravitational force" can be neglected. In this regard it is similar to the corresponding local shell frame around a nonspinning black hole described by the metric [33] on page 2-33.

Equation [33] in Query 23 is limited in many ways: It deals with two adjacent events in the equatorial plane of an extreme-spin black hole. However, this equation is useful for analyzing nearby events that occur along the same ring.

Now (finally!) we can define the reduced circumference R everywhere external to the horizon, even inside the static limit. A ring rider measures the circumference of his freely rotating ring and divides this circumference by 2π.

$$\begin{pmatrix} \text{circumference of} \\ \text{freely rotating ring} \end{pmatrix} \equiv 2\pi R \qquad [34]$$

The result is a formal definition of the **reduced circumference** R for this zero angular momentum (freely rotating!) ring. The value of R, along with the value of r from equation [4], is then stamped on each rotating ring for all to see and for everyone to use. The same values of R and r can also be stamped on each nonrotating ring that coincides with an already measured rotating ring. (Of course, nonrotating rings can exist only outside the static limit.)

In principle this set of rotating rings extends from the horizon to infinite radius. For a pair of events near one another along a given ring, the proper distance $d\sigma$ between them is given by the equation

$$d\sigma = R d\phi_{\text{ring}} \qquad [35.\ dr = dt = 0]$$

QUERY 24 **Speed of light along the ring is unity for ring riders.** From the metric [33], show that the ring rider measures the speed of light along the ring to have the magnitude unity. Is this value the same for motion of the light in both directions along the ring (Figure 6)?

QUERY 25 **Is motion along ring free or locked?** *Hard thought question; optional.* Equation [33] says that the ring rider on every ring can use special relativity in analyzing motion along the ring. So he must be able to move freely back and forth along the ring, even on a ring near the horizon. In contrast, Query 14 asserts that the tangential motion near the horizon is rigidly locked to the rotation of the black hole. Locked or free? What's going on?

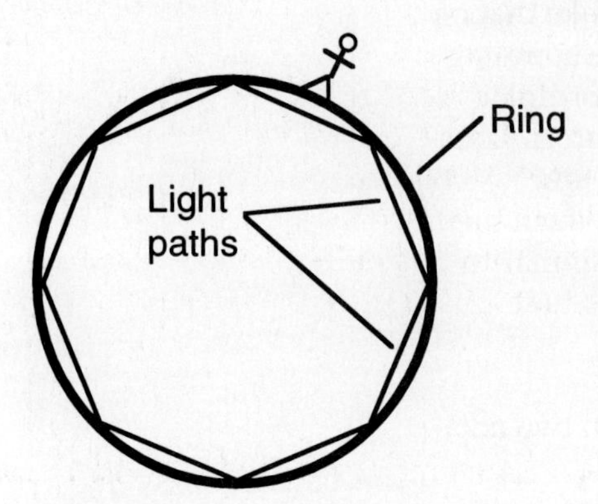

Figure 6 *Silvered inner surface of rotating ring allows signaling at light speed $v = 1$ as measured by observers on the same ring, with synchronization of their clocks, etc. Light-path segments shown as straight will be curved. We assume that each segment is arbitrarily short so that the light skims along close to the ring.*

10 Negative Energy: The Penrose Process

Roger Penrose devised a scheme for milking energy from a spinning black hole. This scheme is called the **Penrose process** (see references). The Penrose process depends on the prediction that in some orbits inside the ergosphere a particle can have *negative* total energy. Before we detail the Penrose process, we need to describe negative total energy.

Negative Total Energy

What can negative total energy possibly mean? Negative energy is nothing new. In Newtonian mechanics the potential energy of a particle at rest far from Sun is usually taken to be zero by convention. Then a particle at rest near Sun has zero kinetic energy and negative potential energy, yielding a total energy less than zero. But in Newtonian mechanics the zero point of potential energy is arbitrary, and all reasonable choices of this zero point lead to the same description of motion. In contrast, special relativity determines the rest energy of a free material particle in flat spacetime, setting its rest energy equal to its mass. So the arbitrary choice of a zero point for energy is lost, and a particle far from a center of gravitational attraction always has an energy that is positive.

For Schwarzschild geometry the physical system differs from Newtonian. A particle at rest near the horizon of a nonspinning black hole has zero total energy (from equation [18] in Sample Problem 1, page 3-12). The meaning? That it takes an energy equal to its rest energy ($= m$) to remove this particle to rest at a large distance from the black hole (where it has the energy m). As a consequence, if the particle drops into the black hole from its stationary position next to the horizon, then the mass of the combined black-hole particle system (measured by a far-away observer) does not change.

For Kerr geometry the physical system differs from that in Schwarzschild geometry. A particle can have a negative energy near a spinning black hole. The meaning? An energy *greater* than its rest energy (greater than m) is required to remove such a particle to rest at a great distance from the black hole. If the particle with negative energy is captured by the spinning black hole, the black hole's mass and angular momentum decrease. (See Section 11.) This process can be repeated until the black hole has zero angular momentum. Then it becomes a "dead" Schwarzschild black hole, from which only Hawking radiation can extract energy (box, page 2-4).

Strategy of the Penrose Process

The strategy of the Penrose process is similar to the following series of unethical financial transactions:

1. You and I decide to share our money. Our combined net worth is positive.

2. I give you all my money, then borrow money from a bank and give that to you as well. My bank debt is a negative entry on my accounting balance sheet, so now my net worth is negative.

3. I declare bankruptcy and the bank is stuck with my debt.

The net result is the transfer of money from me and from the bank to you. The bank provides the mechanism by which I can enter a state of negative net worth.

The Penrose process is similar:

1. Starting at a distant radius, you and I together descend to a position inside the ergosphere.

2. We are moving together tangentially inside the ergosphere in the rotation direction. You push me away *violently* in a direction opposite to the direction of rotation. This push puts you into a new trajectory and puts me into a state of negative energy.

3. I drop into the black hole, which is stuck with my negative energy. You continue in your new trajectory, arriving at a distant radius with augmented energy.

The net result is the transfer of energy from me and from the black hole to you. The spinning black hole provides the mechanism by which I can enter a state of negative energy.

This entire strategy rests on the assumption that an object can achieve a state of negative energy in the space surrounding a spinning black hole. Is this assumption correct? Look again at expression [19] for the energy of a stone near an extreme Kerr black hole:

$$\frac{E}{m} = \left(1 - \frac{2M}{r}\right)\frac{dt}{d\tau} + \frac{2M^2}{r}\frac{d\phi}{d\tau} \qquad [19]$$

Can this energy be negative? Start to answer this question by finding the "critical" condition under which the energy is zero.

QUERY 26 **Conditions for zero energy.** Set $E/m = 0$ in equation [19] and show that the resulting expression for the bookkeeper rate of change of angle is

$$\left(\frac{d\phi}{dt}\right)_{E=0} = \frac{2M - r}{2M^2} \qquad [36]$$

Under what conditions is this angular velocity negative? positive?

QUERY 27 **Bookkeeper tangential velocity for zero energy.** Now assume that the direction of motion is tangential and show that the bookkeeper velocity is given by the expression

$$v_{\text{bkkpr E=0}} = R\left(\frac{d\phi}{dt}\right)_{E=0} = \frac{R(2M - r)}{2M^2} \qquad [37.\ dr = 0]$$

QUERY 28 **Bookkeeper tangential velocities for negative energy.** Now redo the analysis for the circumstance that the particle energy is negative. Show that the condition is

$$v_{\text{bkkpr E=neg}} < \frac{R(2M - r)}{2M^2} \qquad [38.\ dr = 0]$$

Figure 7 shows a plot of equation [37] along with plots of the positive and negative tangential velocities of light from Figure 2. The tangential motion of *any* particle must be bounded by the curves of tangential light motion. (Inside the ergosphere even light moving "in the negative tangential direction" moves forward, in the direction of rotation, according to the remote bookkeeper.) In addition, equation [38] tells us that a particle with negative energy must have a tangential velocity that lies below the heavy line in the Figure 7. The shaded area in that figure conforms to these conditions and shows the range of bookkeeper tangential velocities of a stone for which the stone has negative energy.

Next we turn our attention away from the bookkeeper to what the ring rider sees (Query 29).

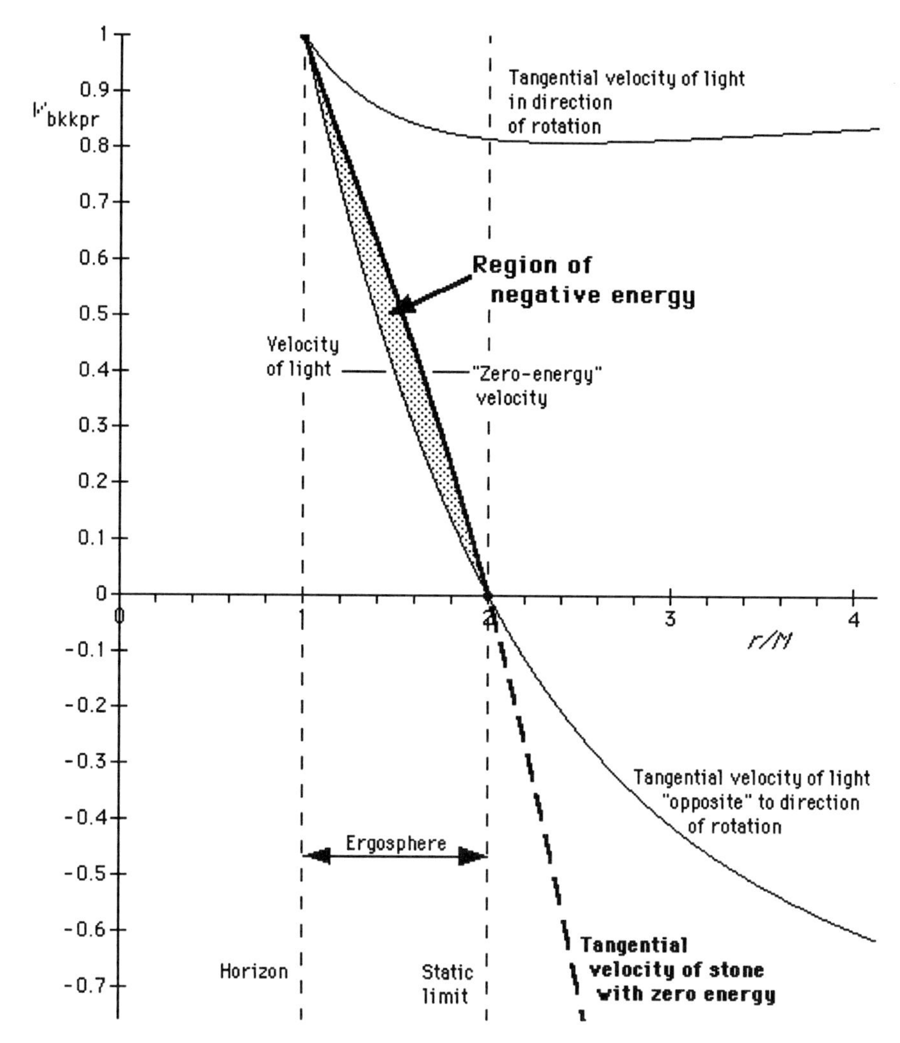

Figure 7 *Computer plot showing bookkeeper tangential velocities of light (thin curves) and tangential velocity of a stone with zero energy (thick curve), calculated using equation [37]. For r greater than 2M, the static limit, the particle cannot have zero energy (or negative energy), because it would have to be moving in a negative tangential direction with a speed greater than that of light in that direction. Only inside the ergosphere is this critical tangential velocity possible. The shaded area shows the range of bookkeeper velocities for which the stone has negative energy.*

QUERY 29 **Ring rider velocity for zero energy.** *Optional—messy algebra!* A stone moves tangentially along a rotating ring. For what values of the ring velocity v_{ring} will the energy measured at infinity be negative? Set $E/m = 0$ in equation [19]. Then make substitutions from equations [29] and [32] to convert variables to $d\phi_{\text{ring}}$ and dt_{ring}. Simplify using equation [4]. Show that the result is

$$v_{\text{ring, E=0}} = R\frac{d\phi_{\text{ring}}}{dt_{\text{ring}}} = -\frac{1}{2}\left(\frac{r}{M} - 1\right)\frac{r}{M} \qquad [39]$$

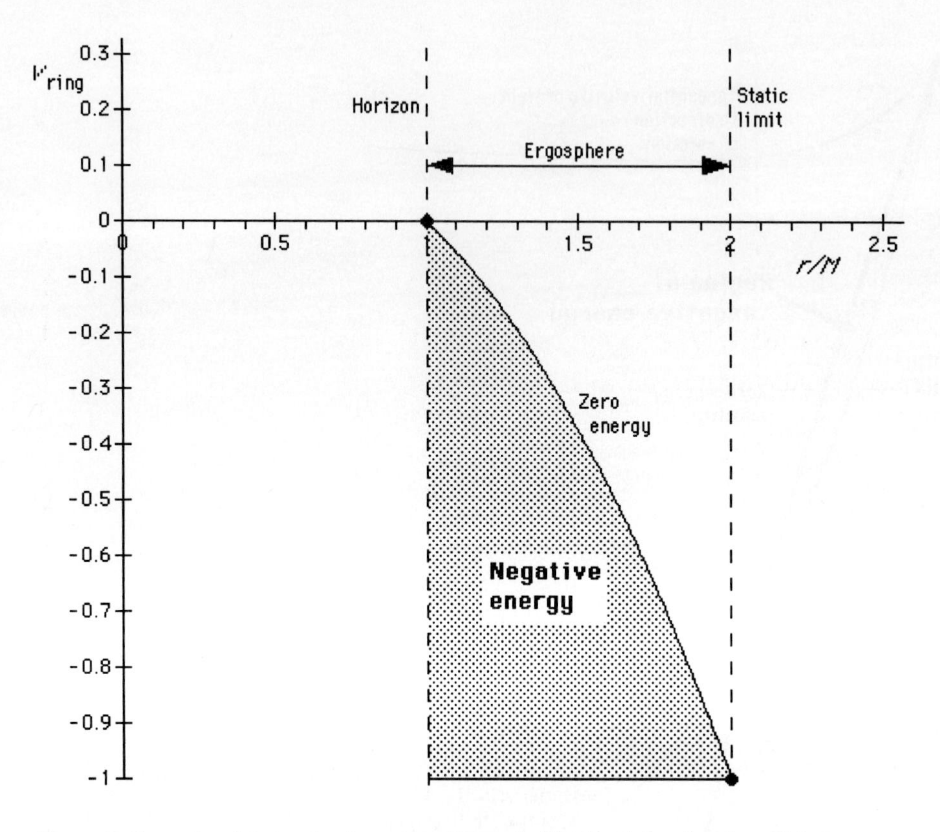

Figure 8 *Computer plot showing the range of ring velocities (shaded region) for which the energy measured at infinity is negative. Negative ring velocity means motion along the ring in a direction opposite to the direction of rotation of the black hole.*

Figure 8 plots equation [39] for ring velocity. Energy measured at infinity E/m will be negative for values of the ring velocity in the shaded region of the plot. The range of ring velocities for which energy is negative depends on the radius of the ring. Limiting cases are interesting: For a ring at the static limit, motion backward along the ring with the speed of light leads to zero energy. In contrast, for a ring near the horizon, any backward velocity, no matter how small, leads to negative energy.

11 Quasar Power

How much total energy can be extracted from a rotating black hole? In general relativity, energy is a seamless whole; we cannot separate the kinetic from the rest energy of a rotating object. Milking energy from a rotating black hole changes its mass M along with its angular momentum J. Analysis has identified a so-called **irreducible mass** M_{irr} that is the smallest residual mass that results when all the angular momentum is milked out of a rotating black hole. This irreducible mass M_{irr} of an uncharged rotating black hole with angular momentum parameter $a = J/M$ is given by the equation

$$M_{\text{irr}}^2 = \frac{1}{2}[M^2 + M(M^2 - a^2)^{1/2}] \qquad [40]$$

or equivalently

$$M^2 = M_{\text{irr}}^2 + \frac{J^2}{4M_{\text{irr}}^2} \qquad [41]$$

(Wald, page 326. The corresponding equation in Misner, Thorne, and Wheeler, page 913, contains misprints.) This result was discovered in Princeton by a 19-year-old Athenian, Demetrios Christodoulou, who never finished high school.

The final state is a nonrotating Schwarzschild black hole of mass M_{irr}. The net result is that a total energy $M - M_{\text{irr}}$ has been extracted from an uncharged rotating black hole.

QUERY 30 **Irreducible mass of extreme Kerr black hole.** What is the irreducible mass of an uncharged extreme Kerr black hole of mass M? What fraction of the mass M of an extreme Kerr black hole can be extracted in the form of energy by an **advanced civilization** (defined as a civilization that can accomplish any engineering feat not forbidden by the laws of Nature)?

QUERY 31 **How much energy is available from the monster in our galaxy?** Imagine that the black hole of mass $M = 2.6 \times 10^6 M_{\text{Sun}}$ thought to exist at the center of our galaxy is an extreme Kerr black hole. How much total energy can be milked from it? Express your answer as a multiple of the mass M_{Sun} of our Sun.

From where do quasars get their power (box, page F-4)? Probably not directly from the Penrose process (Section 10). One set of theories has the quasar radiation coming from the gravitational energy of matter descending toward the black hole as it orbits in an accretion disk. This matter interacts with other matter in the disk in a complicated manner not well understood. As debris in the disk moves toward the center, it is compressed along with its magnetic fields, is heated, and emits radiation copiously. The net result is to convert its gravitational energy into radiation with high efficiency (high compared with nuclear reactions on Earth). Note that the angular momentum of the black hole may actually be increased during this process, depending on the overall angular momentum of the gas and clouds swirling into the black hole. Another theory derives the quasar output from the rotation energy of the black hole itself, employing magnetic field lines to couple black hole rotation energy to the matter swirling around exterior to the horizon of the black hole. Such a model leads to reduction in the rotation rate of the black hole.

QUERY 32 **Quasar output.** How much energy does a quasar put out each second? Suppose that the quasar emits energy at a rate 100 times the rate of our entire galaxy, which contains approximately 10^{11} stars similar to our Sun. How much energy does Sun put out per second? Luminous energy from Sun pours down on the outer atmosphere of Earth at a rate of 1370 watts per square meter (called the **solar constant).** From the solar constant, estimate the energy production rate of our Sun in watts, then of our galaxy, and then of a quasar that emits energy at 100 times the rate of our galaxy. This rate corresponds to the total conversion to energy of how many Sun masses per Earth-year?

The details of the emission of radiation by quasars may be complicated, but the analysis in the present project provides the basis for an estimate of the energy *available* for such processes.

Suppose that each element of the accretion disk circles the black hole at the same rate of rotation as the local ring (an unrealistic assumption, since rotating with the ring does not place the particle in a stable circular orbit). As a given bit of debris moves inward, let it radiate energy sufficient to keep it at rest with respect to the local ring. For a bit of debris riding on the ring, the time $d\tau$ between ticks on its wristwatch is the same as time dt_{ring} between ticks of the ring clocks, since they are relatively at rest. Equation [19] for the energy of this bit of debris then becomes

$$\frac{E}{m} = \left(1 - \frac{2M}{r}\right)\frac{dt}{dt_{ring}} + \frac{2M^2}{r}\frac{d\phi}{dt_{ring}} \qquad [42]$$

Now, the relation between ring time increments and bookkeeper time increments is given by equation [32]:

$$dt_{ring} = \frac{r-M}{R}dt \qquad [32]$$

QUERY 33 **Energy of stone riding on the ring.** Substitute equation [32] into equation [42], use equation [27] for the resulting $d\phi/dt$, and collect terms over a common denominator $R(r - M)$ to obtain

$$\frac{E}{m} = \frac{\left(1 - \frac{2M}{r}\right)R^2 + \frac{4M^4}{r^2}}{R(r-M)} \qquad \text{[43. riding on ring]}$$

For the expression for R^2 in the numerator (only) substitute from equation [4] and simplify to show that, for a stone riding on the ring,

$$\frac{E}{m} = \frac{r-M}{R} \qquad \text{[44. riding on ring]}$$

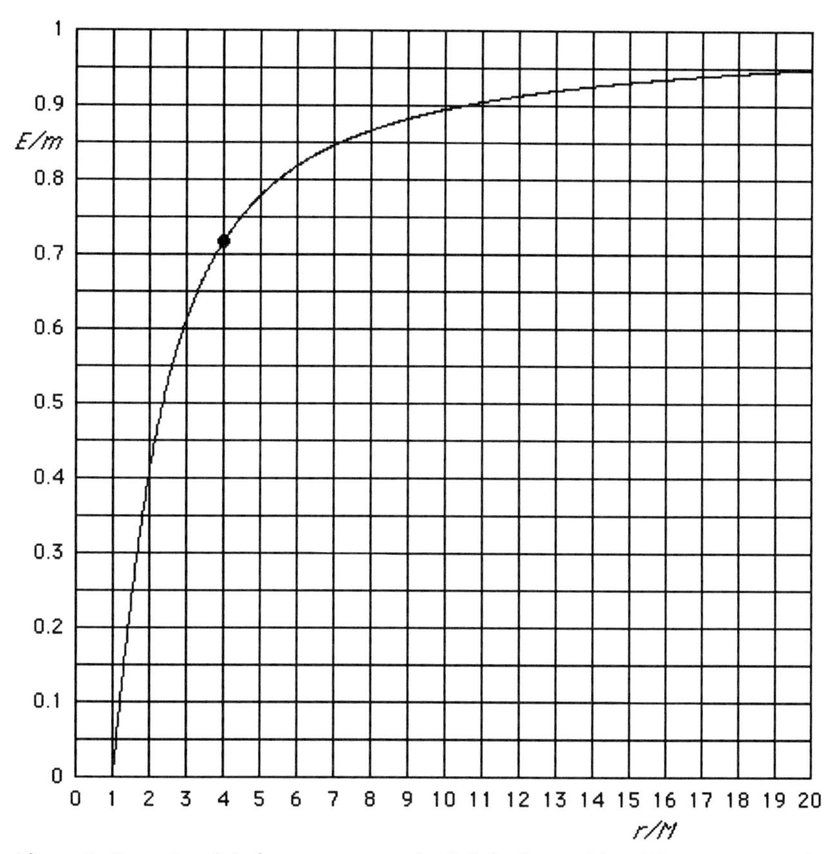

Figure 9 *Computer plot of energy measured at infinity for an object riding at rest on a ring rotating at various radii around an extreme Kerr black hole. Example shown by dot on the diagram: Stone riding at rest on a ring at $r = 4M$ has total energy measured at infinity of $E = 0.72m$.*

Equation [44] is a simple expression but awkward to calculate because R is a function of r (equations [4] and [13]). However, the computer has no difficulty with these complications and plots the result in Figure 9.

> QUERY 34 **Brilliant garbage.** A blob of matter starts at rest at a great distance from a black hole and gradually descends, riding at rest on each local ring and emitting any change of energy as radiation. Now this matter rides on the ring at $r = 2M$, the static limit. From Figure 9, determine what fraction of its original rest energy it has radiated thus far. In principle, what is the *maximum* fraction of its original rest energy that can be radiated before it disappears inward across the horizon of the black hole?

Figure 9 is not to be taken literally. In practice the rings do not rotate at the same rate as the accretion disk, and the accretion disk itself is not a perfectly efficient emitter of radiation. A few percent of the rest energy of swirling particles may be emitted in the form of radiation before they plunge across the horizon. Still, a few percent is far greater than the efficiency of nuclear reactors on Earth.

12 A "Practical" Penrose Process

Using results of Sections 10 and 11, we can devise a "practical" Penrose process by which energy can be milked from an extreme spinning black hole. Actually, this process is "practical" only for an advanced civilization, one that can accomplish any engineering feat not forbidden by the laws of Nature. Outline of the strategy: Equal masses of matter and antimatter (say positrons and electrons united in positronium molecules, in bulk as liquid positronium) are carried down to a rotating ring just outside the horizon of an extreme Kerr black hole. There the matter and antimatter are combined (annihilated) to create two oppositely moving pulses of radiation. One pulse has negative energy and drops into the black hole, robbing the black hole of some of its mass and energy of rotation. The other pulse has positive energy and escapes to a distant observer who uses this energy for practical purposes. Now for the details.

The generalization of equation [44] for a particle moving along a rotating ring is given by the equation

$$\frac{E}{E_{\text{ring}}} = \frac{r-M}{R} + \frac{2M^2}{rR} v_{\text{ring}} \qquad [45]$$

where $v_{\text{ring}} = R d\phi_{\text{ring}}/dt_{\text{ring}}$. Equation [45] comes from applying a boatload of algebra to equations [19], [29], and [32] and simplifying using equation [4]. In addition, the derivation of [45] employs the following results of special relativity:

$$E_{\text{ring}} = m\gamma_{\text{ring}} \qquad [46. \text{ special relativity}]$$

where

$$\gamma_{\text{ring}} \equiv \frac{1}{(1-v_{\text{ring}}^2)^{1/2}} \qquad [47. \text{ special relativity}]$$

A final transformation (time stretching) from special relativity tells us that

$$dt_{\text{ring}} = \gamma_{\text{ring}} d\tau \qquad [48. \text{ special relativity}]$$

where $d\tau$ is the wristwatch time of the stone moving along the ring.

Note that in equation [45], v_{ring} can be positive or negative, corresponding to motion in either direction along the ring. Under some circumstances, the result is negative energy for the particle.

Now apply some simplifying circumstances. First keep constant the value of $E_{\text{ring}} = m\gamma_{\text{ring}}$ in equation [46] while letting m go to zero and v_{ring} go to plus or minus one. The result signifies a pulse of light.

Second, apply equation [45] to a rotating ring very close to the horizon, as a limiting case. In other words $r \longrightarrow M$ and $R \longrightarrow 2M$. Equation [45] becomes

$$\frac{E}{E_{ring}} = \pm 1 \qquad \text{[49. light flash moving along ring as } r \rightarrow M]$$

With these equations we can analyze the following idealized method for milking energy from the black hole. Start with a mass m of matter and an equal mass m of antimatter. Total mass: $2m$.

Phase 1. Take the total load of mass $2m$ down to a position at rest on a ring near to the horizon of an extreme Kerr black hole, milking off the energy as it makes successive moves from rest on one ring to rest on the next lower ring.

QUERY 35 **Energy extracted in Phase 1**. When Phase 1 is completed, how much energy will have been milked off for use at a distant location?

Phase 2. Combine the matter and antimatter at rest on the near-horizon ring and direct the resulting light pulses in opposite directions along the ring at the horizon.

QUERY 36 **Energies of tangential light flashes.** Just after Phase 2 is completed, what is the ring energy E_{ring} of each of the two light pulses moving along the ring as measured locally by a rider on that ring? What is the energy measured at infinity E of each of these flashes?

Now the light flash with negative energy drops across the horizon into the black hole, thereby reducing the angular momentum (and mass) of the spinning hole. In contrast, the light flash with positive energy flies out to a great distance and its energy is employed for useful purposes.

QUERY 37 **Total energy extracted.** In summary, what is the total useful energy made available to distant engineers as a result of this entire procedure? How much mass/energy was the input for this process?

QUERY 38 **Phase 1 reduction of angular momentum?** *Thought question, optional.* Does the energy extracted in Phase 1 by itself reduce the rotation rate of the black hole? In answering, recall the analogous extraction of energy from a nonrotating black hole (Exercise 6, Chapter 3).

13 Challenges

Nothing but algebra stands in the way of completing a full analysis of orbits of stones and light in the equatorial plane of the extreme Kerr black hole. The strategies required are analogous to those that led to similar results for the nonrotating Schwarzschild black hole (Chapters 4 and 5).

- Computing the orbits of a stone from equations that relate dr and $d\phi$ to the passage of wristwatch time $d\tau$ (similar to equations [21] and [22], page 4-9).

- Carrying out qualitative descriptions of different classes of orbits using an effective potential (similar to equation [32], page 4-18).

- Finding stable circular orbits, similar to those analyzed in the exercises of Chapter 4. (Stable orbits allow a more realistic analysis of the behavior and energy of a particle orbiting with the accretion disk.)

- Predicting orbits of light as done in Chapter 5.

- Predicting details of life inside the horizon, comparable to the analysis carried out for the Schwarzschild black hole in Project B, Inside the Black Hole. Such an analysis is probably fantasy, since inside the Cauchy horizon (choosing the minus sign in equation [2]) spacetime appears to be unstable, hence not described by the Kerr metric, and possibly lethal to incautious divers.

- Verifying that an extreme spinning black hole cannot accept additional angular momentum. Can an object moving in the direction of rotation of an extreme black hole cross the horizon and thus increase the angular momentum of this structure which already has maximum angular momentum?

Much of the complicated algebra that lies on the way to these outcomes springs from the relation between the radius r and the reduced circumference R given by equation [4]. Once the algebra is mastered, results can be plotted using a simple computer graphing program.

- For readers with unfettered ambition or for those skilled in the use of computer algebra manipulation programs, the outcomes of this project can be rederived for a black hole that spins with angular momentum parameter $a = J/M$ *less* than its maximum value. Start with the metric [1] and use the more general reduced circumference R_a, defined by the equation (valid in the equatorial plane)

$$R_a^2 = r^2 + a^2 + \frac{2Ma^2}{r} \qquad [50]$$

  The resulting equations are easy to check at the extremes: They go to the Schwarzschild limit when $a \longrightarrow 0$ and to the expressions derived in this project when $a \longrightarrow M$.

- We have studied two important metrics: the Schwarzschild metric for a nonspinning black hole and the Kerr metric for a spinning black hole. You can apply the skills you have now mastered to analyze the consequences of a third metric, the so-called **Riessner-Nordstrøm metric** for an electrically charged nonspinning black hole. For a pair of events that occur near one another on a plane through the center of such a charged black hole, the Riessner-Nordstrøm metric has the form

$$d\tau^2 = \left(1 - \frac{2M}{r} + \frac{Q^2}{r^2}\right)dt^2 - \frac{dr^2}{\left(1 - \frac{2M}{r} + \frac{Q^2}{r^2}\right)} - r^2 d\phi^2 \qquad [51]$$

Here Q is the electric charge of the black hole in units of length.

Good luck!

The essence of newer physics

Of all the entities I have encountered in my life in physics, none approaches the black hole in fascination. And none, I think, is a more important constituent of this universe we call home. The black hole epitomizes the revolution wrought by general relativity. It pushes to an extreme—and therefore tests to the limit—the features of general relativity (the dynamics of curved spacetime) that set it apart from special relativity (the physics of static, "flat" spacetime) and the earlier mechanics of Newton. Spacetime curvature. Geometry as part of physics. Gravitational radiation. All of these things become, with black holes, not tiny corrections to older physics, but the essence of newer physics.

—John Archibald Wheeler

14 Basic References to the Spinning Black Hole

Introductory references to the spinning black hole

For the human and scientific story of the spinning black hole, read Kip S. Thorne, *Black Holes and Time Warps: Einstein's Outrageous Legacy,* W. W. Norton, New York, 1994, pages 46–54 and 286–299.

Bernard F. Schutz has an excellent analytic treatment in *A First Course in General Relativity,* Cambridge University Press, New York, 1985, pages 294–305.

Chapter 33 of Misner, Thorne, and Wheeler's *Gravitation*, W. H. Freeman and Company, San Francisco (now New York), 1973, is very thorough, with wonderful summary boxes, though beset with the mathematics of tensors and differential forms. It is also approximately 30 years old.

Chapter 12 of Robert M. Wald's *General Relativity* (University of Chicago Press, Chicago, 1984) is authoritative and straightforward. The mathematics is deep; you have to "read through the mathematics" to find the physical conclusions, which are clearly stated.

Section 12.7 of *Black Holes, White Dwarfs, and Neutron Stars* by Stuart L. Shapiro and Saul A. Teukolsky (John Wiley, New York, 1983, pages 357–364), covers the spinning black hole, mostly with algebra rather than tensors, and discusses orbits in some detail.

Steven Detweiler, editor, *Black Holes: Selected Reprints*, American Association of Physics Teachers, New York, 1982. This collection is out of print but may be available in some physics libraries.

Original references to the spinning black hole

The first paper: R. P. Kerr, "Gravitational Field of a Spinning Mass as an Example of Algebraically Special Metrics," *Physical Review Letters*, Volume 11, pages 237–238 (1963).

Choice of coordinate system can make thinking about the physics convenient or awkward. Boyer and Lindquist devised the coordinates that illuminate our analysis in this project. Robert H. Boyer and Richard W. Lindquist, "Maximum Analytic Extension of the Kerr Metric," *Journal of Mathematical Physics*, Volume 8, Number 2, pages 265–281 (February 1967). See also Brandon Carter, "Global Structure of the Kerr Family of Gravitational Fields," *Physical Review*, Volume 174, Number 5, pages 1559–1571 (1968).

For completeness, the Newman electrically charged black hole: E. T. Newman, E. Couch, K. Chinnapared, A. Exton, A. Prakash, and R. Torrence, "Metric of a Rotating, Charged Mass," *Journal of Mathematical Physics*; Volume 6, Number 6, pages 918–919 (1965); also E. T. Newman and A. I. Janis, "Note on the Kerr Spinning-Particle Metric," *Journal of Mathematical Physics*, Volume 6, Number 6, pages 915–917 (1965).

The Penrose process, to help you milk the energy of rotation from the spinning black hole: R. Penrose, "Gravitational Collapse: The Role of General Relativity," *Revista del Nuovo Cimento*, Volume 1, pages 252–276 (1969).

15 Further References and Acknowledgments

Initial quote: S. Chandrasekhar, *Truth and Beauty: Aesthetics and Motivations in Science*, University of Chicago Press, 1987, pages 153–154.

Quote in reader objection, page F-3: Stephen Hawking, *Black Holes and Baby Universes*, Bantam Books, New York, 1993, pages 91–92.

E. Poisson and W. Israel, "Inner-Horizon Instability and Mass Inflation in Black Holes," *Physical Review Letters*, Volume 63, Number 16, pages 1663–1666 (16 October 1989).

The value of the angular momentum of Sun (page F-6) was provided by Douglas Gough, private communication.

Stephan Jay Olson suggested using light flashes as part of the practical Penrose process in Section 12.

Final quote: John Archibald Wheeler with Kenneth Ford, *Geons, Black Holes, and Quantum Foam, A Life in Physics*, New York, W. W. Norton & Company, 1998, page 312.

Project G The Friedmann Universe

- *How does the size of the Universe change with time?*
- *Will the Universe recontract—or expand forever?*
- *How can there be galaxies that we have not yet seen?*
- *Do recent observations tell us the final fate of the Universe?*

PROJECT G

The Friedmann Universe

Will we ever penetrate the mystery of creation? There is no more inspiring evidence that the answer will someday be "yes" than our power to predict, and predict correctly, and predict against all expectation, so fantastic a phenomenon as the expansion of the universe.

—Misner, Thorne, and Wheeler

1 "The Biggest Blunder of My Life"

The Friedmann model Universe is the simplest cosmological model based on Einstein's field equations. In 1922 Alexander A. Friedmann idealized the sprinkling of stars through space as a cloud of dust at zero pressure and of uniform density. Depending on the value of this density, the Universe may either expand forever or else expand to a maximum size and recontract to a Big Crunch. In the recontracting case there is a final instant after which there is no "after," as well as an earliest moment before which there is no "before." (John Archibald Wheeler favors the model Universe that recontracts, in part because this model has no difficulties with boundary conditions: There aren't any! There is no spacetime "beyond" either end.)

Friedmann's stunning prediction of a Universe that starts off expanding was unexpected, and at first Einstein did not accept it. Inspired by Benedict de Spinoza, the greatest of his heroes, Einstein had from his youth felt that there could be no moment of creation. He believed that time, physical law, and the Universe stand eternal, from everlasting to everlasting. On this basis he rejected the Friedmann cosmology. Only after Edwin Hubble documented the recession of galaxies—and the faster recession of more distant galaxies—was expansion generally accepted. Then Einstein embraced the Friedmann model and confided to George Gamow that his previous objection was "the biggest blunder of my life."

What follows is an adaptation of material in *Principles of Physical Cosmology* by P. J. E. Peebles (Princeton University Press, 1993) and *Gravitation* by Misner, Thorne, and Wheeler (W. H. Freeman, 1973), abbreviated MTW. Our treatment in this project is looser and more informal than the analyses in earlier pages, a sign that at the end of this book we are running out of physical systems described by simple metrics.

Friedmann's model of the Universe satisfies what is called the **cosmological principle**, which says that on the large scale the Universe is assumed

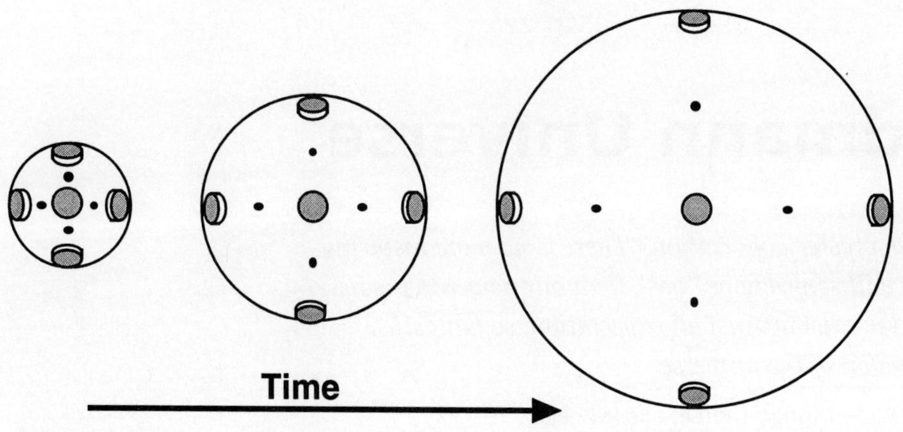

Figure 1 A model of the expansion of the Universe: an expanding balloon with coins glued on it. The rider on each coin is a two-dimensional creature for whom the surface of the balloon is the only space there is. Each rider sees nearby coins moving away uniformly in all directions, more distant coins moving away more rapidly. Thus each observer sees herself at the center of the Universe. By projecting the relative motion backward in time, each coin-rider derives the value of the current time t measured since the expansion began. Coins do not expand as the balloon radius increases. Similarly, our galaxy and our solar system (size determined by spacetime curvature due to local structures) and molecules in our bodies (size determined by quantum mechanics), stay the same size as the Universe expands.

to be everywhere **homogeneous** (for example, has everywhere the same density) and **isotropic** (has the same properties when looking in any direction). In other words, all positions in the Universe are essentially equivalent. Recent observations confirm that these assumptions are roughly correct, but only when density is averaged over regions of space large compared with clusters of galaxies.

2 A One-Dimensional Creature on a Two-Dimensional Circle

We begin with an analogy: coins glued to an expanding spherical balloon (Figure 1). The coins mimic creatures confined to a two-dimensional spherical surface expanding in a three-dimensional space. As the balloon expands, each coin rider sees all other coins moving away. Each rider sees the same picture, so each rider is free to call his or her position the center of the model Universe. (Talk about ego!)

How do the two-dimensional coin creatures analyze the geometry of their balloon? To answer this question, think of an even simpler geometry: a one-dimensional creature living on a two-dimensional circle of radius r (Figure 2). For now, assume that r does not change with time.

From Figure 2, we have

$$r^2 = x^2 + y^2 \qquad [1]$$

Or, rearranging,

$$y^2 = r^2 - x^2 \qquad [2]$$

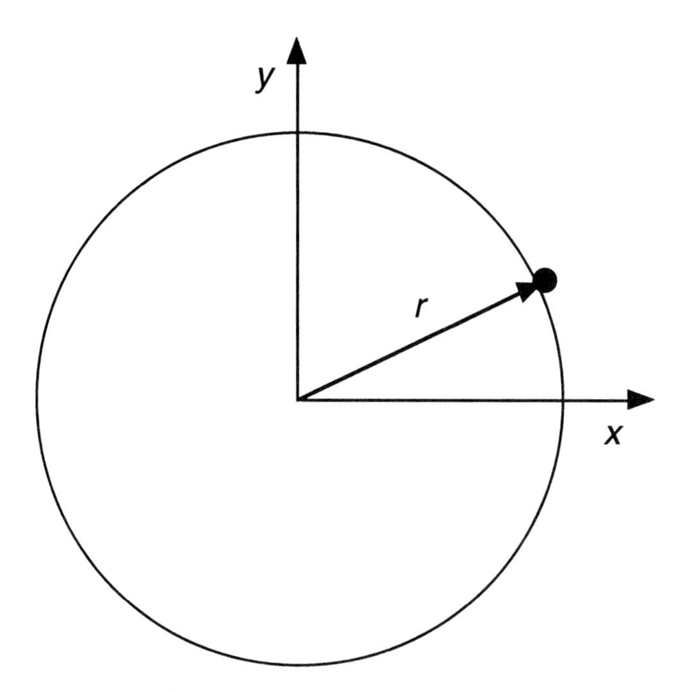

Figure 2 *One-dimensional creature (dot) living on the circumference of a two-dimensional circle.*

The one-dimensional creature wants to measure short lengths along the circle specified by a fixed value of *r*. Take the differential of both sides of equation [2], keeping *r* constant:

$$2ydy = -2xdx \qquad [3.\ r = \text{constant}]$$

or, substituting from equation [2],

$$dy = -\frac{xdx}{y} = -\frac{xdx}{(r^2 - x^2)^{1/2}} \qquad [4.\ r = \text{constant}]$$

The one-dimensional observer lays a short measuring rod of length *dl* along his circle. Substituting from equation [4], we have

$$dl^2 = dx^2 + dy^2 = dx^2 + \frac{x^2dx^2}{(r^2 - x^2)} \qquad [5.\ r = \text{constant}]$$

QUERY 1 **Constant *dl* but different *dx*.** Collect terms in equation [5]. Show that the result can be written

$$dl^2 = \frac{dx^2}{1 - x^2/r^2} \qquad [6.\ r = \text{constant}]$$

Figure 3 shows the relation between *dx* and *dl* for our one-dimensional observer.

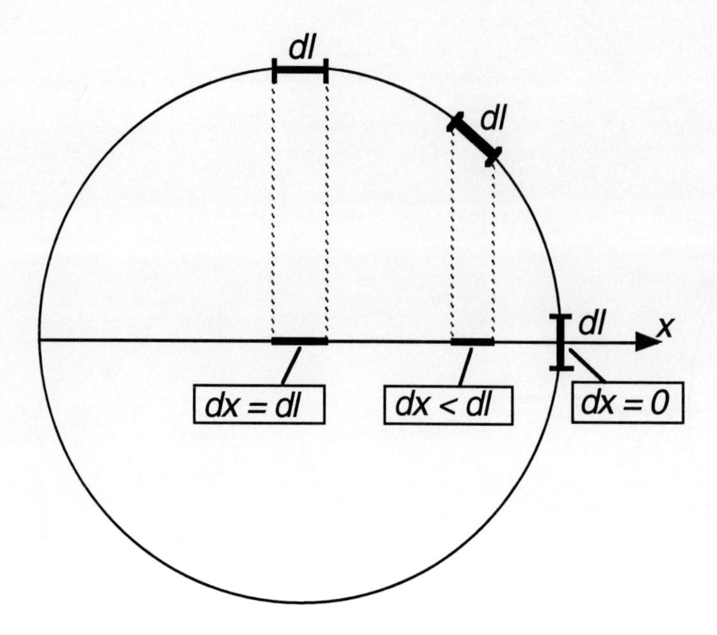

Figure 3 *A one-dimensional creature measures equal lengths dl along the circumference of a circle. Equal values of dl correspond to different values of dx, increments along the horizontal coordinate. (The right side of equation [6] is indeterminate at x = r, so we show here the result of multiplying both sides of [6] by the denominator of the right hand side.)*

3 Three-Dimensional Us on a Four-Dimensional Balloon

Now we jump from a one-dimensional circle in a two-dimensional space (Figures 2 and 3) to a two-dimensional spherical surface in a three-dimensional space (Figure 1) and finally to our first model of the Universe: a three-dimensional hypersphere in a four-dimensional space. (These four dimensions are *spatial,* to which we later add time as a fifth dimension.)

Following Peebles (his page 60) we call the fourth space dimension w and call the four-dimensional distance from the center R. Then "the three-sphere" (hypersphere) of space is the set of points (x, y, z, w) at fixed distance R from the origin:

$$R^2 = x^2 + y^2 + z^2 + w^2 = r^2 + w^2 \qquad [7]$$

where

$$r^2 = x^2 + y^2 + z^2 \qquad [8]$$

Combine [7] and [8]:

$$w^2 = R^2 - r^2 \qquad [9]$$

QUERY 2 **Eliminate fourth space dimension.** We are looking for a metric. A metric employs differentials to describe experiments that take place during times short compared with the age of the Universe. So take the differential of equation [9] at constant R (that is, for a given size of the Universe). Show that

$$dw = -\frac{rdr}{w} = -\frac{rdr}{(R^2 - r^2)^{1/2}} \qquad \text{[10. } R = \text{constant]}$$

The square of the spatial four-dimensional line element dl is

$$dl^2 = dx^2 + dy^2 + dz^2 + dw^2 = dx^2 + dy^2 + dz^2 + \frac{r^2 dr^2}{R^2 - r^2} \qquad \text{[11 R=const]}$$

For simplicity, consider only events that occur on a *single spatial plane* ($z = 0$ and $dz = 0$), as we have done throughout this book. Such a limitation allows us to describe positions on this plane using only two coordinates, r and ϕ. In these polar coordinates equation [11] becomes

$$dl^2 = dr^2 + r^2 d\phi^2 + \frac{r^2 dr^2}{R^2 - r^2} \qquad \text{[12. } z = 0, R = \text{constant]}$$

QUERY 3 **Increments of distance.** Collect dr^2 terms in equation [12]. Show that the result can be written

$$dl^2 = \frac{dr^2}{1 - r^2/R^2} + r^2 d\phi^2 \qquad \text{[13. } z = 0, R = \text{constant]}$$

In equation [13], R is the **hyperradius of the Universe** and we can choose our space origin ($r = 0$) anywhere; the Universe has no unique center.

Let's explore some consequences (Query 4) of the space equation [13] under the assumption that R is constant. Suppose that we lay metersticks of constant length dl along the radial direction ($d\phi = 0$) starting at different distances r from the center at $r = 0$.

QUERY 4 **Constant *dl* but variable *dr*.** For constant length *dl* of the meterstick (and $d\phi = 0$), discuss qualitatively the magnitude of the increment of the radius *dr* for the following conditions.

A. The value of *r* is very much less than *R*.

B. The value of *r* is a significant fraction of *R* but is not equal to *R*.

C. The value of *r* approaches *R*.

Draw an explicit analogy to results in Figure 3.

4 Metric of the Model Universe

We assumed that *R*, the hyper-radius of the Universe, is constant with time in deriving equations [10] – [13]. These equations do not change in any important way if *R* changes so slowly that it remains effectively constant during the time that a particle or light flash moves an incremental distance *dl*. Define a new variable $u \equiv r/R(t)$, so that, for slowly-varying $R(t)$,

$$u \equiv \frac{r}{R(t)} \quad \text{and} \quad du = \frac{dr}{R(t)} \qquad [14]$$

Substitute these expressions into equation [13] and introduce a new index *k*, discussed after equation [15].

$$dl^2 = R^2(t)\left(\frac{du^2}{1 - ku^2} + u^2 d\phi^2\right) \qquad [15]$$

Here the explicit time variation is present only in the factor $R(t)$. Thus far we have been discussing a model Universe that has the geometry of a hypersphere, which we call a **closed geometry**. In Section 6 we describe model Universes with other geometries. The constant *k* has been introduced into equation [15] to generalize it to such models. For the present closed case, $k = +1$. In Section 6 we set $k = -1$ for a model Universe that we will later define as an **open Universe** and $k = 0$ for a model Universe that we will later define as a **flat Universe.**

How can we use expression [15] for *dl* to construct a "metric for the Universe"? The metric must satisfy Einstein's field equations. We naively write down a metric in the same form as those we have met earlier:

$$d\tau^2 = dt^2 - dl^2 = dt^2 - R^2(t)\left(\frac{du^2}{1 - ku^2} + u^2 d\phi^2\right) \qquad [16]$$

What is the physical meaning of the incremental time lapse dt? It is the change in time recorded on a "dust clock" at rest with respect to the uniform dust of which the Universe is composed in this model. And how is the time t established? Each dust particle remains at rest at constant u and ϕ, so $d\tau = dt$ measures wristwatch time (proper time). Thus t is the total time recorded on a dust-clock since the beginning of the Universe.

QUERY 5 **Clock synchronization in a static Universe.** For R constant, describe a method of synchronizing the clocks on dust specks. Does your method differ from that for a latticework in flat spacetime (page 1-17)?

Does the metric [16] satisfy Einstein's field equations? Yes, provided $R(t)$ takes one of a set of particular forms, some of which are described in Section 6. (Had we used equation [13] instead of equation [15] to construct metric [16], the result would *not* satisfy the field equations.)

QUERY 6 **Nearly flat local spacetime.** Assume that $R(t)$ is constant for the duration of a particular short experiment and that the spatial dimensions of this experiment (characterized by the value of r) are small compared with R. Show that under these assumptions, equation [16] is arbitrarily close to a flat spacetime metric in a suitably small region of spacetime.

Metric [16] tells a lot about the closed Universe according to this model. The requirement $u < 1$, that is $r < R$, does not constrain the distance we can travel. We can establish a center, move a distance, say $r = R/2$, establish another center there, move another distance $R/2$, and so on. For the model of the closed Universe, with $k = +1$, this process eventually brings us back to our starting point in the same way a two-dimensional creature moving "straight" (that is, along a great circle) on the balloon of Figure 1 will eventually return to his starting point.

5 Time Development of the Closed Friedmann Model Universe

The analysis thus far has been rather general. Now we focus on the results of Friedmann's solution of Einstein's field equations, which tells us how R changes with time. The Friedmann conditions (uniform dust, zero pressure) inserted into the Einstein field equations yield the following result for a closed model Universe (MTW, page 705):

$$\left(\frac{dR}{dt}\right)^2 + 1 = \frac{8\pi}{3}\rho R^2 \qquad [17]$$

Here ρ (Greek rho) is the local mass density of the Friedmann dust, assumed uniform in space. (Actually, when R was about one ten-thousandth of its present value, radiation was an equal contributor to the mass-energy density. In the very early stages of the Universe, radiation was the dominant contributor to its mass-energy.) The equation of mass conservation for the Friedmann dust-filled model Universe shows that in the following expression M is constant:

$$M = \frac{4}{3}\pi R^3 \rho \qquad [18]$$

It is tempting to read the right side of equation [18] as "volume times density" and call the symbol M on the left "the mass of the Universe." Such a generalization must be treated with care. As MTW point out (page 705), "there is no 'platform' outside the Universe on which to stand to measure its attraction via periods of Keplerian orbits or in any other way." It is true that a closed model Universe, one that has sufficient mass density to recontract on itself, has a finite volume. This volume (at a particular time) multiplied by the given value of mass density ρ (at that time) yields an unambiguous value for mass M. Smaller values of M lead to model Universes (described in Section 6) that are not closed and do not have finite volume. To these model Universes the concept "total mass" cannot be naively applied. For all Friedmann models, the local mass–energy density ρ can be observed directly, but M cannot. For this reason we call the symbol M the **mass parameter** of the model Universe.

QUERY 7 **Rate of change of *R* as a function of *M*.** Use equation [18] to rewrite equation [17] as

$$\left(\frac{dR}{dt}\right)^2 + 1 = \frac{2M}{R} \qquad [19]$$

QUERY 8 **Maximum value of *R*.** When the Universe stops expanding, what is the value of R? Shades of Schwarzschild!

Equation [19] can be rewritten in an evocative form by multiplying through by $m/2$, where m is the mass of a single dust particle:

$$\frac{1}{2}m\left(\frac{dR}{dt}\right)^2 - \frac{mM}{R} = -\frac{m}{2} \qquad [20]$$

Equation [20] has the form of a Newtonian energy equation for a dust particle "at the edge of the Universe," expressed in geometric units. The first term on the left corresponds to the kinetic energy of the dust particle, the second term its potential energy in the "gravitational field of the Universe," and the term $-m/2$ on the right to the total energy of the dust particle. This naive Newtonian interpretation is an analogy only, since such concepts as "the edge of the Universe" and "gravitational field of the Universe" have no meaning in general relativity.

What happens to the radius $R(t)$ of the Friedmann model Universe as time goes by? We shall see that the integrated solution to equation [19] is a **cycloid**, the trajectory of a spot on the rim of a wheel as the wheel rolls without slipping along a horizontal surface (Figure 4 on page G-10). We use the Greek lower-case symbol η (eta) to describe the angle of rotation of the wheel. Introduce the following substitution into equation [19]:

$$d\eta = \frac{dt}{R(t)} \qquad [21]$$

The result is

$$\frac{1}{R^2}\left(\frac{dR}{d\eta}\right)^2 + 1 = \frac{2M}{R} \qquad [22]$$

We will call η the **time parameter.** Eliminate the constant $2M$ (and make the integration a breeze) with the further substitution

$$q = \left(\frac{R}{2M}\right)^{1/2} \qquad [23]$$

QUERY 9 **Integrating the Universe.** Substitute [23] into [22], simplify, and prepare for integration to obtain

$$d\eta = \pm\frac{2dq}{(1-q^2)^{1/2}} \qquad [24]$$

Carry out the integration using a table of integrals, leading to

$$\eta = 2\arcsin q \qquad \text{or} \qquad q = \sin\frac{\eta}{2} \qquad [25]$$

Substitute for q from equation [23]. Finally, use the trigonometric identity

$$2\left(\sin\frac{\eta}{2}\right)^2 = 1 - \cos\eta \qquad [26]$$

Show that the result can be written

$$\frac{R}{M} = 1 - \cos\eta \qquad [27.\ 0 \le \eta \le 2\pi]$$

As the cosmic wheel rolls along, its angle of rotation η increases. The right-hand side of equation [27] increases from zero to the value two at $\eta = \pi$. Then the radius R drops again to zero at $\eta = 2\pi$. In this Friedmann model, the closed Universe explodes, expands to a maximum stage, then recontracts to a final singularity.

Equation [27] is called a **parametric equation** because the physical quantity of interest—the radius R of the model Universe—is expressed in terms of the angle *parameter* η, rather than as a function of the time t directly. The time t satisfies a second parametric equation.

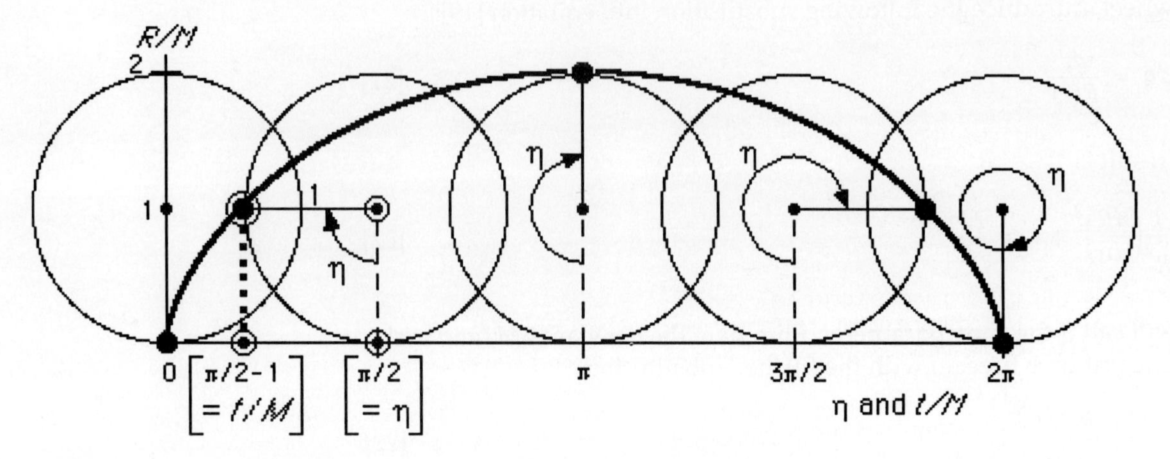

Figure 4 *Computer construction: A point on the rim of a wheel rotating without slipping traces out a cycloid, the time development of the Universe according to the closed Friedmann model. The Universe reaches its maximum size at time $t = \pi M$ and recontracts to the Big Crunch at $t = 2\pi M$.* ***Example:*** *Look at the square at the left side, with circled dots at its corners. The circled dots on the right side of this square have the horizontal coordinate $\eta = \pi/2$. This is the angle through which the wheel has rolled and is also the horizontal coordinate of the center of the wheel. And what are the coordinates for the dot on the rim? Equation [29] tells us that the value of the time is $t/M = \eta - \sin(\eta)$. For $\eta = \pi/2$ we have $\sin(\eta) = 1$ and the value of the time is $t/M = \pi/2 - 1$, as shown for the two circled dots at the left side of the square. The value of R/M (the R-value of the Universe at this time) is given by equation [27], which locates the large dot on the heavy line tracing out the cycloid.*

QUERY 10 **Timing the Universe.** From equation [21] write

$$dt = R(t)d\eta \qquad [28]$$

Substitute for R from equation [27] and integrate. Show that the result is

$$\frac{t}{M} = \eta - \sin\eta \qquad [29.\ 0 \le \eta \le 2\pi]$$

Equations [27] and [29] plot as the cycloid in Figure 4. The time development of the Universe is described by a spot on a wheel that rotates without slipping on a horizontal plane. These equations have no adjustable parameter, no "fudge factor" to explain away an incorrect prediction. For a model Universe with a mass parameter M great enough to lead to recontraction, these equations uniquely predict the radius R as a function of time t.

What is the physical meaning of the time t in these equations? In an expanding or contracting model Universe, clocks on different dust particles are not relatively at rest. Therefore the synchronization process may not be as simple as that developed to answer Query 5. The time on each dust clock must be set individually. There are two idealized ways to

Opinion: The Bang-to-Crunch Universe Too Simple to be Wrong!

Spacetime tells mass how to move, and mass tells spacetime how to curve. If the black hole provides our closest plain-speaking witness to spacetime curvature, the Cosmos itself looks like the one operative on the largest scale of space and time. The stars bear witness to the scales of space and time characteristic of the Cosmos. Sun gets its energy by burning hydrogen to helium and some of that helium to heavier elements, profiting from the difference in mass per nucleon between hydrogen and helium: 1.00783 for hydrogen and 1.00065 for helium (in units that set the mass of the most common isotope of carbon equal to 12).

That space is expanding shows most directly in the red shift of light from distant galaxies, a red shift which is greater the more distant the galaxy. The inflation of a balloon (Figure 1) provides a simple model for such an expansion of the Universe. This model tells its story in the functional dependence of the radius $R(t)$ of the sphere on the time t. In this model the curvature of spacetime in the large has only two components. One is the momentary "intrinsic curvature," fixed by the momentary radius $R(t)$ of the idealized three-geometry: $6/R^2(t)$. The other contribution to the curvature arises from the variation of this radius with the cosmological time, t: $(6/R^2)(dR/dt)^2$. Einstein, considering the matter in the days before Hubble had seen and measured the expansion of the Universe, found it natural to think of a closed Universe with an essentially constant radius R.

At this point one comes hard up against the second part of gravitation theory: mass tells spacetime how to curve. In Einstein's time there did not seem to be enough mass around to curve up the Universe into closure. Therefore Einstein postulated an additional source of curvature, a so-called "cosmological constant." Going into the doorway of the Institute for Advanced Study's Fuld Hall with Einstein and George Gamow, I heard Einstein say to Gamow about the cosmological constant, "That was the biggest blunder of my life." If we drop that term, then the equation in which matter tells spacetime how to curve becomes

$$\left(\frac{1}{R}\frac{dR}{dt}\right)^2 + \frac{1}{R^2} = \frac{8\pi}{3}\rho$$

This equation forecasts the connection between radius and time depicted by the cycloid in Figure 4 (explanation in the caption to that figure). The Universe begins with a Big Bang. In a phase of gradually slowing expansion, it reaches a maximum radius and recontracts to zero radius—a "big bang" to "big crunch" history.

An article by John Noble Wilford in the Science Times section of the *New York Times* for Tuesday March 3, 1998, reports observations by two separate groups of investigators which they interpret as showing that today the expansion of the Universe is speeding up rather than undergoing the slowdown expected for any approach to maximum expansion. [See box page G-20.] Later that day I encountered a hard-bitten veteran gravitation physics colleague in the elevator of the Princeton physics building and asked him if he believed the purported evidence of accelerating expansion. "No," he replied. Neither do I. Why not? Two reasons: (1) Because the speed-up argument relies too trustingly on the supernovas being standard candles.
(2) Because such an expansion would, it seems to me, contradict a view of cosmology too simple to be wrong. Such clashes between theory and experiment have often triggered decisive advances in physics. We can hope that some decisive advance is in the offing.

— John Archibald Wheeler

do this. First, each clock can be set to zero at the instant of the Big Bang. Carrying out this procedure presents technical difficulties. Second, each clock can be set at any time using the given model of the Universe. The clocks on every dust speck are assumed to be identical by symmetry. Each clock observer knows that she lives in a closed Friedmann model Universe. From the parameters of her Universe and the recession rate of nearby dust particles, she derives the location of the Universe along the curve of Figure 4 (same as the lower curve in Figure 5) and hence the time t since the Big Bang.

Will the wheel roll round more than once, according to this model Universe? Will the Universe emerge again and again from the singularity and disappear into another singularity? General relativity cannot answer these questions. In this model, time itself begins with the Big Bang and ends with the Big Crunch. Without time there are no laws of physics as we know them and therefore no scientific predictions that apply beyond the boundaries at either end of the curve in Figure 4.

While the closed Friedmann model Universe is the one favored by John Wheeler (box on page G-11), current observations do not appear to be consistent with a closed model of the Universe (see box, page G-20). Indeed, these observations are not consistent with *any* of the model Universes discussed in this project.

6 Open and Flat Model Universes

Section 5 described a closed model Universe, a Universe of finite three-dimensional volume whose value of the mass parameter M is large enough so that the Universe first expands but ultimately recontracts to a final crunch. If the mass parameter of the Universe is not large enough to lead to this recontraction, then the Cosmos will expand forever. The model Universe that expands forever is called an **open Universe**. For eternal expansion of the Friedmann pressure-free dust model, the field equations yield (compare with equation [19] on page G-8)

$$\left(\frac{dR}{dt}\right)^2 - 1 = \frac{2M}{R} \qquad [30]$$

QUERY 11 **Expansion rate for an open Universe.** Examine what equation [30] says qualitatively about the time development of the Universe. What happens to the expansion rate *dR/dt* for *R* very small and *R* very large? Is there any value (or limiting value) of *R* for which the expansion rate goes to zero?

Write equation [30] in a form similar to that of equation [20]. Interpret the resulting equation in terms of Newtonian energy of a dust particle. In this case is the total energy of the dust particle positive, negative, or zero?

QUERY 12 **Integration of *dR/dt*.** Prepare for integration of equation [30] using the change of variables in equation [23]. Show that the result is

$$\frac{dt}{M} = \pm\frac{4q^2dq}{(1+q^2)^{1/2}} \qquad [31]$$

Integrating equation [31] with a table of integrals leads to a physical outcome radically different from that for the recontracting model Universe:

$$\frac{t}{M} = 2\left(\frac{R}{2M}\right)^{1/2}\left(\frac{R}{2M}+1\right)^{1/2} - 2\ln\left[\left(\frac{R}{2M}\right)^{1/2} + \left(\frac{R}{2M}+1\right)^{1/2}\right] \qquad [32]$$

This curve is plotted in Figure 5, along with curves for the other cases.

The third model of the Friedmann model Universe lies between the cases of contraction and eternal expansion. This third model describes expansion at a steadily decreasing rate that tends to zero in the limit of large radius. If the mass parameter takes on a critical value, which we call M_{crit}, the result obeys the equation

$$\left(\frac{dR}{dt}\right)^2 = \frac{2M_{crit}}{R} \qquad [33]$$

QUERY 13 **Expansion rates for a flat Universe.** Examine what equation [33] says qualitatively about the time development of the Universe. What happens to the expansion rate *dR/dt* for *R* very small and *R* very large? Is there any value (or limiting value) of *R* for which the expansion rate goes to zero? Does this model Universe expand forever?

Write equation [33] in a form similar to that of equation [20]. Interpret the resulting equation in terms of Newtonian energy. In this case is the total energy of the dust particle positive, negative, or zero?

QUERY 14 **Single equation for all three Friedmann models**. Write down a single equation for *dR/dt* for all three Friedmann models of the Universe. Use the parameter *k*, defined in the paragraph following equation [15] on page G-6.

QUERY 15 **Radius *R(t)* for flat Universe.** Integrate equation [33] from radius 0 to *R* and from time 0 to *t*. Show that the result can be written

$$\frac{t}{M_{crit}} = \frac{4}{3}\left(\frac{R}{2M_{crit}}\right)^{3/2} \qquad [34]$$

The resulting curve is plotted in Figure 5, along with curves for the other two cases.

The three cases we have been describing are often distinguished using the parameter Ω (capital omega, the last letter of the Greek alphabet).

$$\Omega \equiv \frac{M}{M_{crit}} \qquad [35]$$

Then the three models are characterized by different values of Ω as follows:

$\Omega > 1$. The mass parameter is large enough to recontract the Universe. The result is called a **closed Universe** and is said to be a Universe with overall **positive spatial curvature**. The constant *k* has the value +1 in the metric [16].

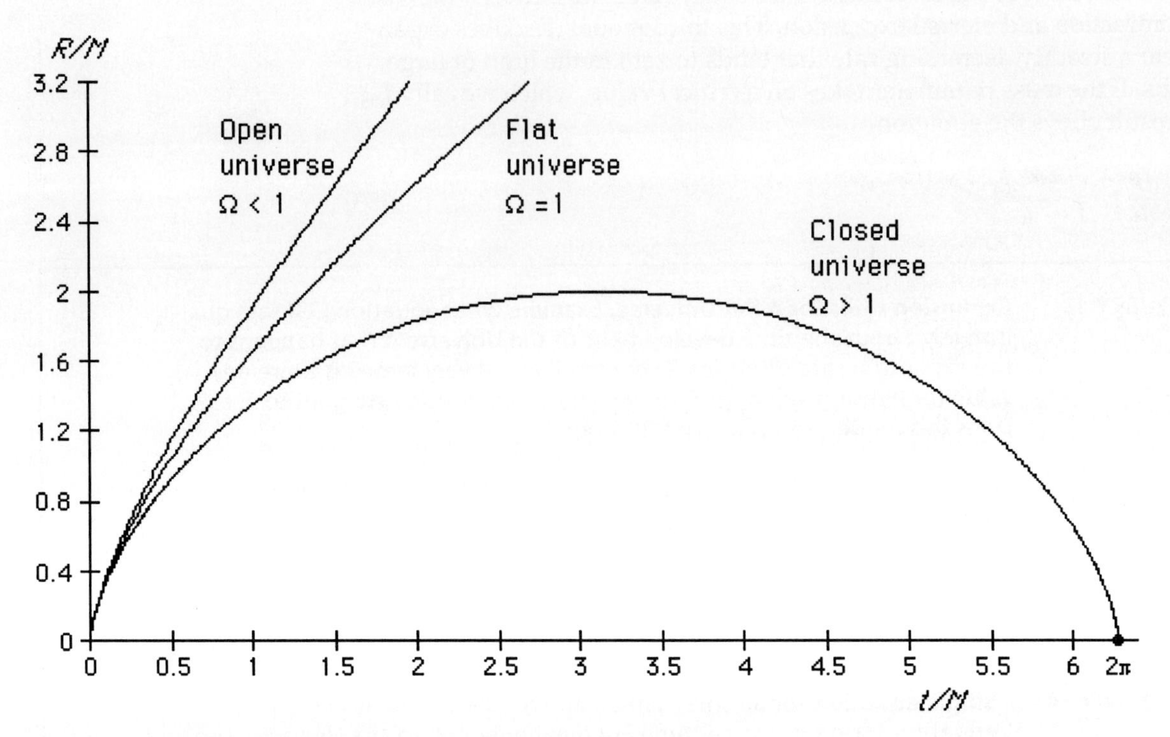

Figure 5 *Computer plot: Time development of three versions of the Friedmann model Universe composed of pressure-free dust: the open Universe, the flat Universe, and the closed Universe. The open Universe continues to expand forever. The flat Universe also continues to expand forever, but the rate of expansion gradually approaches zero. The closed Universe expands to a maximum size, then contracts to a final crunch. The parameter omega* Ω *is the ratio of the mass parameter M of each of these models to the critical mass parameter* M_{crit} *that leads to a flat Universe. When the mass parameter has a value equal to* M_{crit}, *then* $\Omega = 1$. *A value of* Ω *less than unity corresponds to the open, ever-expanding Universe, whereas a value of* Ω *greater than unity leads to a closed Universe.*

$\Omega = 1$. The mass parameter has the critical value M_{crit} such that the Universe continues to expand, but the rate of expansion approaches zero. This result is called a **flat Universe** and is said to be a Universe with overall **zero spatial curvature.** The constant k has the value 0 in the metric [16], yielding a space part that is flat. (It is *only* the space part that is flat.)

$\Omega < 1$. The mass parameter is small enough that the rate of expansion of the Universe decreases but never reaches zero. This result is called an **open Universe** and is said to be a Universe with overall **negative spatial curvature.** The constant k has the value -1 in the metric [16].

Do any of these three models of expansion describe the Universe we live in? There are many other models of the Universe, some more realistic than our Friedmann pressure-free dust models. Cosmology has made great strides in recent years, but as yet no one model of the Universe has won general acceptance. All matter in the Universe for which we have direct evidence adds up to only 10^{-2} to 10^{-1} of the critical mass M_{crit}. Until recently the most popular cosmological theories assumed, nevertheless, that the total mass was equal to M_{crit}, which led theorists to look for so-called dark matter, matter of new and exotic kinds, to account for the needed mass (so far observed only indirectly). In the meantime, recent observations have been interpreted as evidence that the rate of expansion of the Universe is actually *increasing* with time, as described in the box on page G-20.

7 Simplifications for the Closed Model Universe

How much of the Universe can we see right now? Are some galaxies so far away that light from them has not had time to reach us since the Big Bang? And what can these questions possibly mean, since all parts of the Universe were together at the Big Bang itself?

We cannot answer these questions until we have decided what model Universe correctly represents the one in which we live. In the meantime, we can answer the questions for one of our simplified model universes. We choose the closed model Universe, for which the parameter $k = 1$ in the metric [16]. This metric (for two spatial dimensions) becomes

$$d\tau^2 = dt^2 - R^2(t)\left(\frac{du^2}{1-u^2} + u^2 d\phi^2\right) \qquad [36]$$

This equation can be simplified still further by converting from time increment dt to the angle increment $d\eta$ using equation [21].

QUERY 16 **Time as an angle.** For dt in equation [36] substitute from equation [21] to show that the metric can be written

$$d\tau^2 = R^2(t)\left[d\eta^2 - \left(\frac{du^2}{1-u^2} + u^2 d\phi^2\right)\right] \qquad [37]$$

The quantity in the square bracket of equation [37] is independent of $R(t)$ and will have the same form for the entire history of the Universe—according to this closed-Universe model.

We can simplify metric [37] even further when we model the space part in the rounded brackets by a unit sphere (Figure 6).

The angle ϕ (the direction in the spatial plane we are considering) is measured around a latitude of this unit sphere from some arbitrary initial direction, and the variable u is the perpendicular distance from the axis to the point being described. The variable u can be replaced by the angle θ measured from the upward vertical axis (Figure 7). For a unit sphere, the *distance* from the north pole along a longitude is also measured by θ in radians. From Figure 7, we have a relation between u and the new angle θ

$$u = \sin\theta \qquad [38]$$

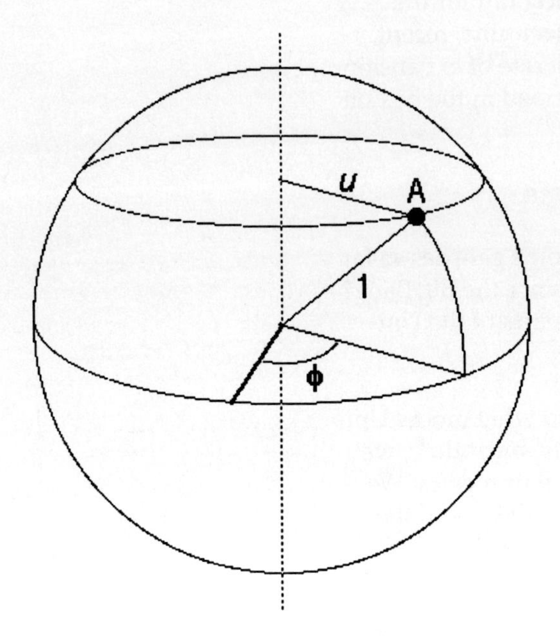

Figure 6 *Location of an event A on the unit sphere using the coordinates u and ϕ.*

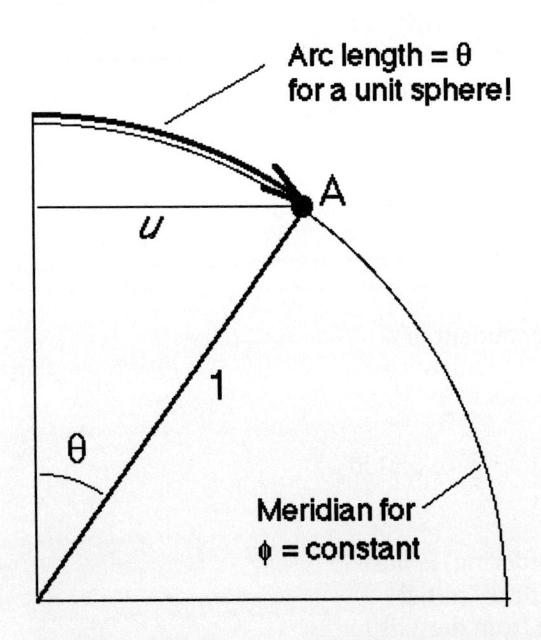

Figure 7 *Conversion from variable u to θ using a unit sphere that represents the time-independent geometry of the closed Universe model.*

from which

$$du = \cos\theta \; d\theta \qquad [39]$$

and we can use a familiar trigonometric identity to write

$$d\theta = \frac{du}{\cos\theta} = \frac{du}{(1 - \sin^2\theta)^{1/2}} = \frac{du}{(1 - u^2)^{1/2}} \qquad [40]$$

QUERY 17 **Metric for the closed Universe model.** Substitute recent expressions into [37] to show that the metric can be simplified to the form

$$d\tau^2 = R^2(t)\,[d\eta^2 - (d\theta^2 + \sin^2\theta \; d\phi^2)] \qquad [41]$$

The metric [41] carries a powerful and simple description of the closed model Universe with our usual restriction to two dimensions on a spatial plane. Here ϕ is the usual direction on that plane with respect to some arbitrary direction of zero angle, and r is coded in the variable θ through equations [14] and [38]. The size of the Universe is $R(t)$, given by equation [27], and the corresponding "angle" η is related to t by equation [29]. Every location in two spatial dimensions in this model Universe at a fixed time is described by a position on a unit sphere multiplied by that radius. The metric [41] relates adjacent events in the model closed Universe. The wristwatch time $d\tau$ between this pair of events is a product of the current size $R(t)$ of the Universe and the spacetime separation between the two events given by the expression in square brackets in equation [41].

According to this model (refer to Figures 4 and 5), when η reaches the value π then the Universe reaches its point of maximum expansion, and when η reaches 2π the Universe has again contracted to the Big Crunch.

Now we can analyze how light moves in our closed model Universe. For light $d\tau = 0$ and the metric [41] collapses to the expression

$$d\eta = \pm(d\theta^2 + \sin^2\theta \; d\phi^2)^{1/2} \qquad [42.\ \text{light}]$$

The right side of this equation has a simple interpretation: It is the distance between two nearby points on the unit sphere. Equation [42] says that in an increment of time as measured by η, light moves an equal increment of distance on the unit sphere as measured by the angles θ and ϕ.

8 Seeing the Big Bang

Our everyday language trips us up when we apply it to the Universe as a whole. Space and time were created at the Big Bang. Equation [41] tells us that at the beginning, when the radius $R(t)$ is zero, the proper time $d\tau$ (as well as the proper distance $d\sigma$) between any two events is zero. We are present at the Big Bang *wherever we are* in the spatial plane represented by the unit sphere (Figure 6). The Big Bang occurs *everywhere* on the unit sphere at $t = \eta = 0$.

Equation [42] helps us to describe how we continue to see the big bang at later times. Take our position to be at the north pole of the unit sphere and ask what we see as we look outward. It takes time for light from the Big Bang to reach us from other portions of the unit sphere. The light we see will move along longitudes of the unit sphere, namely with $d\phi = 0$ in equation [42]. As the time parameter increases, the available angle over which the light travels also increases, so that

$$d\eta = d\theta \qquad \text{[43. light]}$$

This is an easy equation to integrate! Figure 8 shows the ring on the unit sphere representing the source of light from the Big Bang that we see at the time parameter η.

Let dt_1 be the period of a light wave emitted at time t_1 when the size of the Universe is described by $R(t_1)$. Let this light wave propagate a great distance, such as that from S to O in Figure 8. What will be the observed period dt_2 at reception, when the Universe has expanded to $R(t_2)$? Equation [42], page G-17, says that at all stages of expansion light moves equal distances along the unit sphere in equal units of the time parameter η, so that $d\eta$ remains constant as the wave propagates. Equation [21] on page G-9 says that $dt = R(t)d\eta$. Hence for a larger $R(t)$, we have a larger dt, a longer period, a redder light signal. This **cosmic red shift** is a generalization in expanding curved spacetime of the Doppler shift of light in flat spacetime.

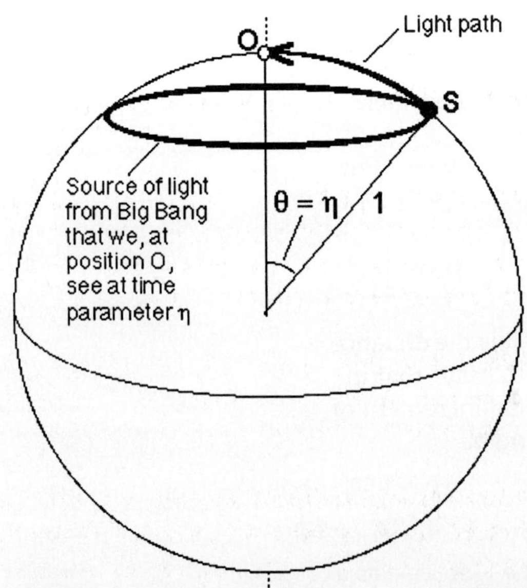

Figure 8 *Unit sphere representing the location of all events that occur on a plane in space. The Big Bang occurs everywhere on this unit sphere at t = 0. The heavy ring shows the source of light from the Big Bang that we at location O see at a time given by the time parameter η. The light path from S to O represents all light paths along fixed longitudes from the source ring to our location.*

QUERY 18 **Seeing the whole Universe.** Continue predictions implied by the construction in Figure 8. Pay attention to light which travels directly to us at position O along lines of longitude, for which $d\phi = 0$.

A. How long after the Big Bang will light have reached us from all parts of the Universe? Express your answer in the "time angle" η.

B. At the time found in part A, what is the state of expansion of the Universe?

C. Suppose that it takes some time after the Big Bang for galaxies to condense out of the primordial gases. At the time found in part A, will we have seen every galaxy in the Universe that lies on the spatial plane represented by the unit sphere? Why or why not?

QUERY 19 **When will we see it all?** How many billion years after the Big Bang will we have been able to receive light from the entire Universe, according to this closed-Universe Friedmann model? From part B of Query 18, this occurs when the Universe has reached it maximum expansion. Assume that the mass density is twice the critical mass density of five hydrogen atoms (effectively five protons) per cubic meter *at the time of maximum expansion*. In brief, the mass density at maximum expansion corresponds to 10 protons/meters^3. Assume this rough approximation is numerically exact for purposes of the following calculations.

A. Express the characteristic radius R of the Universe in terms of its mass parameter M at the moment of maximum expansion, at which time we will have received light from all of the universe.

B. Use the result of part A plus equation [18], page G-8 to find an expression for the mass parameter M as a function of the mass density ρ at the moment of maximum expansion.

C. Use equation [29], page G-10 and the value of η at maximum expansion to derive an expression for the time t at this state of expansion. Verify that the geometric units are the same on both sides of this equation.

D. Compute the value of the density ρ_{conv} in conventional units kilogram/meter^3. Convert to geometric units using the conversion factor G/c^2 (page 2-14 and inside the back cover). The result to one significant digit is 1×10^{-53} meter^{-2}. Find the result to three significant digits.

E. Find the value of M from your result of part B and substitute your value from part D into the expression for t in part C. The result to one digit is 2×10^{26} meters of time. Find the result to three significant digits.

F. Convert your result in part E to seconds and then to years (conversion factors inside the back cover). According to this model of the universe (and our assumed value of the mass density at maximum expansion), how many billion years after the Big Bang will we have received light from the entire Universe?

Accelerating Expansion of the Universe?

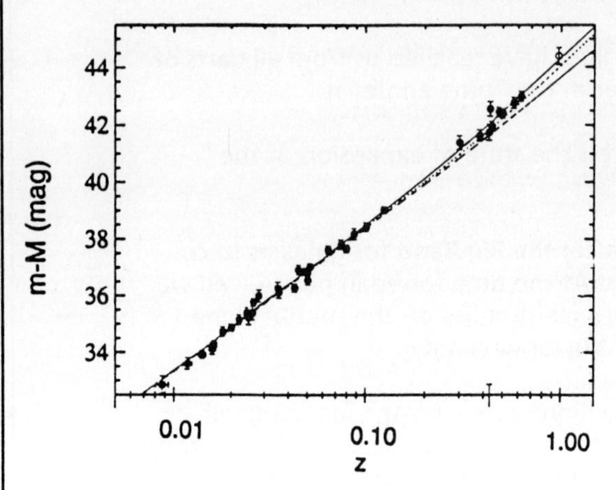

The results of Riess, Filippenko, and their coworkers.

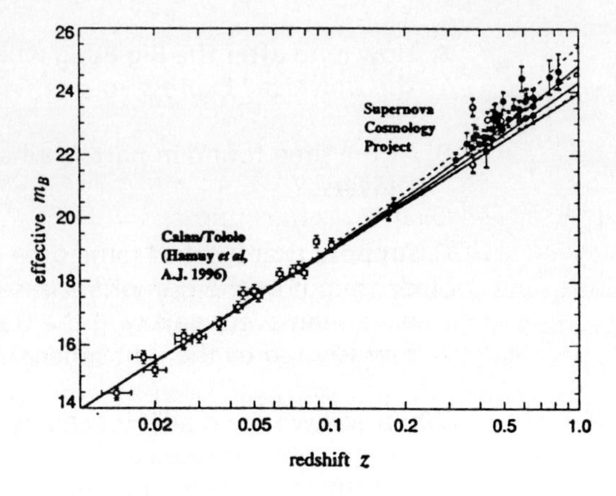

The results of Perlmutter and his coworkers.

Is the rate at which the Universe expands actually increasing with time? That is the conclusion of two separate groups, one group including Adam Riess, Alexei V. Filippenko, and eighteen other people, the other group including S. Perlmutter and thirty-one other people.

These observers use light from an exploding supernova of a particular kind, the so-called Type Ia supernova. They believe that the Type Ia supernova results when a white dwarf gradually accretes mass from a large binary companion, finally reaching a mass at which the white dwarf becomes unstable and explodes into a supernova. The "slow fuse" on the gradual accretion process may lead to almost the same size explosion on each such occasion, giving us a "standard candle" of the same intrinsic brightness, provided that nuclear burning has left every white dwarf with the same nuclear composition. If so, the brightness of the explosion as seen from Earth provides a measure of the distance to the supernova. The cosmic red shift of the light tells us how fast the supernova is receding. Because supernovas are so bright, they can be seen at a very great distance, which brings us information about the Universe much of the way back to the Big Bang.

The figures display observed brightness on the vertical scale versus red shift on the horizontal scale. Calibration of the vertical scale is in **stellar magnitude**, a logarithmic measure of observed brightness. The larger the magnitude number, the *dimmer* is the observed star. (The vertical scales of the two graphs differ by a standard magnitude, which moves the curve up and down without changing its shape.) The horizontal scale, also logarithmic, is calibrated in what is called the **red-shift factor *z***, defined implicitly in the following equation

$$\lambda_{\text{observed}} \equiv (1 + z)\, \lambda_{\text{emitted}} \qquad [44]$$

Here $\lambda_{\text{observed}}$ is the wavelength of light observed from Earth, while λ_{emitted} is the wavelength of the light emitted from the source. The emitted wavelength is known if one knows the emitting atom, identified from the pattern of different wavelengths.

If the data points in the figure lie along a straight line, then the expansion velocity is proportional to distance, as one would expect if nothing slowed down matter blasted out of the Big Bang. But the data points in the figure appear to lie slightly *above* the straight line for the most distant supernovas (upper right on the diagrams). This could mean that the most distant supernovas (highest on the vertical axis) are moving away slower than expected (farther to the left on the horizontal axis than expected). We are watching motions of these most distant supernovas as they were long ago, because it takes light such a long time to reach us. Could this mean that the expansion rate long ago was slower than it is now? And how significant is the apparent deviation from the straight line? You be the judge.

Whatever you think about the claimed change in expansion rate, the use of supernovas as standard candles for small and intermediate distances leads to a new and improved value for the average expansion rate of the Universe. From this average expansion rate, Riess, Filippenko, and their colleagues derive a time of $(14.2 \pm 1.7) \times 10^9$ years for the age of the Universe. The result derived by Perlmutter and company is $(14.5 \pm 1.0) \times 10^9$ years.

The results on previous pages illustrate the predictions for our closed-model Universe. Unfortunately they cannot correspond to reality. The mass-energy density in the early Universe was dominated by radiation and extreme pressure, whereas our simple Friedmann models consist of pressure-free dust. Moreover, in the plasma soup (swarm of uncombined electrons and nuclei) in the early Universe, light was continually absorbed and scattered and could not propagate in a straight line. Because of these effects, light could not move directly to us from the Big Bang. At present the so-called **cosmic background radiation** provides our earliest view of the Universe. The cosmic background radiation is microwave radiation that permeates the Universe, emitted just before the moment at which electrons in the cooling soup combined with protons to form atoms. With the formation of atoms, the Universe suddenly became transparent to electromagnetic radiation. This occurred about 300 000 years after the Big Bang. The cosmic red shift reduces the characteristic temperature of the background radiation that we observe from thousands of degrees Kelvin to 2.73 degrees Kelvin.

So at present we could not see the Big Bang directly, through the blanket of early plasma, whatever the predictions of our model. Neutrinos and gravity waves are much less affected by plasma than is light. So future observations made using neutrinos or gravity waves may penetrate the wall of early plasma, letting us peek farther back toward our ultimate origin. *Selah.*

How can physics live up to its true greatness except by a new revolution in outlook which dwarfs all its past revolutions? And when it comes, will we not say to each other, "Oh, how beautiful and simple it all is! How could we ever have missed it so long!"

—John Archibald Wheeler

When shall I cease from wondering?

—Galileo Galilei

9 References and Acknowledgments

Edmund Bertschinger made many important suggestions for this project.

Initial quote, Charles W. Misner, Kip S. Thorne, and John Archibald Wheeler, *Gravitation*, W. H. Freeman and Company, San Francisco (now New York), 1971, page 707. Slightly edited by JAW.

P. J. E. Peebles, *Principles of Physical Cosmology*, Princeton University Press, 1993.

"Observational Evidence from Supernovae for an Accelerating Universe and a Cosmological Constant," Adam G. Riess, Alexei V. Filippenko, Peter Challis, Alejandro Clocchiatti, Alan Diercks, Peter M. Garnavich, Ron L. Gilliland, Craig H. Hogan, Saurabh Jha, Robert P. Kirshner, B. Leibudgut, M. M. Phillips, David Riess, Brian P. Schmidt, Robert A. Schommer, Nicholas B. Suntzeff, and John Tonry, *Astronomical Journal*, Volume 116, pages 1009–1038 (September 1998).

"Measurements of Ω and Λ from 42 High-Redshift Supernovae," S. Perlmutter, G. Aldering, G. Goldhaber, R. A. Knop, P. Nugent, P. G. Castro, S. Deustua, S. Fabro, A. Goodbar, D. E. Groom, I. M. Hook, A. G. Kim, M. Y. Kim, J. C. Lee, N. J. Nunes, R. Pain, C. R. Pennypacker, R. Quimby, C. Lidman, R. S. Ellis, M. Irwin, R. G. McMahon, P. Ruiz-LaPuente, N. Walton, B. Schaefer, B. J. Boyle, A. V. Filippenko, T. Matheson, A. S. Fruchter, N. Panagia, H. J. M. Newberg, W. J. Couch, *Astrophysical Journal*, Volume 517, Number 2, pages 565–586 (1999).

Selah Used many places in the Biblical Psalms. Exact meaning uncertain. Our favorite definition is from *Bucer's Psalms* in the year 1530: "This word Selah signifyeth ye sentence before to be pond'red with a depe affecte, long to be rested upon and the voyce there to be exalted." *Oxford English Dictionary*, 2nd edition, Clarendon Press, Oxford, 1989.

Final quote by Wheeler from John Archibald Wheeler, "On Recognizing 'Law Without Law'," *American Journal of Physics*, Volume 51, pages 398–404 (1983).

Readings in General Relativity

Note: ISBNs are for paperback editions, when available.

Popular Books

Black Holes and Time Warps: Einstein's Outrageous Legacy, Kip S. Thorne, W. W. Norton & Co., New York, 1994, ISBN 0-393-31276-3. Our favorite popular treatment, as described in the front matter of this book.

Black Holes, J.-P. Luminet, Cambridge University Press, 1992, ISBN 0-521 - 40906-3. Superb, brief book. Similar to Thorne without describing the personalities.

Relativity: The Special and General Theory, Albert Einstein, Crown Publishers, New York, 1961, ISBN 0-517-025302. A popular treatment by the Old Master himself. Enjoyable for the depth of physics, the humane viewpoint, and the charm of old-fashioned trains racing past embankments. Published originally in 1916.

Was Einstein Right? Putting General Relativity to the Test, Second Edition, Clifford M. Will, Basic Books/Perseus Group, New York, 1993, ISBN 0-465-09086-9. A careful description of the experimental and observational demonstrations verifying Einstein's theory. Beautifully and engagingly written.

A Journey into Gravity and Spacetime, John Archibald Wheeler, Scientific American Library, New York, 1990, distributed by W. H. Freeman and Company, ISBN 0-7167-5016-3. One of the authors of this book examines the physical bases of Einstein's equations, their inner workings, and their astonishing predictions.

Gravity's Fatal Attraction: Black Holes in the Universe, Mitchell Begelman and Martin Rees, Scientific American Library, New York, 1996, distributed by W. H. Freeman and Company, ISSN 1040-3213, ISBN 0-7187-5074-0. Gorgeously illustrated and artfully written glossy book full of current topics in astrophysics.

Black Holes and Baby Universes and Other Essays, Stephen Hawking, New York, Bantam Books, 1993, ISBN 0-553-09523-4. Part autobiography, part science popularization, part speculations by perhaps the foremost general relativist of our time.

General Relativity from A to B, Robert Geroch, University of Chicago Press, 1978, ISBN 0-226-28864-1. Developing ideas of spacetime, starting with a single-observer description of spacetime and ending with the black hole.

Geons, Black Holes, and Quantum Foam, John Archibald Wheeler with Kenneth Ford, W. W. Norton & Company, 1998, ISBN 0-393-04642-7. Autobiography of one author of this book, including the story of his contributions to general relativity.

Collections of Original Papers

The Principle of Relativity, A. Einstein, H. A. Lorentz, H. Weyl, H. Minkowski, Dover Publications, Inc., New York, 1952, Standard Book Number 486-60081-5. A wonderful, inexpensive set of translations of many of the original papers in special and general relativity.

Black Holes, Selected Reprints, edited by Steven Detweiler, American Association of Physics Teachers, 1982, no ISBN. Here are the classic black hole papers by Michell, Schwarzschild, Oppenheimer and Snyder, Kerr, Boyer and Lindquist, and so on. Unfortunately out of print, but may be in some libraries.

Introductory Treatments of General Relativity

Flat and Curved Space-Times, George F. R. Ellis and Ruth M. Williams, Clarendon Press, Oxford, 1988, ISBN 0-19-851169-8. The only introductory treatment we know (besides the present book) that introduces general relativity without tensors (Ellis and Williams have an appendix on tensors.) A leisurely and informative trip through special relativity is followed by treatment of curved spacetime, the Schwarzschild metric, and simple cosmological models.

Note: The following books all introduce and use tensors.

A First Course in General Relativity, Bernard Schutz, Cambridge University Press, 1985 (reprinted with corrections to 1999), ISBN 0-521-27703-5. A wonderful introductory treatment. Gracefully intertwines mathematics, experiment, and a magisterial overview.

Essential Relativity, Special, General and Cosmological, Revised Second Edition, Wolfgang Rindler, Springer-Verlag, 1977, ISBN 0-387-10090-3. Rindler (who coined the term *horizon* for the black hole) is the dean of relativity authors. Wide overview with carefully motivated mathematics.

General Relativity, I. R. Kenyon, Oxford University Press, New York, 1990, ISBN 0-19-851996-6. The briefest of these introductory treatments; physically well motivated.

Advanced Texts in General Relativity and Cosmology

Gravitation, Charles W. Misner, Kip S. Thorne and John Archibald Wheeler, W. H. Freeman and Company, San Francisco (now New York),1973, ISBN 0-7167-0344-0. The beloved "telephone book," nearly the size of the Manhattan telephone directory, dear to the heart of every general relativist. A clear, comprehensive, and complete text that addresses most of the important aspects of general relativity from the geometrical point of view—as of the 1970s.

Gravitation and Spacetime, Second Edition, Ohanian and Ruffini, W. W. Norton & Co., New York, 1994, ISBN 0-393-96501-5. A physical approach, invoking many experiments, starting with the linear (weak field) approximation to general relativity and gradually introducing the geometric formalism as the applications demand it.

Black Holes, White Dwarfs, and Neutron Stars, Stuart L. Shapiro and Saul A. Teukolsky, John Wiley, New York, 1983, ISBN 0-471-87316-0. A careful treatment of what these condensed objects are actually made of and how they form. Nice treatment of particle and light orbits around both nonspinning and spinning black holes.

General Relativity, Robert M. Wald, University of Chicago Press, 1984, ISBN 0-226-87033-2) Full-blast mathematical treatment of general relativity from the beginning; comprehensive and authoritative. Final chapter on quantum effects in strong gravitational fields.

Gravitation and Cosmology, Steven Weinberg, John Wiley, New York, 1972, ISBN 0-471-92567-5. Most treatments of general relativity emphasize the geometric. Weinberg's treatment is an analytic one in which the geometrical content of general relativity is minimized.

Principles of Physical Cosmology, P. J. E. Peebles, Princeton University Press, 1993, ISBN 0-691-01933-9. From one end of the Universe to the other, through all its components, and from the beginning of the theory to its modern developments.

Cosmological Physics, John A. Peacock, Cambridge University Press, 1999, ISBN 0-521-42270-1. The most up-to-date of these advanced texts, includes the basics of both general relativity and quantum field theory as needed to engage the latest astrophysical observations. Lots of physical insights about the latest developments when one "reads around" the mathematics.

Eccentric and Absorbing

Relativity Visualized, Lewis Carroll Epstein, Insight Press, San Francisco, 1997, ISBN 0-953218-05-X. An enjoyable and eccentric presentation of special and general relativity, done primarily with figures and graphics. Available in some bookstores, or send $19.95 plus $2 handling to Insight Press, 614 Vermont Street, San Francisco, CA 94107-2636, USA.

Faster Than Light: Superluminal Loopholes in Physics, Nick Herbert, New American Library, a Division of Penguin Books, Markham, Ontario, 1988, ISBN 0-452-26317-4. Herbert tries every trick he can think of in flat and curved spacetime to figure out how to go faster than light (and thus backward in time). You be the judge of whether or not he succeeds.

Time Machines: Time Travel in Physics, Metaphysics, and Science Fiction, Paul J. Nahin, American Institute of Physics, New York, 1993, ISBN 0-88318-935-6. Nahin uses an extensive survey of the science fiction literature about time travel to illustrate and engage modern speculations on the subject. Tech Notes appendices go more deeply into the theory. A wonderful read.

GLOSSARIES OF SYMBOLS AND TERMS

Symbols

There is no abbreviation of units in this book. For example, the words meter *and* second *are spelled out, so the symbol* m *always means mass and the symbol* s *always means a distance in space. The one exception to the no-abbreviation rule is subscripts. For example, the subscript* conv *means "in conventional units."*

Hyphenated numbers are page numbers in the book.

GREEK LETTERS

γ (gamma) Stretch factor of special relativity: $\gamma = (1 - v^2)^{-1/2}$. No units. 3-25

η (eta) Angle of rotation of the "cosmic wheel" in the Friedmann model of the closed Universe. No units. G-9, G-17–19

θ (theta) (1) Angle measured from a point on a shell with respect to the radially outward direction (4-22, 5-10, 5-18, 5-21). (2) Angle and arc length along the unit circle describing the closed Friedmann model Universe (G-16–19). No units.

ρ (rho) The average density of the Universe. Unit: meter/meter^3 = meter^{-2}. G-7–8

σ (sigma) Proper distance between two events. Unit: meter. 1-4

$d\sigma$ ("dee sigma") Increment of proper distance between two adjacent events. Unit: meter. 2-19

τ (tau) Wristwatch time (proper time) between two events. Unit: meter. 1-2

$d\tau$ ("dee tau") Increment of wristwatch time (proper time) between two adjacent events. Unit: meter. 2-17, 2-19–20, F-2

ϕ (phi) Measure of angle in a plane through a center of gravitational attraction. ϕ has the same meaning in general relativity as in Euclidean geometry. No units. 2-16

ψ (psi) Direction in which observer looks to see incoming light. B-22–24

ω (omega) (1) The angular rate of rotation around a center according to some observer. Unit: radians/second (C-5). (2) Angular velocity of radial harmonic oscillation of Mercury in its orbit. Unit: radians/second (C-4). (3) The rate of angular rotation of the ring riders (zero angular momentum observers or ZAMOs) around a spinning black hole, as reckoned by the remote observer. Unit: meter^{-1} (F-16).

$\Omega = M/M_{crit}$ (capital omega) Parameter whose value describes the different Friedmann model Universes. M is the mass parameter of the Universe and M_{crit} is the critical value that leads to a flat-space Universe. No units. G-13–14

ROMAN LETTERS

$a = J/M$ Ratio of angular momentum to mass of a spinning black hole. Unit: meter. F-2

b Impact parameter of an object near a center of attraction. Unit: meters. 4-6, 5-6

c Speed of light. In flat spacetime *defined* to have the value given inside the back cover. Unit: meters/second. 1-2

dl Four-dimensional increment of distance used in the metric for the Universe in the Friedmann model. Unit: meter. G-3

dr Increment of radial separation, where r may be the Euclidean radius or the reduced circumference (2-7–11) or the Boyer-Lindquist radius (F-2). Unit: meter.

dt Increment of time between two adjacent events, measured by the far-away observer. Unit: meter. 2-19, F-2

E Energy of a particle. In special relativity, given by $E = m\,dt/d\tau$ (1-11). In general relativity, the energy measured at infinity (3-9, F-13). Units: Typically we use the dimensionless expression E/m, allowing the reader to choose any unit desired for energy, as long as it is the same unit as for mass m (1-11, 3-9).

g Local acceleration of gravity, either g_{conv} in conventional units, meters/second^2 or g in geometric units, meters^{-1}. 2-47, 3-31, B-19–20

h Planck's constant. Unit: joule-second. 5-34

J Angular momentum of a spinning black hole. Unit: meter^2. F-2

k Parameter used in Friedmann models of the Universe that distinguishes the closed Universe ($k = 1$) from the flat Universe ($k = 0$) and from the open Universe ($k = -1$). No unit. G-6

k_B Boltzmann constant. Unit: joule/degree Kelvin. 5-27

$L = mr^2 d\phi/d\tau$ Angular momentum of a particle around a center of attraction. Unit: kilogram-meter. 4-3

$L^* = L/(mM)$ where m is the mass of the orbiting object and M is mass of center of attraction. Unitless measure of angular momentum. 4-28

m mass of a test particle. Unit: kilogram or any other measure chosen by the reader. (*See* definition of *E*.) 1-12–13, 3-9

M (1) Mass of a center of gravitational attraction (2-13–15). (2) Mass parameter of the Universe (G-8). Unit: meter.

M_{crit} "Critical" mass parameter of the flat Friedmann model Universe. Unit: meter. G-13

$p = mds/d\tau$ Momentum in special relativity. Typically we use the expression p/m, allowing the reader to choose any unit desired for momentum, as long as it is the same unit as for mass m. 1-12

r (1) Radius from a center in Euclidean geometry. (2) Reduced circumference in Schwarzschild spacetime (2-7–11, B-8). (3) Boyer-Lindquist radius in Kerr spacetime (F-2). Unit: meter.

$r^* = r/M$, where M is the mass of a center of attraction. Unitless measure of reduced circumference. B-16, 4-28

R (1) Radius of Sun (D-2, E-3). (2) Reduced circumference in Kerr spacetime (F-6, F-11, F-19). (3) Radius of the Universe (G-4).

s Spatial (frame) separation between two events in a given reference frame. Unit: meter, 1-2

t Frame time in special relativity (1-2). Far-away time in Schwarzschild spacetime (2-27–30) and Kerr spacetime (F-2–3). Unit: meter.

$t^* = t/M$ where M is the mass of the center of attraction. Unitless measure of far-away time. B-16

U Effective potential used in relativistic analysis of orbit of the planet Mercury. Unit: kilogram. C-6

v Speed (meters of distance per meter of light-travel time—fraction of the speed of light) as measured in some reference frame. No units. 1-3

V Effective potential for orbiting particle or light pulse. Unit: meter. 4-12, 4-18, 5-11

$V^* = V/m$ where m is the mass of the orbiting object. Unitless measure of effective potential. 4-28

w Fourth space dimension in Friedmann model of the Universe. Unit: meter. G-4

z cosmic red shift factor. No units. G-20

Terms

Peter M. Brown

Theory, experiment, observation, and definition are born into the world together. Therefore a definition cannot stand alone and often seems circular. The purpose of this glossary of terms is not to teach physics but to clarify the terms in which physical experiments are described. Listed here are "workhorse" terms, those central to the story line of this book. For other terms, such as Hawking radiation or ZAMOs, consult the index.

Hyphenated numbers are page numbers in the book.

accretion disk Disk of material swirling around a spinning black hole. F-2, F-4, F-25

aging Elapsed time (proper time) recorded on the wristwatch of an object that moves along a given worldline between two fixed events. 1-5, 3-4

Aging, Principle of Extremal *See* **Principle of Extremal Aging**

angular momentum Second constant of the motion (the first is energy) for angular motion of test particles near spinning and nonspinning centers of attraction. We generally use the ratio angular momentum over mass, for which the unit is meter. 4-2–5, F-13

angular momentum parameter, $a = J/M$ Ratio of the angular momentum to the mass of a spinning black hole. Unit: meter. F-2

azimuthal angle Angle ϕ that (along with radius and time) locates an event in a plane. The azimuthal angle has the same meaning in general relativity as in Euclidean geometry. 2-16

black hole In this book, any structure with a horizon and surrounded at a great distance by flat spacetime.

blue shift *See* **gravitational blue shift.**

bookkeeper coordinates Three coordinates of an event recorded by a remote observer: r-coordinate, azimuthal angle ϕ, and far-away time t. The r-coordinate has a different meaning for Schwarzschild spacetime (2-7–11, 2-19) than for Kerr spacetime (F-2, F-19). *See also* **Boyer-Lindquist coordinates** and **reduced circumference.**

Boyer-Lindquist coordinates (Kerr coordinates) Three coordinates r, azimuthal angle ϕ, and far-away time t for events around a spinning, uncharged black hole. Boyer-Lindquist coordinates ϕ and t are similar to Schwarzschild coordinates, but the radial coordinate r has a different meaning (F-2, F-19). *See also* **reduced circumference.**

conv Subscript meaning "in conventional units." 1-13

coordinate radius *See* **reduced circumference**.

cosmological principle Assumption for many models of the Universe that the Universe is homogeneous and isotropic. In other words, the Universe looks the same wherever you are located in it and in whichever direction you look. G-1–2

critical mass According to the Friedmann model, if the mass parameter of the Universe has a critical value M_{crit}, then the rate of change of the radius R of the Universe approaches zero for large times. G-13

curvature factor In this book, the coefficient $(1 - 2M/r)$ that appears in the Schwarzschild metric as the coefficient of the dt^2 term and as the denominator in the dr^2 term. 2-21

curvature of spacetime Property of spacetime evidenced by tidal accelerations (relative accelerations) of free test particles. 1-16, 2-6

distance, proper *See* **proper distance**

effective potential Expression used to analyze orbits in terms of radial motion. 4-11–20, C-6–7, 5-10–13

Einstein ring Circular image of a distant light source due to light deflected around all sides of an intermediate astronomical object. 5-25, D-9–11

embedding diagrams Figures on a two-dimensional page that help us to visualize curved-space geometry. 2-25

energy The term *energy* is used in two ways in this book, in its special-relativity form and in the general-relativistic form called energy measured at infinity. Units: We usually write energy in a ratio E/m. This allows energy E and mass m to be expressed in the same units, chosen by the user, for example kilograms for both numerator and denominator—or million electron-volts or joules. 1-11, 3-9

energy in special relativity In a free-float frame the energy of a particle is defined by $E/m = dt/d\tau$. Here dt is the time between two nearby events on the trajectory of the particle as measured in the local frame and $d\tau$ is the wristwatch time of the particle between the same two events (1-7–12). The shell observer in Schwarzschild spacetime uses the special relativity expression in his local measurement of energy (3-17–19).

energy in general relativity The energy of a free test particle is a constant of the motion as the particle orbits or plunges into the center of gravitation. (3-6–10) The energy of the test particle is measured in principle from the mass of the combined system (test particle plus center of gravitation) derived from the gravitational attraction it has for a remote satellite minus the mass of the center of gravitational attraction alone. *See* Figure 4, page 3-11. Energy is described by a different formula for a particle near a nonspinning black hole (3-9) than for a particle near a spinning black hole (F-13).

equatorial plane Flat plane passing through the equator and the center of a spinning black hole (F-2). For a non-spinning black hole, by symmetry, the "equatorial" plane passing through the center can have any orientation (2-15–16).

ergosphere Region of space around a spinning black hole between the static limit and the horizon (F-7). The ergosphere is the region where the energy-extraction process of Roger Penrose can operate (F-20–25, F-28–29).

event A location in space at an instant of time; that is, a point in spacetime. A firecracker explosion is an example of an event. The explosion fixes both a position in space and a time. 1-2, 2-1

event horizon *See* **horizon**

extremal aging *See* **Principle of Extremal Aging**

extreme Kerr black hole A spinning black hole with the maximum possible angular momentum. F-5–7

far-away time (ephemeris time, *t*-coordinate) Time *t* measured by clocks remote from and stationary relative to a center of gravitation. 2-19, 2-27–30, F-3

flat spacetime Region of spacetime in which it is possible to set up a free-float (inertial) reference frame. 1-14–16, 2-4–7

frame distance Distance between two events in spacetime as measured by an observer in a particular frame of reference. 1-2

frame dragging Phenomenon that requires a rocket ship to fire its rockets tangentially (as well as radially) to keep it from being swept along in the direction of rotation of a spinning black hole (F-7–10). A test particle can circulate around a spinning black hole and still have zero angular momentum (F-13–16).

frame time Time between two events as measured by an observer in a given frame of reference. 1-2

free-float frame (inertial frame) Generally, a reference frame in which a free test particle initially at rest remains at rest. More technically, a reference frame with respect to which relative (tidal) accelerations of test particles can be neglected for the purposes of a given experiment, 1-14–16, 2-4–7.

Friedmann Universe Model that idealizes the Universe as composed of dust of zero pressure and uniform density, Project G

general relativity Einstein's theory of gravitation that describes the curvature of spacetime resulting from the presence of mass and pressure. General relativity also describes the motion of test particles in the resulting curved spacetime, gravitational waves, and the structure and development of the Universe (1-1, Project G). General relativity is a classical theory, one that does not describe quantum effects. A quantum theory of gravitation does not yet exist. 2-24

geodesic The worldline in spacetime followed by a free particle. 2-4, 3-4

Global Positioning System (GPS) System for locating position on Earth by means of timing signals sent from atomic clocks in orbiting satellites to a specialized receiver. Project A

gravitation Effect of mass-energy on spacetime, evidenced by the relative or tidal accelerations of free test particles. 1-14–16, 2-5

gravitational blue shift Decrease in the period of light as it moves toward the center of gravitational attraction. 2-13, 2-30

gravitational red shift Increase in the period of light as it moves away from the center of gravitational attraction. 2-12–13

horizon One-way surface surrounding a black hole, defined by the property that anything may pass inward through the horizon, but (in the non-quantum description) nothing, not even light, may pass outward. 2-21, B-13–16, 5-34, F-4

inertial frame *See* **free-float frame**

invariant Property of any quantity whose value is independent of the frame of reference. 1-3, 1-13

irreducible mass Mass to which an originally spinning black hole is reduced when all its energy of rotation has been extracted or lost. F-24–25

Kerr metric Metric that describes spacetime around a spinning, uncharged center of gravitational attraction. F-2, F-6

light, speed of Conversion factor used to express space and time in the same units (1-2), or to convert mass in kilograms to meters (2-14). Different values of light speed in curved spacetime (5-2–5, 5-7–9, Project E).

mass In special relativity the invariant property of an object related to its energy E and linear momentum p by the equation $m^2 = E^2 - p^2$ (1-12–13). In general relativity, the mass of a test particle is assumed to be the same as that derived in special relativity and to be small enough that its effect on spacetime in its vicinity can be neglected. In this book the unit of m is the same as the unit of energy E, chosen by the reader, so the ratio E/m has no units. The much greater mass M of an astronomical object curves spacetime in its vicinity; the value of M can be measured using the motion of a beacon in distant orbit around it (Figure 4, page 3-11). In this book the mass M of an astronomical object is measured in meters (2-13-15). *See also* **mass parameter.**

mass parameter of the Universe Quantity M that characterizes the amount of matter in the Universe. In the Friedmann model (Project G) the value of the mass parameter is equal to the local density of dust (assumed uniform throughout space) multiplied by the Euclidean volume of a sphere of radius R, where R is the radius of the Universe. Unit: meter. G-8

metric In its timelike (or spacelike) form, the metric is an expression that gives the interval of proper time (or proper distance) between two adjacent events in terms of the incremental separation in coordinate time between the events and the coordinate distance between them. 1-2, 1-4, 2-19–24, 2-32, B-12–13, F-2, F-6, G-6, G-15, G-17

microlensing Focusing of light from a distant star or galaxy by the gravitation of an object between that distant object and the observer. D-11–12

momentum (linear momentum) In this book, we use the linear momentum p of special relativity, defined for a free-float frame by $p = m ds/d\tau$, where ds is the incremental distance traveled as measured in that free-float frame and $d\tau$ is the time to move this distance as recorded on the wristwatch carried by the particle. 1-12

observer Collection of rods and recording clocks associated with a given frame of reference. 1-16–18, 2-35, 2-38

Penrose process Proposed procedure by which energy can be extracted from a spinning black hole. F-20–25, F-28–29

perihelion of Mercury, advance of Project C

potential, effective *See* **effective potential**

Principle of Extremal Aging A free object takes the worldline between two events for which the time lapse between these events recorded on its wristwatch is a maximum or minimum (an extremum). 1-5–7, 3-1–5, F-13

proper distance The distance between two events measured in a frame in which they occur at the same time (1-4, 2-22). Measured distance between spherical shells (2-22, 2-45).

proper length The length of an object as observed in a frame in which it is at rest. 1-5

proper time *See* **wristwatch time**

pulsar A rotating neutron star that sends out a sweeping searchlight beam of radiation as it spins. F-1

quantum gravity A theory, not yet developed, that would combine general relativity and quantum mechanics. 2-24, B-5

quasar Extremely bright source of light and radiation, thought to be spinning black hole with accretion disk from which radiation is emitted. F-4, F-24–27

***r*-coordinate** *See* **reduced circumference.**

rain frame A coordinate frame based on clocks and objects in free fall toward and into a black hole starting from rest at a great distance. B-4–6

recording clocks The entire collection of clocks in a lattice of clocks, each of which records the time, spatial location, and nature of any event occurring nearby. 1-17, 2-38

red shift *See* **gravitational red shift**

reduced circumference Radial location r of an event or object. Near a nonspinning center of attraction the value of r is determined by measuring the circumference of a great circle that passes through the event or object and centered on the point of attraction, then dividing this circumference by 2π (2-7–11). Inside the horizon another method of measuring r is required (B-8). Near a spinning center of attraction, the reduced circumference R is measured either on a nonrotating ring (F-6, F-9, F-11) or on a freely rotating ring which has zero angular momentum (F-19).

reference clock Clock used to synchronize all other clocks in a free-float frame. 1-17

ring, ring rider, ring coordinates, etc. These terms refer to a set of concentric rings in the equatorial plane of a rotating black hole, each ring with zero angular momentum, that is at rest with respect to the "current of space" swirling around the black hole. F-16–20

Schwarzschild bookkeeper Accountant who records events using the Schwarzschild coordinates r, ϕ, and t . 2-35 *See also* **Schwarzschild lattice.**

Schwarzschild lattice Collection of shell markings and clocks reading far-away time that allows the Schwarzschild coordinates r, ϕ, and t to be read off directly next to any event that occurs outside the horizon of a Schwarzschild black hole. 2-38

Schwarzschild map Diagram on which are plotted events and orbits in terms of the Schwarzschild coordinates r, ϕ, and t. 4-8, 5-13–19

Schwarzschild metric Metric used to describe spacetime surrounding a spherically symmetric, uncharged, nonrotating body. 2-19–24

Schwarzschild radius Radius (reduced circumference) of the horizon at $r = 2M$ for a spherically symmetric, uncharged, nonrotating body. 2-7–11, 2-19, 2-21–22

Schwarzschild sphere *(Schwarzschild surface, Schwarzschild horizon, event horizon, horizon)* Horizon of a Schwarzschild black hole. 2-21

shell In Schwarzschild spacetime, one of a set of latticelike stationary spherical surfaces concentric to the center of attraction (2-9). In Kerr spacetime, the term for a nonrotating ring in the equatorial plane (F-13).

simultaneous Occurring at the same time as recorded in a given frame of reference. 1-4

spacetime Arena in which events take place. Around steady-state structures, the geometry of spacetime is described by the metric. 2-1

spacetime interval Collective name given to the timelike spacetime interval and the spacelike spacetime interval. 1-5

speed Frame distance s covered by a particle in a given reference frame divided by the corresponding frame time t, i.e., $v = s/t$. In this book s and t are measured in the same units, so v has no units and is a fraction of the speed of light. 1-3

special relativity Study of the laws of physics expressed with respect to free-float (inertial) frames and comparison of observations in overlapping free-float frames in uniform relative motion. Special relativity correctly predicts results of experiments with light and with particles (particles are limited to speeds less than the speed of light). Special relativity cannot analyze phenomena that take place throughout a large region surrounding a center of gravitational attraction, because a single free-float frame cannot be defined in such a large region. General relativity must be used to describe such phenomena correctly. 1-1, 1-14

spherical shell *See* **shell**

static limit Surface surrounding a spinning black hole at which the dragging of space becomes so strong that everything is swept along in the direction of rotation. At the static limit light launched tangentially opposite to the direction of motion initially stands still. F-7–10

synchronization of clocks Process by which clocks in a given reference frame are set to read the same time as the reference clock. 1-17, G-7, G-10–11

time, proper *See* **wristwatch time**

time, wristwatch *See* **wristwatch time**

tidal acceleration Relative acceleration of two free test particles located in different parts of a reference frame. Tidal acceleration is a true indicator of gravitation and spacetime curvature. When tidal acceleration affects the result of a given experiment, then a free-float (inertial) reference frame cannot be defined for purposes of that experiment and general relativity must be used to analyze it. 1-14, 2-6

Universe, closed In this book, the Friedmann model Universe that expands and later contracts to a Big Crunch. Project G

Universe, flat In this book, the Friedmann model Universe that expands at a rate that tends toward zero in the limit of large times. G-12–15

Universe, Friedmann Model that idealizes the Universe as composed of dust of zero pressure and uniform density. Project G

Universe, open In this book, the Friedmann model Universe that continues expanding forever and at a constant rate for large times. G-12–15

worldline (1) Path in space and time taken by a particle or light flash. (2) Curve on the spacetime diagram representing the motion of the particle or light flash. If the particle is free, its worldline is a *geodesic*. 1-9, 3-3

wristwatch time The term wristwatch time between two events is used in two related ways in this book. (1) As proper time, τ, it is the total time (aging) recorded by a clock carried along *any* possible worldline between the two events, events that are not necessarily near one another (3-3–5). (2) As timelike spacetime interval, it is the incremental time $d\tau$ between two nearby events as recorded by a clock that follows a geodesic between them (1-1). The wristwatch time is zero along the worldline of a light flash (5-2, 5-6).

INDEX

F

G

H

I

K

L

M

N

O

P

Q

R

S

T

U

V

W

Z

Selected Formulas

The **Schwarzschild metric** describes the separation between two neighboring events in the vicinity of a spherically symmetric, nonrotating center of gravitational attraction. This is equation [10] on page 2-19.

$$d\tau^2 = \left(1-\frac{2M}{r}\right)dt^2 - \frac{dr^2}{\left(1-\frac{2M}{r}\right)} - r^2 d\phi^2 \qquad \text{[A]}$$

- $d\tau$ is the *wristwatch time* between the two events as measured on a wristwatch that moves directly from one event to the other.
- dt is the time between the events measured on a clock far from the center (page 2-27)
- r is the *reduced circumference:* circumference divided by 2π (page 2-7).
- M is the mass of the center of attraction measured in units of meters (page 2-13).

Equation [A] is called the *timelike version* of the Schwarzschild metric, useful when a clock can be carried between the two events at less than the speed of light. If this is not possible, then we use the *spacelike version,* equation [11] on page 2-19.

$$d\sigma^2 = -\left(1-\frac{2M}{r}\right)dt^2 + \frac{dr^2}{\left(1-\frac{2M}{r}\right)} + r^2 d\phi^2 \qquad \text{[B]}$$

Here $d\sigma$ is the *proper distance* between the two events: the distance between them recorded by a measuring rod moving in such a way that the two events occur at the same time in its rest frame.

Equation [C] relates the time lapse dt_{shell} between two ticks of a *shell clock* (a clock at rest on a stationary spherical shell of radius r concentric to the center of attraction) to the time lapse dt between the same two ticks measured by a far-away clock. This is equation [19] on page 2-23.

$$dt_{\text{shell}} = \left(1-\frac{2M}{r}\right)^{1/2} dt \qquad \text{[C]}$$

Equation [D] relates the radial distance dr_{shell} measured directly by the shell observer between two events that lie along the same radial direction and the radial separation dr calculated by a far-away observer. This is equation [12] on page 2-22.

$$dr_{\text{shell}} = \frac{dr}{\left(1-\frac{2M}{r}\right)^{1/2}} \qquad \text{[D]}$$

The following approximation is used often in this book. The approximation is accurate for positive, negative, or fractional values of n.

$$(1 + d)^n \approx 1 + nd \quad \text{provided} \quad |d| \ll 1 \quad \text{and} \quad |nd| \ll 1 \qquad \text{[E]}$$

Selected Formulas

The **Schwarzschild metric** describes the separation between two neighboring events in the vicinity of a spherically symmetric, nonrotating center of gravitational attraction. This is equation [10] on page 2-19.

$$d\tau^2 = \left(1-\frac{2M}{r}\right)dt^2 - \frac{dr^2}{\left(1-\frac{2M}{r}\right)} - r^2 d\phi^2 \qquad \text{[A]}$$

- $d\tau$ is the *wristwatch time* between the two events as measured on a wristwatch that moves directly from one event to the other.
- dt is the time between the events measured on a clock far from the center (page 2-27)
- r is the *reduced circumference:* circumference divided by 2π (page 2-7).
- M is the mass of the center of attraction measured in units of meters (page 2-13).

Equation [A] is called the *timelike version* of the Schwarzschild metric, useful when a clock can be carried between the two events at less than the speed of light. If this is not possible, then we use the *spacelike version,* equation [11] on page 2-19.

$$d\sigma^2 = -\left(1-\frac{2M}{r}\right)dt^2 + \frac{dr^2}{\left(1-\frac{2M}{r}\right)} + r^2 d\phi^2 \qquad \text{[B]}$$

Here $d\sigma$ is the *proper distance* between the two events: the distance between them recorded by a measuring rod moving in such a way that the two events occur at the same time in its rest frame.

Equation [C] relates the time lapse dt_{shell} between two ticks of a *shell clock* (a clock at rest on a stationary spherical shell of radius r concentric to the center of attraction) to the time lapse dt between the same two ticks measured by a far-away clock. This is equation [19] on page 2-23.

$$dr_{\text{shell}} = \left(1-\frac{2M}{r}\right)^{1/2} dt \qquad \text{[C]}$$

Equation [D] relates the radial distance dr_{shell} measured directly by the shell observer between two events that lie along the same radial direction and the radial separation dr calculated by a far-away observer. This is equation [12] on page 2-22.

$$dr_{\text{shell}} = \frac{dr}{\left(1-\frac{2M}{r}\right)^{1/2}} \qquad \text{[D]}$$

The following approximation is used often in this book. The approximation is accurate for positive, negative, or fractional values of n.

$$(1 + d)^n \approx 1 + nd \quad \text{provided} \quad |d| \ll 1 \quad \text{and} \quad |nd| \ll 1 \qquad \text{[E]}$$